Decision Processes in Dynamic Probabilistic Systems

Mathematics and Its Applications (*East European Series*)

Volume 42

Decision Processes in Dynamic Probabilistic Systems

by

Adrian V. Gheorghe

Polytechnic Institute,
Bucharest, Romania

KLUWER ACADEMIC PUBLISHERS

DORDRECHT / BOSTON / LONDON

Library of Congress Cataloging in Publication Data

Gheorghe, Adrian V.
    Decision processes in dynamic probabilistic systems / Adrian
V. Gheorghe.
        p.   cm. -- (Mathematics and its applications. East European
series ; v. 42)
    Includes bibliographical references.
    Includes index.
    ISBN 0-7923-0544-2 (alk. paper)
    1. Decision-making. 2. Markov processes.   I. Title. II. Series:
Mathematics and its applications (Kluwer Academic Publishers). East
European series ; v. 42.
T57.95.G49 1990
003'.56--dc20   90-4766

ISBN 0-7923-0544-2

Published by Kluwer Academic Publishers,
P.O. Box 17, 3300 AA Dordrecht, The Netherlands.

Kluwer Academic Publishers incorporates
the publishing programmes of
D. Reidel, Martinus Nijhoff, Dr W. Junk and MTP Press.

Sold and distributed in the U.S.A. and Canada
by Kluwer Academic Publishers,
101 Philip Drive, Norwell, MA 02061, U.S.A.

In all other countries, sold and distributed
by Kluwer Academic Publishers Group,
P.O. Box 322, 3300 AH Dordrecht, The Netherlands.

Printed on acid-free paper

To my children, Alexandra and Paul

'Et moi, ..., si j'avait su comment en revenir,
je n'y serais point allé.'

Jules Verne

The series is divergent; therefore we may be
able to do something with it.

O. Heaviside

One service mathematics has rendered the
human race. It has put common sense back
where it belongs, on the topmost shelf next
to the dusty canister labelled 'discarded non-
sense'.

Eric T. Bell

Mathematics is a tool for thought. A highly necessary tool in a world where both feedback and non-linearities abound. Similarly, all kinds of parts of mathematics serve as tools for other parts and for other sciences.

Applying a simple rewriting rule to the quote on the right above one finds such statements as: 'One service topology has rendered mathematical physics ...'; 'One service logic has rendered computer science ...'; 'One service category theory has rendered mathematics ...'. All arguably true. And all statements obtainable this way form part of the raison d'être of this series.

This series, *Mathematics and Its Applications*, started in 1977. Now that over one hundred volumes have appeared it seems opportune to reexamine its scope. At the time I wrote

"Growing specialization and diversification have brought a host of monographs and textbooks on increasingly specialized topics. However, the 'tree' of knowledge of mathematics and related fields does not grow only by putting forth new branches. It also happens, quite often in fact, that branches which were thought to be completely disparate are suddenly seen to be related. Further, the kind and level of sophistication of mathematics applied in various sciences has changed drastically in recent years: measure theory is used (non-trivially) in regional and theoretical economics; algebraic geometry interacts with physics; the Minkowsky lemma, coding theory and the structure of water meet one another in packing and covering theory; quantum fields, crystal defects and mathematical programming profit from homotopy theory; Lie algebras are relevant to filtering; and prediction and electrical engineering can use Stein spaces. And in addition to this there are such new emerging subdisciplines as 'experimental mathematics', 'CFD', 'completely integrable systems', 'chaos, synergetics and large-scale order', which are almost impossible to fit into the existing classification schemes. They draw upon widely different sections of mathematics."

By and large, all this still applies today. It is still true that at first sight mathematics seems rather fragmented and that to find, see, and exploit the deeper underlying interrelations more effort is needed and so are books that can help mathematicians and scientists do so. Accordingly MIA will continue to try to make such books available.

If anything, the description I gave in 1977 is now an understatement. To the examples of interaction areas one should add string theory where Riemann surfaces, algebraic geometry, modular functions, knots, quantum field theory, Kac-Moody algebras, monstrous moonshine (and more) all come together. And to the examples of things which can be usefully applied let me add the topic 'finite geometry'; a combination of words which sounds like it might not even exist, let alone be applicable. And yet it is being applied: to statistics via designs, to radar/sonar detection arrays (via finite projective planes), and to bus connections of VLSI chips (via difference sets). There seems to be no part of (so-called pure) mathematics that is not in immediate danger of being applied. And, accordingly, the applied mathematician needs to be aware of much more. Besides analysis and numerics, the traditional workhorses, he may need all kinds of combinatorics, algebra, probability, and so on.

In addition, the applied scientist needs to cope increasingly with the nonlinear world and the extra mathematical sophistication that this requires. For that is where the rewards are. Linear models are honest and a bit sad and depressing: proportional efforts and results. It is in the nonlinear world that infinitesimal inputs may result in macroscopic outputs (or vice versa). To appreciate what I am hinting at: if electronics were linear we would have no fun with transistors and computers; we would have no TV; in fact you would not be reading these lines.

There is also no safety in ignoring such outlandish things as nonstandard analysis, superspace and anticommuting integration, p-adic and ultrametric space. All three have applications in both electrical engineering and physics. Once, complex numbers were equally outlandish, but they frequently proved the shortest path between 'real' results. Similarly, the first two topics named have already provided a number of 'wormhole' paths. There is no telling where all this is leading - fortunately.

Thus the original scope of the series, which for various (sound) reasons now comprises five subseries: white (Japan), yellow (China), red (USSR), blue (Eastern Europe), and green (everything else), still applies. It has been enlarged a bit to include books treating of the tools from one subdiscipline which are used in others. Thus the series still aims at books dealing with:

- a central concept which plays an important role in several different mathematical and/or scientific specialization areas;
- new applications of the results and ideas from one area of scientific endeavour into another;
- influences which the results, problems and concepts of one field of enquiry have, and have had, on the development of another.

There is much in the world that can be fruitfully modelled mathematically. One exceedingly important goal of such modelling is to use these models as a central component of a decision support system. Here are two instances of (decision) problems which occur and which are extensively treated in the present volume: emergency repairs of electrical power systems; clinical medical diagnosis and treatment (of the respiratory system).

A good class of model for use in decision making should take a number of operators into account:
- partial observability; the 'state' of the system being modelled is not completely nor directly observable
- the attitudes of the decision makers towards risk
- uncertainty and random components in the environment and system.
This leads to the very general class of Markovian decision models which are studied systematically in this book. They are quite general and, hence, applicable to a multitude of situations; they are also structured strongly enough so that a great deal of theory is available. Both these aspects will become clear to the attentive reader of this volume.

The shortest path between two truths in the real domain passes through the complex domain.

J. Hadamard

Never lend books, for no one ever returns them; the only books I have in my library are books that other folk have lent me.

Anatole France

La physique ne nous donne pas seulement l'occasion de résoudre des problèmes ... elle nous fait pressentir la solution.

H. Poincaré

The function of an expert is not to be more right than other people, but to be wrong for more sophisticated reasons.

David Butler

Amsterdam, 25 April 1990   Michiel Hazewinkel

CONTENTS

Introduction

Dealing with uncertainty in a complex environment is treated analytically with different models from probabilistic to heuristic ones. And it is still going on in a large scale effort to identify means of solving, in an analytical way, specific problems by using dynamic probabilistic systems.

A special class of such models is known as Markovian processes and it mainly includes Markov and semi-Markov systems. If we add a decision process we obtain a Markovian decision process.

The present work unifies, in an original manner, three distinct dimensions in decision processes in probabilistic systems (see diagram).

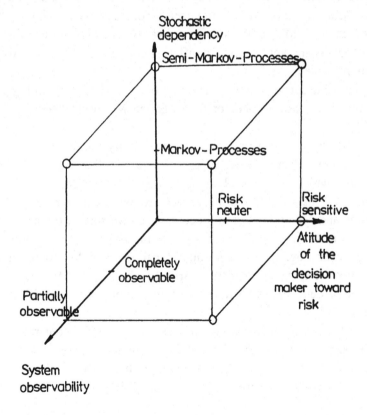

Classification of Markovian Decision Processes.

Three parameters were taken into consideration:

— stochastic dependency (dealing with Markov and semi-Markov processes);

— system observability (identification of completely or partially observable Markovian processes);

— attitude towards risk of a decision maker (dealing with risk-sensitive – averse or prone – or risk neutral persons).

Within this universe there are a finite number of processes which can model the real world under different circumstances. The discounting process can also be taken into consideration and adequate optimization models must be used accordingly.

The present book is written using first the most complicated Markovian decision processes (e.g. Risk-Sensitive Partially Observable Semi-Markov Decision Processes – RSMDP) which later degenerate into simpler cases; as was mentioned, the discounted effect has also been taken into consideration.

The first chapter of the work deals with semi-Markov and Markov chains. It introduces definitions and basic properties for these stochastic processes as well as algebraic and analytical methods of study of such systems. The chapter also covers the investigation of transient and recurrent processes, Markovian populations, partially observable Markovian chains, rewards and discounting for decision processes. Models and applications in different engineering fields and for hospital planning are presented.

Chapter 2 introduces the basic notions for dynamic programming and linear programming which are needed for investigating Markovian decision processes (MDP).

The next chapter investigates the risk sensitivity phenomenon as applied to decision models in a dynamic environment. Special attention is given to exponential utility functions which obey special properties (e.g. delta property).

Chapter 4 deals with Markovian decision processes with complete information. Semi-Markov and Markov decision processes are investigated by means of a large variety of optimization algorithms such as value iteration, policy iteration with discounting, linear programming, and sub-optimal actions in Markovian processes.

The next chapter investigates partially observable Markovian decision processes using dedicated algorithms such as the one-pass algorithm with, and without, discounting.

In Chapter 6 special Markovian processes with policy constraints when the decision maker can be either neutral or risk-averse are investigated.

The last chapter is dedicated to applied studies of stochastic models to engineering, learning processes, and to medical diagnosis and planning under uncertainty.

The history of this book is very complex and because decision processes in dynamic probabilistic systems deals with uncertain processes, the future of the present work was always almost uncertain. Only the help and continuous encouragement which I have received from Dr. D.J. Larner and Ms. Victoria Mol of Kluwer Academic Pulishers made it possible to realize its publication. I also wish to express my great thanks to Ms. Joyce Andrew, who undertook the wordprocessing form of this book; she really had a formidable task navigating through my handwritten manuscript.

Many friends and colleagues were pushing me into the right direction for the final form of the manuscript. I must give many thanks to Professor Edmond Nicolau (Bucharest Polytechnic Institute) and Professor Solomon Marcus (University of Bucharest) who helped me during the turbulent history of the book.

During initial efforts to have the book written, faculty advice and encouragement came from Professor Ronald Howard (Stanford University), Dr. Edward J. Sondik (Stanford University), as well as from many colleagues and friends from the Department of Engineering Economic Systems – Stanford University, and Department of Systems Science, The City University, London (U.K.).

Special words of appreciation and love are devoted to my wife Aurora and my children, Alexandra and Paul, who understood me and were next to me all the time; I took a lot of time from their Sundays and holidays but they were very understanding of my effort among the many University duties that I had to attend to all the time.

I would like to express my consideration and appreciation for the excellent cooperation I developed with Kluwer Academic Publishers, as well as to all my colleagues, friends and enemies who really made me determined to finalize this work.

Adrian V. Gheorghe,
Department of Management Science,
Bucharest Polytechnic Institute.

SEMI-MARKOV AND MARKOV CHAINS

1.1 Definitions and basic properties

Markovian processes (semi-Markov and Markov) are processes included in a wider class of processes where one has an explicit *time* dependence (the *dynamic* aspect) as well as the *stochastic* character of the states evolution (therefore *probabilistic*). They are part of *dynamic probabilistic systems*. In practical situations the importance of this type of process is very great. Dynamic probabilistic systems are characterized by *states*, *holding times* in any given state and *transitions* among states. The present work deals almost exclusively with this important case of dynamic probabilistic systems. A physical, biological, technological economic system can be described by values of some variables considered as fundamental by those who observe a specified system. According to Howard, the state of a system "represents all we need to know to describe the system at any instant" [9]. The systems under consideration do change in time and in an indetermininistic way — therefore, probabilistic. In this work we shall consider that the systems can occupy a *finite* number of states. A change in the system's state is described by the *waiting time* and the *transitions* among the different states.

We shall consider that waiting time and transition probabilities obey some predetermined regularities, and, among these, the most important is that the system evolves, conditioned only by the immediate past. This includes the whole history in its last state and in its last "waiting" chance in that state (or in a finite number of states and associated waiting times). This, apparent particular, form of the "past memory" encoding is sometimes called the "Markov assumption", or the "Markov property" and this mainly justifies the probabilistic schemes introduced in the present work.

The reader could estimate the field of practical applications of Markovian models described in this book.

The class of Markovian decision models is included in the field of operations research as the application of rational tools of analysis to decision-making, either in engineering, biology or economics and computer sciences [15].

Applying Markovian decision processes as rational tools of analysis to decision-making is included in the typical characteristics of a decision problem that is presented to a systems analyst.

Next, two important properties of a complex decision problem are presented [71]:

(a) they involve the commitment of a considerable amount of resources;

(b) there is a great deal of uncertainty, as well as fuzziness about the alternatives available to the decision-maker, the external factors influencing the decision, as well as the goals, the objectives as well as the constraints of the decision-maker.

The phases of applying systems analysis to a real class of situations are problem formulation, modelling analysis and optimization and implementation. Among these phases many important inter-relationships are involved so that we must keep all of the phases in view at all times, constantly appraising the influence of the present activity on the other phases of the systems analysis process.

The systems analyst must understand the decision problem well enough, so that, in arriving at an appropriate mathematical representation, he should be able to capture the true essence of the problem.

When a problem exists, most decision-makers find it difficult to be explicit about the actual details of the decision problem.

Within the framework of Markovian decision models, we have to identify the range of alternatives or decision variables that are available to the decision-maker. As was already emphasized in the literature, "it has been truly said that one good creative alternative can render a fantastic amount of mathematical analysis irrelevant" [71]. The main role of a systems analyst is to provide the language, and analyse the relative value of the alternatives available.

For using complex models, as a given in the present book, it is necessasry to understand the relationships between the alternatives, the outcomes and the uncertain environment within which the decision process must be carried out. It is of prime importance to identidfy the pertinent parameters of the state variables of the Markovian process that define the relationships between decision actions, environment and outcomes.

The next stage, following problem formulation, is to construct the appropriate mathematical model (e.g. completely or partially observable, risk averse or risk preference decision processes) that describes the relationships between the alternatives and the outcomes.

The modelling via the Markov assumption is a highly individualized activity in which the systems analyst applies his talents and *a priori* experience and knowledge "to painting an abstract picture of the physical world" [106] to [103].

According to Smallwood, "when we consider the complete formulation of the decision problem there is a great pressure to include every relevant aspect of the problem in the model of the process; on the other hand, we must remember that the capabilites of our tools as well as our computational resources are limited. ... It will do us very little good at all to propose a mathematical model for the relevant relationships in the decision problem if we are required to estimate parameters of the model that deal with relationships for which there is no data or for which the collection of the data would consume a very large amount of our resources" [71].

This is the case of the evaluation of probabilities for Markovian decision models. We can use subjective probabilities with special practical results in complex decision situations.

In the implementation stage of the modelling process we attempt to translate the optimum policy calculated (e.g. repair versus replacement) in the model world into a set of actions in the real world.

In using a complex systems analysis procedure we must always have in mind what models are available to us and thus allow our formulation of the decision problem (e.g. finite or infinite horizon, Markov or semi-Markov processes, etc.) to be guided by the eventual mathematical structure that we shall use to quantify and compute the important relationships.

It is clear that we must "keep in mind just what policies are implementable and guard against the omission of those factors in the problem that will render our analyis impractical, ... we must ensure that we adequately encode the important relationships, at the same time eliminating that complexity from the model that will render our analysis and optimization phase infeasible from a practical point of view" [71].

We shall remark that Markovian processes play a central role in dynamic probabilistic schemes and, therefore, in associated decision processes [25], [28], [33], [39], [40], [44], [62], [64], [95].

1.1.1. DISCRETE TIME SEMI-MARKOV AND MARKOV BEHAVIOUR OF SYSTEMS

Let us consider a system which can be described by a finite number of states, given in this work by the set $\{1, 2, ... , N\}$. If we consider time t which can take only integer values $t = 0, 1, 2, ... $, we can, in this case, talk about a discrete time behaviour of the system. $S(t)$ represents the system's state at time t. We shall assume that, if at time t the system was in state i (we shall write this as $S(t) = i$) before making a transition to state j, the system was waiting in state i a random time τ_{ij}, which can also take integer values $(0, 1, 2, ...)$.

What is the evolutionary "history" of the system up to time t? This will be described by the past set of occupied states up to time t — let them be $i_1, i_2, ... i_t$ and by the past set of waiting times in those states given by $m_{i_1 i_2}, m_{i_2 i_3}, ... , m_{i_{t-1} i_t}$. It is natural to consider that the future evolution of the process — i.e. the future state of the system and the associated waiting time — will be conditioned (probabilistically determined) by that history. A semi-Markov process (and in particular, a Markov process) is an intuitive formulation of a process whereby the history is reduced to the "previous moment". A semi-Markov process is defined by the transition probabilities among states:

$$\Pr\{S(t+1) = j \,|\, S(t) = i\} = p_{ij}(t); \quad i,j = 1, \ldots N$$

and the waiting times in a given state:

$$\Pr\{\tau_{ij} = m\} = h_{ij}(m).$$

In a particular case, when the system is not waiting for a random time in the state where it enters, and the transitions take place at every time moment — the system could stay for a longer period of time in the same state without this being prescribed by an *a priori* distribution function — we say the system behaves in accordance with a Markov chain with a finite number of states. It is easy to observe that, if:

$$h_{ij}(m) = \partial(m-1), \quad m = 1, 2, \ldots ; \quad i,j = 1, \ldots , N$$

with

$$\partial(m) = \begin{cases} 0 & \text{if } m = 0, \\ 1 & \text{otherwise,} \end{cases}$$

then the system evolution will be completely defined by the transition probabilities among the states $\{1, 2, \ldots N\}$. According to the "Markov property", we shall have:

$$\Pr\{S(t+1) = j | S(t) = i_k, \quad k = 1, 2, \ldots , t\}$$

$$= \Pr\{S(t+1) = j | S(t) = i_t\} \tag{1.1}$$

for simple Markov chains [1]. We shall write this as:

$$p_{ij}(t) = \Pr\{S(t+1) = j | S(t) = i\}, \quad i\,j = 1, 2, \ldots , N. \tag{1.2}$$

Quantities $p_{ij}(t)$ are probabilities and, in addition, from any state i, the system will have to make a transition to one of the states $1, 2, 3, \ldots , N$, these alternatives being exclusive (i.e. the system cannot be, at time $t+1$ in both state k and state l).

From the above we can write the following relations:

---

[1] The case when the future evolution of a system is conditioned not by the last state in which the system was, but by a finite number (two, three, etc.) of states, the associated Markov chain is called a multiple chain. In this work we shall only investigate the case of simple chains.

$$0 \leq p_{ij}(t) \leq 1; \qquad i, j = 1, 2, \dots, N,$$

$$\sum_{j=1}^{N} p_{ij}(t) = 1; \qquad i = 1, 2, \dots, N,$$

for every $t = 0, 1, 2, \dots$.

At any moment t, the Markov process is characterized by the matrix $P(t)$, $N \times N$ dimensions, with $p_{ij}(t)$ as elements:

$$P(t) = [p_{ij}(t)] = \begin{bmatrix} p_{11}(t) & \cdot & \cdot & \cdot & p_{1N}(t) \\ \cdot & & & & \cdot \\ \cdot & & & & \cdot \\ \cdot & & & & \cdot \\ p_{N1}(t) & \cdot & \cdot & \cdot & p_{NN}(t) \end{bmatrix} \qquad (1.3)$$

The evolution of the system will be determined by a set of matrices $\{P(t), \ t = 0, 1, \dots\}$.

In the particular case when the transition probabilities $p_{ij}(t)$, $i, j = 1, \dots, N$ do not depend on t — i.e. the transition probabilities among the system's states are the same for the whole time evolution of the process — we shall say the Markov chain is homogeneous (or stationary). In this case the system evolution is determined by a single matrix P.

Let $\pi_i(0) = \Pr\{S(0) = i\}$, $i = 1, 2, \dots, N$ be the initial probabilities for the system at time zero for any of its states $i = 1, 2, \dots, N$.

They form an N-dimensional vector in addition $\sum_{i=1}^{N} \pi_i(0) = 1$ and we shall use the notation $\pi_i(0) = \pi_i$, $i = 1, 2, \dots, N$.

It can be easily shown that by using these probabilities and the probability transition matrix, $P(t)$ (for the stationary case, P), we can compute the absolute probabilities so that, at any moment, the system could be in a given state. If we use the notation:

$$\pi_i(t) = \Pr\{S(t) = i\}, \qquad i = 1, 2, \dots, N; \ t = 0, 1, 2, \dots ; \qquad (1.4)$$

with the obvious condition

$$\sum_{i=1}^{N} \pi_i(t) = 1, \qquad (1.5)$$

it is clear that for any $\pi_i(t)$, $i = 1, 2, \dots, N$, $t = 0, 1, 2, \dots$, Markov processes are perfectly defined by π_i, $i = 1, 2, \dots, N$ and $P(t)$ (P, respectively). In this respect, we can say that the initial probability vector and the transition probability matrix completely determine the Markovian evolution.

In Figure 1.1, a Markov chain with a special emphasis on the states and the transitions among them is presented in a diagrammatic form.

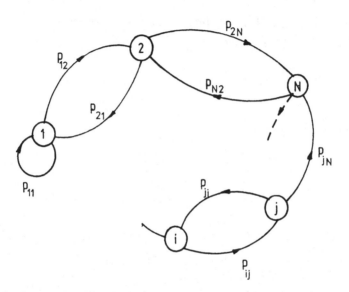

Figure 1.1. Markovian chain with special emphasis on states and transitions.

1.1.2. MULTI-STEP TRANSITION PROBABILITY

Now we shall investigate the case of homogeneous (regular) Markov chains (i.e. $p_{ij}(t) = p_{ij}$).

Let $\phi_{ij}(t)$ be the probability that the process will occupy state j at time t, conditional on that it occupied state i at time 0:

$$\phi_{ij}(t) = \Pr\{S(t) = j | S(0) = i\}; \quad i, j = 1, 2, \dots, N; \quad t \geq 0. \tag{1.6}$$

By definition, this is the multi-step transition probability, or the t-step transition probability, of the Markov process from state i to state j. It is easy to show that:

$$\phi_{ij}(t + 1) = \sum_{k=1}^{N} \phi_{ik}(t) p_{kj}; \quad t = 1, 2, \dots, \tag{1.7}$$

relation, also known as the Chapman-Kolmogorov relation, with the additional property that:

$$\phi_{ij}(0) = \partial_{ij} = \begin{cases} 1 & i = j, \\ 0 & i \neq j, \end{cases} \qquad (1.8)$$

for $i, j = 1, 2, \ldots, N$.

The t-step transition probabilities also satisfy the following properties:

$$0 \leq \phi_{ij}(t) \leq 1; \quad i, j = 1, 2, \ldots, N; \quad t \geq 0$$

$$\sum_{i=1}^{N} \phi_{ij}(t) = 1; \quad i = 1, 2, \ldots, N; \quad t \geq 0. \qquad (1.9)$$

For regular Markov chains, in a matrix form, the multi-step transition probability can be written as:

$$\Phi(t+1) = \Phi(t)P; \quad t \geq 0 \qquad (1.10)$$

$$\Phi(0) = I$$

and, in general,

$$\Phi(t) = P^t; \quad t \geq 0,$$

where:

$$\Phi(t) = [\phi_{ij}(t)] = \begin{bmatrix} \phi_{11}(t) & \cdot & \cdot & \cdot & \phi_{1N}(t) \\ \cdot & & & & \cdot \\ \cdot & & & & \cdot \\ \phi_{N1}(t) & \cdot & \cdot & \cdot & \phi_{NN}(t) \end{bmatrix} \qquad (1.11)$$

Between the state probability and the t-step transition probabilities, we can write the relation:

$$\pi_j(t) = \sum_{i=1}^{N} \pi_i(0) \, \phi_{ij}(t); \qquad j = 1, 2, \ldots, N; \quad t \geq 0. \qquad (1.12)$$

If $\pi(t)$ is the state probability vector ($\pi(t) = [\pi_1(t), \ldots, \pi_N(t)]$), then:

$$\pi(t) = \pi(0) \, \Phi(t); \qquad t \geq 0 \qquad (1.13)$$

or

$$\pi(t) = \pi(0) \, P^t; \quad t \geq 0. \qquad (1.14)$$

From the above relation, we can write:

$$\pi(t + 1) = \pi(0) \, P^{t+1} = \pi(0) P^t P$$

or

$$\pi(t + 1) = \pi(t) \, P^t; \qquad t \geq 0. \tag{1.15}$$

1.1.3. SEMI-MARKOV PROCESSES

As was already mentioned, a more general class within that of Markov processes is that of semi-Markov processes [9], [27], [34], [35], [42], [65], [66].

We shall again mention that a semi-Markov process is a process where the successive occupation of states is governed by the corresponding transition probabilities of a regular Markov process as well as by the waiting probabilities of the system in a state before making a transition to a different state. The waiting period of the process in a state before making a transition to a different state is described by a random variable which takes integer values. Its distribution depends on the present state as well as on the state where the next transition will take place. The parameters characterizing a semi-Markov process are the transition probabilities p_{ij} as well as the probability mass function $h_{ij}(\cdot)$.

Transition probabilities p_{ij}, $i, j = 1, 2, \ldots, N$ satisfy, in a similar way to the case of Markov chains, the following conditions:

$$p_{ij} \geq 0; \quad i, j = 1, 2, \ldots N$$

$$\sum_{j=1}^{N} p_{ij} = 1; \quad i = 1, 2, \ldots N.$$

The τ_{ij} represents the holding time for the transition from state i to state j, the time the semi-Markov process will spend in a state i before making a transition to state j and is governed by the probability mass function $h_{ij}(\cdot)$:

$$h_{ij}(m) = \Pr\{\tau_{ij} = m\}; \quad m \geq 1; \quad i, j = 1, 2, \ldots, N$$

$$h_{ij}(0) = 0. \tag{1.16}$$

A possible trajectory for a semi-Markov process is represented in Figure 1.2.

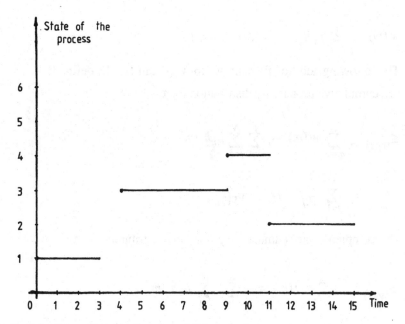

Figure 1.2. Possible trajectory for a semi-Markov process.

As was already mentioned, a Markov process can be obtained from a semi-Markov one, iff:

$$h_{ij}(m) = \partial(m-1); \quad m \geq 0; \quad i,j = 1, 2, \dots, N$$

and

$$\partial(m) = \begin{cases} 0 & \text{if } m = 0 \\ 1 & \text{otherwise.} \end{cases}$$

In connection with the distribution functions for the holding times, we shall consider the following additional quantities:

(a) cumulative probability distribution for τ_{ij}:

$$^{\leq}h_{ij}(t) = \sum_{m=0}^{t} h_{ij}(m) = \Pr\{\tau_{ij} \leq t\}; \tag{1.17}$$

(b) complementary cumulative probability distribution of τ_{ij}:

$$^{>}h_{ij}(t) = \sum_{m=t+1}^{\infty} h_{ij}(m) = 1 - {}^{\leq}h_{ij}(t) = \Pr\{\tau_{ij} > t\}. \tag{1.18}$$

It τ_i is the time spent by the system in state i, then the associated probability mass function is:

$$w_i(m) = \sum_{j=1}^{N} p_{ij} h_{ij}(m) = \Pr\{\tau_i = m\}. \tag{1.19}$$

The following additional quantities for $w_i(\cdot)$ can also be defined:

(a) cumulative probability distribution for τ_i:

$$^{\le}w_i(t) = \sum_{m=0}^{t} w_i(m) = \sum_{m=0}^{t} \sum_{j=1}^{N} p_{ij} h_{ij}(m)$$

$$= \sum_{j=1}^{N} p_{ij} {}^{\le}h_{ij}(t) = \Pr\{\tau_i \le t\} \tag{1.20}$$

(b) complementary cumulative probability distribution for τ_i:

$$^{>}w_i(t) = \sum_{m=t+1}^{\infty} w_i(m) = \sum_{m=t+1}^{\infty} \sum_{j=1}^{N} p_{ij} h_{ij}(m)$$

$$= \sum_{j=1}^{N} p_{ij} {}^{>}h_{ij}(t) = \Pr\{\tau_i > t\} \tag{1.21}$$

As far as the quantity τ_i is concerned, we can easily calculate the mean value $(\bar{\tau}_i)$, the second order moment $(\bar{\tau}_i^2)$ and the variance $\overset{v}{\tau}_i$:

$$\bar{\tau}_i = \sum_{j=1}^{N} p_{ij} \bar{\tau}_{ij} ,$$

$$\bar{\tau}_i^2 = \sum_{j=1}^{N} p_{ij} \bar{\tau}_{ij}^2 ,$$

$$\overset{v}{\tau}_i = \bar{\tau}_i^2 - (\bar{\tau}_i)^2.$$

1.1.4. STATE OCCUPANCY AND WAITING TIME STATISTICS

Next we shall define the basic quantities which characterize the class of dynamic probabilistic systems investigated in the present work (also see Howard [9]).

DEFINITION 1. The probability $\phi_{ij}(t)$ that a semi-Markov process starting from state i at time zero will be in a state j at time t, is called the *interval transition probability*. Its analytical expression is given by:

$$\phi_{ij}(t) = \partial_{ij} {}^{>}w_i(t) + \sum_{k=1}^{N} p_{ik} \sum_{m=0}^{t} h_{ik}(m) \, \phi_{kj}(t-m)$$

$$i, j = 1, 2, \dots, N; \; t \geq 0 \qquad (1.22)$$

where

$$\partial_{ij} = \begin{cases} 1 & \text{if } i = j, \\ 0 & \text{otherwise,} \end{cases}$$

and in a matrix form, we can write:

$$\Phi(t) = {}^{>}W(t) + \sum_{m=0}^{t} (P \; \Box \; H(m)) \, \Phi(t-m), \qquad T \geq 0 \qquad (1.23)$$

$$\Phi(0) = I$$

where ${}^{>}W(t) = [\partial_{ij} \, {}^{>}w_i(t)]$ and I is the identity matrix ($I = [\partial_{ij}]$).

In (1.22), the quantity ${}^{>}w_i(t)$ can be computed as follows:

$$\,^{>}w_i(t) = \sum_{m=t+1}^{\infty} \text{ sum of elements of the } i\text{-th row of} (P \; \Box \; H(m)).$$

(The symbol $\Box$ is for congruent multiplication of P and $H(m)$).

For the case of semi-Markov processes, the interval transition probability plays a similar role to that of the t-step transition probabilitiee as described for the Markov processes.

DEFINITION 1.2. The probability, $e_{ij}(k|t)$, that a semi-Markov process enters state i at time zero, and transition zero enters state j at time t on its k-th transition, is called the *entrance probability* in state j. We may remark that this quantity differs from the interval transition probability; the latter case is concerned with the presence of the process in a state. We can show that $e_{ij}(k|t)$ is expressed by the recurrent relations:

$$e_{ij}(k|t) = \partial_{ij} \, \partial(t) \, \partial(k) + \sum_{r=1}^{N} \sum_{m=0}^{t} p_{ir} \, h_{ir}(m) \, e_{rj}(k-1|t-m),$$

$$i, j = 1, 2, \dots, N; \; k = 0, 1, 2, \dots ; \; t \geq 0, \qquad (1.24)$$

where $e_{ij}(0|t) = \partial_{ij} \, \partial(t)$.

When, in (1.24), we do not take the number of transitions k into account, then the entrance probability for a semi-Markov process is given by:

$$e_{ij}(t) = \partial_{ij}\,\partial(t) + \sum_{r=1}^{N}\sum_{m=0}^{t} p_{ir}\,h_{ir}(m)\,e_{rj}(t-m),$$

$$i, j = 1, 2, \dots, N; \quad t \geq 0, \qquad (1.25)$$

$$e_{ij}(0) = \partial_{ij}\,.$$

The relationship between the interval transition probabilities, $\phi_{ij}(t)$, and the entrance probability in state j, starting from i, is given by the equation:

$$\phi_{ij}(k|t) = \sum_{m=0}^{t} e_{ij}(k|m)\,^{>}w_{j}(t-m) \qquad (1.26)$$

and:

$$\phi_{ij}(t) = \sum_{m=0}^{t} e_{ij}(m)\,^{>}w_{j}(t-m). \qquad (1.27)$$

DEFINITION 1.3. The probability, $\gamma_{ijq}(k|t)$, that a process which started in state i at time zero and transition zero be in state j at time n, having made k transitions by that time, and is to make the next transition to state q, is called the *destination probability*, and is expressed by the relation:

$$\gamma_{ijq}(k|t) = \sum_{m=0}^{t} e_{ij}(k|m)p_{jq}\,^{>}h_{jq}(t-m)$$

$$i, j, q = 1, 2, \dots, N; \quad k, t > 0. \qquad (1.28)$$

DEFINITION 1.4. The transition number k and the length t of the time interval necessary to a semi-Markov process after its entrance into state i, to leave it making a real transition to a different state, determines the so-called duration of a state; the associated probability is denoted by $d_i(k, t)$ [1]:

$$d_i(k,t) = \sum_{m=0}^{t} p_{ii}\,h_{ii}(m)\,d_i(k-1|t-m) + \partial(k-1)\,[w_i(t)-p_{ii}\,h_{ii}(t)] \qquad (1.29)$$

---

[1] $d_i(k, t) = \Pr\{l(k) = t, S(t) \neq i, S(t-1) = S(t-2) \dots = S(1) = i | S(0) = i, l(0) = 0\}$,
where $l(.)$ indicates the system transition.

DEFINITION 1.5. The probability $f_{ij}(k, t)$ that k transitions and time t be required for the semi-Markov process for the first passage from state i to state j [(1)] is called *the first passage time probability*. The *recursive* equation for this quantity is given by:

$$f_{ij}(k, t) = \sum_{\substack{r=1 \\ r \neq j}}^{N} \sum_{m=0}^{t} p_{ir} h_{ir}(m) f_{rj}(k-1, t-m) + \partial(k-1) p_{ij} h_{ij}(t)$$

$$i, j = 1, 2, \ldots, N; \quad k > 0; \quad t > 0. \tag{1.30}$$

In the case that we are only interested in time, rather than in both time and transitions, the quantity $f_{ij}(t)$ is defined as the probability that the first passage from state i to state j requires time t so that

$$f_{ij}(t) = \sum_{\substack{r=1 \\ r \neq j}}^{N} \sum_{m=0}^{t} p_{ir} h_{ir}(m) f_{rj}(t-m) + p_{ij} h_{ij}(t),$$

$$i, j = 1, 2, \ldots, N; \quad k > 0; \quad t > 0. \tag{1.31}$$

The quantity $f_{ij}(. \mid .)$ can be computed by using the interval transition probabilities $\phi_{ij}(. \mid .)$, so that we can write the *recursive* equation:

$$f_{ij}(k, t) = \begin{cases} 0, & \text{for } n = 0 \\ \phi_{ij}(k|t) - \sum_{m=0}^{t-1} \sum_{n=0}^{k} f_{ij}(u, m) \phi_{jj}(k - u|t - m) - \partial_{ij} \partial(k)^{>} w_i(t) & \text{for } n \geq 1. \end{cases}$$

$$\tag{1.32}$$

DEFINITION 1.6. $w_{ij}(k|t)$ is the probability that a semi-Markov process, entering a state i at time zero, enters state j on k occasions in the time period $(0, t)$. If $\gamma_{ij}(t)$ represents the number of times the process enters state j through time t given it entered state time zero then we can write:

[(1)]

$$f_{ij}(k, t) = \begin{cases} \Pr\{l(k) = t, S(t) = j, S(l(k-1)) \neq j, S(l-2) \neq j, \ldots, S(l(i)) = j | S(0) = i, l(0) = 0\} \\ \quad k = 2, 3, 4, \ldots; \quad t = k, k+1, \ldots \\ \Pr\{l(1) = t, S(t) = j | S(0) = i, l(0) = 0\} \\ \quad k = 1, t = 1, 2, 3, \ldots; \quad k = 0 \text{ or } t > k \\ 0, \quad i, j = 1, 2, \ldots, N. \end{cases}$$

$$w_{ij}(k|t) = \Pr\{\gamma_{ij}(t) = k\}.$$

It can be shown that:

$$w_{ij}(k|t) = \sum_{\substack{r=1 \\ r \neq j}}^{N} \sum_{m=0}^{n} p_{ir} \, h_{ir}(m) \, w_{rj}(k|t-m) +$$

$$+ \sum_{m=0}^{t} p_{ij} \, h_{ij}(m) \, w_{jj}(k-1|t-1) + \partial(k) \, {}^{>}w_i(t)$$

$$i, j = 1, 2, \dots, N; \quad k \geq 0; \quad t > 0 \tag{1.33}$$

where $w_{ij}(\,.\,|\,.\,)$ indicates the *state occupancy probability*.

The relationship between the state occupancy probability and the first passage time probability is:

$$w_{ij}(k|t) = \sum_{m=0}^{t} f_{ij}(m) \, w_{jj}(k-1|t-m) + \partial(k) \, {}^{>}f_{ij}(t)$$

$$i, j = 1, 2, \dots, N; \quad k \geq 0; \quad t > 0. \tag{1.34}$$

1.1.5. NON-HOMOGENEOUS MARKOV PROCESSES

There are many practical situations when the homogeneous probability property is not entirely fulfilled for many cases of Markov processes. Therefore, the transition probability $p_{ij}(\,.\,)$ can change in value form one transition to the other. The modelling of seasonal situations referring to the market problem of technology transfer or the penetration of new technologies are examples of such processes.

Let $\phi_{ij}(u, t)$ be the transition probability after t steps from moment u to moment t, equal, by definition, with $\phi_i(u, t) = \Pr\{S_{tj}|S_{ui}\}$ where $S_{\alpha\beta}$ indicates the state β at the moment α of the system.

It can be easily shown that:

$$\phi_{ij}(u, t) = \sum_{k=1}^{N} \phi_{ik}(u, s) \, \phi_{kj}(s, t),$$

$$\phi_{ij}(t) = \phi_{ij}(0, t), \qquad i, j = 1, 2, \dots N; \quad u \leq s \leq t. \tag{1.35}$$

This is called the Chapman-Kolmogorov equation, by extension from the case of homogeneous Markov processes.

In matrix form the Chapman-Kolmogorov equation for the non-homogeneous process is written as:

$$\Phi(u, t) = \Phi(u, s)\,\Phi(s, t); \quad u \leq s \leq t \tag{1.36}$$

and

$$\Phi(t, t) = I \quad \text{(the identity matrix)}.$$

For a non-stationary Markov process, the relationship between $\Phi(\,.\,,\,.\,)$ and transition probabilities $P(n)$ is given by:

$$\Phi(t, t+1) = P(t). \tag{1.37}$$

A special class of processes of practical interest is that of inverse Markov processes. These introduce a special kind of state dependence which has to answer the question: "What is the probability that the system was in each state of the process at some time in the past if we know its present state?" [9]. A special tool in dealing with such a process is represented by the inference theorems (of the Bayesian type) from the classical theory of probability.

We can easily write:

$$\Pr\{S_{ui}|S_{tj}\} = \Pr\{S_{ui}\,S_{tj}\}/\Pr\{S_{tj}\} = \Pr\{S_{ui}\}\,\Pr\{S_{tj}|S_{ui}\}/\Pr\{S_{tj}\} \tag{1.38a}$$

where:

$$\Pr\{S_{tj}\} = \sum_{i=1}^{N} \Pr\{S_{ui}S_{tj}\} = \sum_{i=1}^{N} \Pr\{S_{ui}\}\,\Pr\{S_{tj}|S_{ui}\}. \tag{1.39a}$$

If we denote $\pi_j(t) = \Pr\{S_{tj}\}$, then the relations (1.38a) and (1.39a) become

$$\phi_{ji}(t, u) = \frac{\pi_i(u)\,\phi_{ij}(u, t)}{\pi_j(t)}; \quad u \leq t \tag{1.38b}$$

and;

$$\pi_j(t) = \sum_{i=1}^{N} \pi_i(u)\,\phi_{ij}(u, t); \quad u \leq t. \tag{1.39b}$$

The Chapman-Kolmogorov equation for inverse Markov processes can be written as:

$$\phi_{ji}(t, u) = \sum_{k=1}^{N} \phi_{jk}(t, s) \, \phi_{ki}(s, u); \qquad u \le s \le t \,, \tag{1.40}$$

and in a matrix for, we have the relation:

$$\Phi_{ji}(t, u) = \sum_{k=1}^{N} \Phi(t, s) \, \Phi(s, u); \qquad u \le s \le t \,, \tag{1.41}$$

1.1.6. THE LIMIT THEOREM

By the asymptotic behaviour of a Markov process, we mean its behaviour in the case when the process is allowed a large number of transitions [30], [31], [41], [73].

The limit theorems are propositions which describe, in a concise way, the behaviour of such processes. Let us consider the relation given for homogeneous Markov chains:

$$\pi(t) = \pi(0) \, P^t,$$

where $\pi(t)$ is the limiting probability vector at time t and P is the transition probability matrix. If the set of matrices $\{P^t, \; t = 0, 1, \ldots \}$ has a limit (e.g. $\lim_{t \to \infty} P^t = P^\infty = \Phi$), then we have:

$$\lim_{t \to \infty} \pi(t) = \pi(0) \, P^\infty = \pi. \tag{1.42}$$

Matrix Φ is defined as the multi-step transition probability matrix and is given by the relation:

$$\Phi = \Phi(\infty) = P^\infty. \tag{1.43}$$

By introducing the relation (1.43) into (1.42) we have:

$$\pi = \pi(0)\Phi. \tag{1.44}$$

Next, the state probabilities for the semi-Markov process will be given.

Let $\varphi = (\varphi_1, \ldots, \varphi_N)$ be the interval probability vector (see Definition 1.1). Considering the π-vector of limiting probabilitites associated with the Markov process embedded in a semi-Markov process, the φ-vector has the form (see Howard [8], [9]):

$$\varphi = \frac{1}{\bar{\tau}} \pi M \tag{1.45}$$

where:

$$\bar{\tau} = \sum_{j=1}^{N} \pi_j \bar{\tau}_j \quad \text{and} \quad \pi = [\pi_1, \pi_2, \dots, \pi_N],$$

and M represents the diagonal matrix of mean waiting times in any of the system states.

We can consider the asymptotic behaviour for entrance probabilities (see Definition 1.2) only for the case when all $e_{ij}(t)$, $i = 1, 2, \dots, N$ reach a limit for $t \to \infty$ in the same way as $\phi_{ij}(t)$.

If this limit is denoted by e, then it is demonstrated [9], that

$$\varphi = eM \tag{1.46}$$

and by substitution, taking into account relation (1.45), (1.46) becomes:

$$e = \frac{1}{\bar{\tau}} \pi . \tag{1.47}$$

In addition, destination probabilities (see Definition 1.3) have limit asymptotic values, when $t \to \infty$. In this case, we can write $\gamma_{jq} = \lim_{t \to \infty} \gamma_{ijq}(t)$. The computational equation for γ_{jq} is given by:

$$\gamma_{jq} = \frac{\pi_j p_{jq} \bar{\tau}_q}{\bar{\tau}} ; \qquad j, q = 1, 2, \dots, N. \tag{1.48}$$

Therefore, the quantity γ_{jq} is defined in the state space as (asymptotic!) probability at a given moment, if the process is in state j and its next transition will be to state q. It is obvious that:

$$\sum_{q=1}^{N} \gamma_{jq} = \frac{\pi_j}{\bar{\tau}} \sum_{q=1}^{N} p_{jq} \bar{\tau}_{jq} = \frac{\pi_j \bar{\tau}_j}{\bar{\tau}} = \phi_j . \tag{1.49}$$

1.1.7. THE EFFECT OF SMALL DEVIATIONS IN THE TRANSITION PROBABILITY MATRIX

The transition probability matrix of a stationary Markov chain C is given by $P = [p_{ij}]$; the number of states in C is N. The limiting vector, representing the stationary distribution in the Markov chain is given by $\pi = [\pi_i, \; i = 1, 2, \dots, N]$. Let us consider another stationary Markov chain, C', with the probability transition matrix $P' = [p'_{ij}]$ and the limiting vector $\pi' = [\pi'_i]$. In the case where P' is close to P, we can ask ourselves how close π' is to π. When P and P' are known numerically, the problem can be easily solved by calculating both π and π'. Of great interest is the case where P' is not known exactly but we have information that P' is close to P.

We shall call this the case of small deviations [21].

Let us consider the case when an α exists, for which:

$$(1 + \alpha)^{-1} p_{ij} \leq p'_{ij} \leq (1 + \alpha) p_{ij}, \qquad i, j = 1, 2, \dots, N; \quad i \neq j, \tag{1.50}$$

where α is a positive constant. In calculating the results given below, we must assume that $p'_{ii} (i = 1, 2, \dots, N)$ satisfy the inequalitites given in (1.50).

THEOREM 1.1. *If* (1.50) *holds, then*:

$$\frac{\pi_k}{\pi_k + (1+\alpha)^{2N-2}(1-\pi_k)} \leq \pi'_k \leq \frac{\pi_k}{\pi_k + (1+\alpha)^{-2N+2}(1-\pi_k)} \tag{1.51}$$

for every k, $k = 1, 2, \dots, N$.

The proof is given in [21].

COROLLARY 1.1. *If* (1.50) *holds and α is small enough, then we can have*:

$$|\pi'_k - \pi_k| \leq 2(N-1)(1 - \pi_k)\pi_k \alpha + \mathcal{O}(\alpha^2) \tag{1.52}$$

where $\mathcal{O}(\alpha)$ is a term such that $\mathcal{O}(\alpha)/\alpha$ is bounded in a neighbourhood of the origin.

The problem can be generalized when α depends on state i of the system. If P' differs from P only in one row, h, then we can write:

$$(1 + \alpha)^{-1} p_{hj} \leq p'_{hj} \leq (1 + \alpha) p_{hj}, \tag{1.53}$$

$$j = 1, 2, \dots, h-1, h+1, \dots, N.$$

and

$$p'_{ij} = p_{ij}; \qquad i, j = 1, 2, \dots, N; \quad i \neq h. \tag{1.54}$$

THEOREM 1.2. *If* (1.53) *and* (1.54) *hold, then:*

$$\frac{\pi_k}{\pi_k + (1+\alpha)(1-\pi_k)} \leq \pi'_k \leq \frac{\pi_k}{\pi_k + (1+\alpha)^{-1}(1-\pi_k)}$$

and

$$\frac{\pi_k}{\pi_k + (1+\alpha)\pi_h + (1+\alpha)^2(1-\pi_k-\pi_h)} \leq \pi'_k \leq \frac{\pi_k}{\pi_k + (1+\alpha)^{-1}\pi_h + (1+\alpha)^{-2}(1-\pi_k-\pi_h)}$$

for each k, $k = 1, 2, \dots, h-1, h+1, \dots, N$.

COROLLARY 1.2. *If* (1.53) *and* (1.54) *hold for* α *sufficiently small, then:*

$$|\pi'_h - \pi_h| \leq (1 - \pi_h)\pi_k \,\alpha + \mathcal{O}(\alpha^2)$$

$$|\pi'_k - \pi_k| \leq (2 - 2\pi_k - \pi_h)\,\pi_k\alpha + \mathcal{O}(\alpha^2)$$

for every k, $k \neq h$.

The following results will be of use and interest in the remainder of this paragraph.

LEMMA 1.1. *If* P_i, $i = 1, 2, \dots, N$ *is a cofactor* [1] *of the* (i, i)-*th entry of the matrix* $[I - P]$, *where* I *is a* $N \times N$ *identity matrix and if* $U = \sum_{i=1}^{N} \tilde{P}_i$, *then*

(a) $\pi_k = \dfrac{P_k}{U}$; $k = 1, 2, \dots, N$,

(b) *for every* k, $k = 1, 2, \dots, N$, $\tilde{P}_k$ *can be written as*

$$\tilde{P}_k = \sum_{J_k} P_{1j_1} P_{2j_2} \cdots P_{k-1j_{k-1}} P_{k+1j_{k+1}} \cdots P_{Nj_N} \tag{1.55}$$

---

[1] The cofactor represents the determinant of $[I - P]$ matrix formed by eliminating the i-th row and the i-th column where I is the identity matrix.

where the summation is taken over the set $J_k = \{j_1, j_2, \dots, j_{k-1}, j_k, \dots, j_N\}$ *so that* $j_i \neq i$;

(c) *if* (1.50) *holds, then for every* k, k = 1, 2, ... N, *we have*:

$$(1 + \alpha)^{-N+1} \tilde{P}_k \leq \tilde{P}_k' \leq (1 + \alpha)^{N-1} \tilde{P}_k,$$

$$(1 + \alpha)^{-N+1} (U - \tilde{P}_k) \leq (U - \tilde{P}_k') \leq (1 + \alpha)^{N-1} (u - \tilde{P}_k);$$

(d) *if* (1.53) *and* (1.54) *hold, then* :

$$\tilde{P}_h' = \tilde{P}_h \text{ and } (1 + \alpha)^{-1} (U - \tilde{P}_h) \leq (U' - \tilde{P}_h') \leq (1 + \alpha)(U - \tilde{P}_h),$$

$$(1 + \alpha)^{-1} \tilde{P}_k \leq \tilde{P}_k' \leq (1 + \alpha) \tilde{P}_k,$$

$$(1 + \alpha)^{-1} (U - \tilde{P}_k - \tilde{P}_h) \leq (U_k' - \tilde{P}_k' - \tilde{P}_h') \leq (1 + \alpha)(U - \tilde{P}_k - \tilde{P}_h),$$

for every k, k ≠ h.

Next, in order to obtain the difference between π_k and π_k' we shall use a Taylor's expansion and introduce some additional notations :

(a) $U^{ij} = \sum_{k=1}^{N} \tilde{P}_k^{ij}$,

where $\tilde{P}_k^{ij}$ is the sum of terms in the right hand side of (1.55) which contains p_{ij}, so that finally we can write:

(b) $\tilde{P}_k^{ij} = p_{ij} \dfrac{\partial}{\partial p_{ij}} \tilde{P}_k$ ($i, j, k = 1, 2, \dots, N$; $i \neq j$)

and assume that $|\alpha_{ij}| \leq \alpha$ for sufficiently small α, so that $p_{ij}' = p_{ij}(1 + \alpha_{ij})$, $i, j = 1, 2, \dots, N$; $i \neq j$).

Using the Taylor's expansion formula we can have:

THEOREM 1.3.

$$\pi_k' = \pi_k + \pi_k \sum_{i=1}^{N} \sum_{\substack{j=1 \\ j \neq i}}^{N} \left(\frac{\tilde{P}_k^{ij}}{\tilde{P}_k} - \frac{U^{ij}}{U} \right) \alpha_{ij} + \mathcal{O}(\alpha^2)$$

for every k, k = 1, 2, ... , N.

Proof. Applying Lemma 1.1 to the Markov chain C' we can easily write that

$$\pi'_k = \tilde{P}'_k/U', \quad \text{and by using Taylor's formula:}$$

$$\pi'_k = \frac{\tilde{P}'_k}{U'}\Big|_{p'_{ij}=p_{ij}} + \sum_{i=1}^{N} \sum_{\substack{j=1 \\ j \neq i}}^{N} \left\{ \frac{\partial}{\partial p'_{ij}} \frac{\tilde{P}'_k}{U'}\Big|_{p'_{ij}=p_{ij}} \right\}$$

$$(p'_{ij} - p_{ij}) + \mathcal{R},$$

where $\mathcal{R}$ is a residual term ($\mathcal{R} \cong \mathcal{O}(\alpha^2)$). Using the above relations we can write:

$$\left\{ \frac{\partial}{\partial p'_{ij}} \frac{\tilde{P}'_k}{U'}\Big|_{p'_{ij}=p_{ij}} \right\} (p'_{ij} = p_{ij}) =$$

$$= \left\{ \frac{1}{U'} \frac{\partial \tilde{P}'_k}{\partial p_{ij}}\Big|_{p'_{ij}=p_{ij}} \right\} p'_{ii} \, \alpha_{ij} -$$

$$- \left\{ \frac{\tilde{P}'_k}{U'^2} \frac{\partial U'}{\partial p'_{ij}}\Big|_{p'_{ij}=p_{ij}} \right\} p'_{ij} \, \alpha_{ij} =$$

$$= \frac{1}{U} \tilde{P}_k^{ij} \alpha_{ij} - \frac{\tilde{P}_k}{U^2} U^{ij} \alpha_{ij} =$$

$$= \frac{\tilde{P}_k}{U} \left(\frac{\tilde{P}_k^{ij}}{\tilde{P}_k} - \frac{U^{ij}}{U} \right) \alpha_{ij},$$

where

$$U^{ij} = p_{ij} \frac{\partial}{\partial p_{ij}} U \quad (i, j = 1, 2, \dots, N; \; i \neq j).$$

Finally, we can write

$$\pi'_k = \frac{\tilde{P}_k}{U} + \frac{\tilde{P}_k}{U} \sum_{i=1}^{N} \sum_{j=1}^{N} \left(\frac{\tilde{P}_k^{ij}}{\tilde{P}_k} - \frac{U^{ij}}{U} \right) \alpha_{ij} + \mathcal{O}(\alpha^2),$$

and by using Lemma 1.1 we obtain the result given in Theorem 1.3.

1.1.7.1. *Limits of some important characteristics for Markov chains.* In this chapter, by *important characteristics*, we understand a set of parameters which describe useful aspects in the operation of the Markovian scheme. The importance and relevance of such parameters depend on the structure of the Markov process. We shall open this paragraph by a classification of Markov processes based on some basic properties of their states. In the remainder of the paragraph we shall refer to homogenous chains.

We shall identify two types of states. This will be related to the possibility of the process returning to the state a finite, or infinite, number of times. Their definition requires a few general considerations.

The state of a Markov chain can be grouped according to the *communication* criterion. We shall say that state j is accessible from i if t exists such that $\Pr\{S(t) = j|S(0) = i\} > 0$. States i and j communicate if j is accessible from i and i is accessible from j. A *class of states* is a maximal sub-set of the state space so that any two elements communicate among themselves. A property of states in a Markov chain is called a *class property* if, as soon as a state has this property, it is true for all the states from the class which contains this state.

Let τ_j be the time of the first entrance in state j,

$$\tau_j = \min (t > 1|S(t) = j).$$

Next, we shall use the notation:

$$f(t, i, j) = \Pr\{\tau_j = t|S(0) = i\}$$

and

$$f(i, j) = \sum_{t=1}^{\infty} f(t, i, j) = \Pr\{\tau_j < \infty |S(0) = i\}.$$

We can see that the process makes a transition from state i to state j if, and only if, $f(i, j) > 0$, and makes a transition from state j to state i and back if $f(i, j) \cdot f(j, i) > 0$. It can be proved that for the Markov chain that starts from i the probability of coming back to i at least k times is $[f(i, i)]^k$; in addition if $f(i, i) = 1$, then the probability of revisiting state i an infinite number of times is equal to one; if $f(i, i) < 1$, then the probability of revisiting state i an infinite number of times is equal to zero.

The above considerations justify the following definition: a state i is a *recurrent* (or persistent) one, if $f(i, i) = 1$, and *non-recurrent* (or transient) if $f(i, i) < 1$. It can be easily shown that recurrency and non -recurrency are class properties. A state i

with the property that $\Pr\{S(t+1) = j|S(t) = i\} = 0$ for any j is called an *absorbant state*.

A Markov chain with non-recurrent states is called an *absorbant* one. The name is justified by the fact that such a chain, if started from a non-recurrent state, then, after an evolution in the set of non-recurrent states, will be trapped in a set of recurrent states (this has an *absorbant* character). A Markov chain without non-recurrent states will be called *recurrent..* A recurrent chain whose states create a single class is called an *irreducible* or *ergodic* chain.

a) *Absorbant Markov chains.* Now we shall investigate the variation intervals for some important characteristics which occur within finite Markov chains. We shall consider, first, the case of absorbant Markov chains. It can be shown, in this case, that the probability transition matrix, P, has the form:

$$
P = \left[\begin{array}{c:c}
Q & R \\
\hdashline
0 & I
\end{array}\right]
\begin{array}{l}
s \\
\\
N{-}s
\end{array}
$$

$$
\begin{array}{cc}
s & N-s
\end{array}
$$

and $\mathcal{T} = \{S_2, \dots, S_s\}$ is the set of non-recurrent or transient states $\mathcal{T} = \{S_{s+1}, \dots, S_N\}$, is the set of absorbant states of the Markov chain being considered.

Let $Q = |I - Q|$ and $\tilde{Q}(i, j)$ be a cofactor of the (i, j)-th entrance in the $\tilde{Q}$ matrix.

LEMMA 1.2. *Mean and variance of the random variables* n_i *(which show how many times the system was in state i before making a transtition in an absorbant state) and t (number of transitions before the system arrives in an absorbant state) are given by the relations :*

(a) $\langle n_j \rangle_i = n_{ij} = \tilde{Q}(j, i) / |\tilde{Q}|$; *where i is the starting state of the system ;*

(b) $^v\langle n_j \rangle_i = 2n_{ij}\, n_{jj} - n_{ij} - n^2_{ij} = \dfrac{\tilde{Q}(j, i)}{|\tilde{Q}|^2}\left\{(\tilde{Q}(j,j) - |\tilde{Q}|) + (\tilde{Q}(j,j) - \tilde{Q}(j, i))\right\}$;

(c) $\langle t \rangle_i = \sum\limits_{k=1}^{s} n_{ik} = \dfrac{1}{|\tilde{Q}|}\sum\limits_{k=1}^{s} \tilde{Q}(k, i)$;

(d) $\;{}^v\langle t \rangle_i \;=\; 2 \sum_{k=1}^{s} \sum_{j=1}^{s} n_{ij}\, n_{ik} - \sum_{k=1}^{s} n_{ik} - \left\{ \sum_{k=1}^{s} n_{ij} \right\}^2 ;$

(e) *if m indicates the total number of transient states experienced by the system before arriving in an absorbant state, then, the mean value for m, when the system starts from state i is* :

$$\langle m \rangle_i \;=\; \sum_{k=1}^{s} n_{ik}/n_{kk} \;=\; \sum_{k=1}^{s} \frac{\tilde{Q}(k, i)}{Q(k, k)} \, .$$

If b_{ij} is the probability starting in state $i \in \mathcal{T}$ that the chain is absorbed in state $j \in \mathcal{T}$, then:

$$b_{ij} \;=\; \sum_{k=1}^{s} n_{ik} p_{kj} \;+\; \frac{1}{|\tilde{Q}|} \sum_{k=1}^{s} p_{kj}\, \tilde{Q}(k, i).$$

In a matrix form, if $B = [b_{ij}]$ and $N = [I - Q]^{-1} = \tilde{Q}^{-1}$, then $B = NR$.

If h_{ij} represents the probability starting in state $i \in \mathcal{T}$ that the chain is ever in state $j \notin \mathcal{T}$, then:

$$h_{ij} \;=\; (n_{ii} - 1)/n_{ii} \;=\; \frac{\tilde{Q}(i, i) - |\tilde{Q}|}{\tilde{Q}(i, i)} \qquad\qquad (1.56)$$

and for $j \neq i$, $h_{ij} = \tilde{Q}(j, i)/\tilde{Q}(j, j)$.

In a matrix form, if $H = [h_{ij}]$ and X_{diag} is a diagonal matrix with x_{ii} the i-th diagonal entrance of X, then $H = [N - I]_{\text{di ag}}^{-1}$.

If we consider two Markov chains with P and P', respectively, as transition probability matrices for which:

$$(1 + \alpha)^{-1} p_{ij} \;\leq\; p'_{ij} \;\leq\; (1 + \alpha) p_{ij} \qquad\qquad (1.57)$$

$$i = 1, 2, \ldots, s \,;\; j = 1, 2, \ldots, N \,;\;\; i \neq j,$$

then:

LEMMA 1.3. *If (1.57) holds, then* :

(a) $(1 + \alpha)^{-s} |\tilde{Q}| \leq |\tilde{Q}'| \leq (1 + \alpha)^{s} |\tilde{Q}|$,

(b) $(1 + \alpha)^{-s+1} \tilde{Q}(i, j) \leq \tilde{Q}'(i, j) \leq (1 + \alpha)^{s-1} \tilde{Q}(i, j)$,

(c) $(1 + \alpha)^{-s+1} [\tilde{Q}(i, i) - \tilde{Q}(i, j)] \leq \tilde{Q}'(i, i) - \tilde{Q}'(i, j) \leq (1 + \alpha)^{s-1} [\tilde{Q}(i, i) - \tilde{Q}(i, j)]$,

(d) *if the inequality*

$$(1 + \alpha)^{-1} p_{ii} \leq p'_{ii} \leq (1 + \alpha) p_{ii} ; \quad i = 1, 2, \ldots, s \qquad (1.58)$$

holds, then :

$$(1 + \alpha)^{-s} [\tilde{Q}(i, i) - |\tilde{Q}|] \leq \tilde{Q}'(i, i) - |\tilde{Q}'| \leq (1 + \alpha)^{s} [\tilde{Q}(i, i) - |\tilde{Q}|].$$

This lemma is proved by Takahashi in his work [21].

Using the general results from Lemma 1.1, we can write:

(a) $\langle n'_j \rangle_i \leq (1 + \alpha)^{2s-1} \langle n_j \rangle_i$

(b) $^{v}\langle n'_j \rangle_i \leq (1 + \alpha)^{4s-1} \langle n_j \rangle_i$

(c) $\langle t' \rangle_i \leq (1 + \alpha)^{2s-1} \langle t \rangle_i$

(d) $\langle m' \rangle_i \leq (1 + \alpha)^{2s-2} \langle m \rangle_i$

(e) $b'_{ij} \leq (1 + \alpha)^{2s} b_{ij}$

(f) $h'_{ii} \leq (1 + \alpha)^{2s-1} h_{ii}$

(g) $h'_{ij} \leq (1 + \alpha)^{2s-2} h_{ij} \ (i \neq j)$.

For the above values given by (b) and (f) they also have to fulfill the constraints given by relations (1.57) and (1.58).

When we use relations of the type:

$$\tilde{Q}(i, i) = \tilde{Q}(i, j) + [\tilde{Q}(i, i) - \tilde{Q}(i, j)]$$

or:

$$|\tilde{Q}| = \sum_{\substack{k=1 \\ k \neq i}}^{s} P_{ik} [\tilde{Q}(i, i) - \tilde{Q}(i, k)] + \left(\sum_{k=s+1}^{N} P_{ik} \right) \tilde{Q}(i, i),$$

the bounds can be improved as:

(g') $\quad h'_{ij} \leq \dfrac{h_{ij}}{h_{ij} + (1 + \alpha)^{-2s+2} (1 - h_{ij})}$

(a') $\quad \langle n'_j \rangle_i \leq \dfrac{(1 + \alpha)\, n_{ii}}{\left(\sum\limits_{k=s+1}^{N} P_{ik} \right) n_{ii} + (1 + \alpha)^{-2s+2} \sum\limits_{k=1}^{N} P_{ik} (n_{ii} - n_{ki})}$

b) *General homogeneous Markov chains.* Next, we shall investigate the general case of a non-absorbant Markov chain (i.e. without non-recurrent states). Let us consider a regular (homogeneous) Markov chain with N states and its $P = [p_{ij}]$ matrix; $\pi = [\pi_i\, ; \; i = 1, 2, \dots, N]$ is the limiting vector (stationary distribution of the chain). A is a matrix with each row defined by π, $Z = [z_{ij}] = [I - P + A]^{-1}$ is the fundamental matrix, $M = [m_{ij}]$ is the matrix of mean number of steps required to reach state j for the first time, starting in i and $Y = [y_{ij}]$ the matrix of variances for the number of steps required to reach j starting in i.

$\tilde{P}_k(i, j), \quad i, j, k = 1, 2, \dots, N\,; \; i \neq k, \; j \neq k$ is the cofactor of the entry $-p_{ij}$

(or $1 - p_{ii}$, if $i \neq j$) of the matrix formed by deleting the k-th row and the k-th column from $[I - P]$. This measure can be represented as the sum of products of transition probabilities with plus signs [21].

LEMMA 1.4. *If* $\pi_k = \tilde{P}_k/U$ *and considering that*

$$z_{kk} = \frac{\tilde{P}_k}{U} + \frac{1}{U^2} \sum_{\substack{j=1 \\ j \neq k}}^{N} \sum_{\substack{i=1 \\ i \neq k}}^{N} \tilde{P}_j\, \tilde{P}_k (i, j)$$

and since $m_{kk} = 1/\pi_k$ *and* $y_{kk} = \dfrac{2 z_{kk} - \pi_k}{\pi_k^2}$ *then :*

(a) $\quad m_{kk} = U/\tilde{P}_k$

(b) $m_{jk} = \dfrac{1}{\bar{P}_k} \displaystyle\sum_{\substack{i=1 \\ i \neq k}}^{N} \tilde{P}_k(i,j), \ j \neq k$

(c) $y_{kk} = \dfrac{U}{\bar{P}_k} + \dfrac{2}{\bar{P}_k^2} \displaystyle\sum_{\substack{j=1 \\ j \neq k}}^{N} \sum_{\substack{i=1 \\ i \neq k}}^{N} \tilde{P}_j \, \tilde{P}_k)i,j) \ .$

The following lemma also holds.

LEMMA 1.5. *If* (1.57) *holds, then*

(a) $(1+\alpha)^{-N+2} \, \tilde{P}_k(i,j) < \tilde{P}_k'(i,j) \leq (1+\alpha)^{N-2} \tilde{P}_k(i,j)$

(b) $\pi_k' \leq (1+\alpha)^{2N-2} \pi_k$

(c) $z_{kk}' \leq (1+\alpha)^{4N-5} z_{kk}$

(d) $m_{kk}' \leq (1+\alpha)^{2N-2} m_{kk}$

(e) $m_{jk}' \leq (1+\alpha)^{2N-3} m_{jk} ; \ j \neq k$

(f) $y_{kk}' \leq (1+\alpha)^{4N-5} y_{kk} \ .$

c) *Upper bound of the varince of* π_k. Next, if we consider the case when $p_{ij}' - s$ are mutally independent random variables distributed about p_{ij}; $i, j = 1, 2, \ldots , N$, we shall write:

$$\langle p_{ij}' \rangle = p_{ij} ,$$

$$^v\langle p_{ij}' \rangle = p_{ij}^2 \, \sigma_{ij}^2 \, \sigma^2$$

$$\langle \, |p_{ij}' - p_{ij}|^3 \, \rangle = \mathcal{O}(\sigma^3). \tag{1.59}$$

The following properties of these estimators can be proved (see [21]):

THEOREM 1.4. *If* p_{ij}', $i, j = 1, 2, \ldots , N$ *are mutually independent random variables, satisfying the three conditions given in* (1.59), *then the mean and variance for the limiting vector* π_k' *are:*

$$\langle \pi'_k \rangle = \pi_k + \mathcal{O}(\sigma^2) \tag{1.59a}$$

$$^v \langle \pi'_k \rangle = \pi_k^2 \sum_{\substack{i=1 \\ j \neq i}}^{N} \sum_{j=1}^{N} \left(\frac{\tilde{P}_k^{ij}}{\tilde{P}_k} - \frac{U^{ij}}{U} \right)^2 \sigma_{ij}^2 + \mathcal{O}(\sigma^3) . \tag{1.59b}$$

Proof. Relation (1.59a) is an immediate consequence of Theorem 1.3 with an appropriate change in the residual term. Relation (1.59b) can be proved by taking the expectation of the square of the relation given by Theorem 1.3.

THEOREM 1.5. *If p'_{ij}, $i, j = 1, 2, \ldots , N$, $i \neq j$ are mutually independent random variables which satisfy the conditions (1.59), then :*

(1) $\langle \pi'_k \rangle \leq 2(N-1)\pi_k^2 (1 - \pi_k)^2 \sigma^2 + \mathcal{O}(\sigma^3)$,

(2) *if $p'_{hj}(j = 1, 2, \ldots , h-1, h+1, \ldots , N)$ are mutually independent random variables and $p'_{ij} = p_{ij} (i, j = 1, 2, \ldots , N ; i \neq h)$, then when $k \neq h$,*

$$^v \langle \pi'_k \rangle \leq \pi_k^2 \left\{ (1 - \pi_k)^2 + (1 - \pi_h - \pi_k)^2 \right\} \sigma^2 + \mathcal{O}(\sigma^3),$$

and when $k = h$, then :

$$^v \langle \pi'_k \rangle \leq \pi_k^2 (1 - \pi_h)^2 \sigma^2 + \mathcal{O}(\sigma^3).$$

Proof. We shall give the full proof of part (1) of the above theorem; the proof of part (2) is essentially contained in the following proof.

By Theorem 1.4, we have

$$^v \langle \pi'_k \rangle \leq \sigma^2 \pi_k^2 \sum_{i=1}^{N} \left\{ \sum_{\substack{j=1 \\ j \neq i}}^{N} \left(\frac{\tilde{P}_k^{ij}}{\tilde{P}_k} - \frac{U^{ij}}{U} \right)^2 \right\} + \mathcal{O}(\sigma^3) . \tag{1.59c}$$

The sums in the braces in (1.59c) are influenced by quadratic functions of π_k and π_i as follows. When $i \neq k$,

$$\sum_{j \neq i}^{N} \left(\frac{\tilde{P}_k^{ij}}{\tilde{P}_k} - \frac{U^{ij}}{U} \right)^2 = \frac{1}{(\tilde{P}_k U)^2} \sum_{j \neq i} \left\{ \tilde{P}_k^{ij}(U - \tilde{P}_k) - \tilde{P}_k(U^{ij} - \tilde{P}_k^{ij}) \right\}^2 \leq$$

$$\leq \frac{1}{(\tilde{P}_k U)^2} \left[\sum_{j\neq i} \{\tilde{P}_k^{ij}(U - \tilde{P}_k)\}^2 + \sum_{j\neq i} \{\tilde{P}_k(U_k^{ij} - \tilde{P}_k^{ij})\}^2 \right] \leq$$

$$\leq \frac{1}{(\tilde{P}_k U)^2} \left[\{\sum_{j\neq i} \tilde{P}_k^{ij}(U - \tilde{P}_k)\}^2 + \{\sum_{j\neq i} \tilde{P}_k(U_k^{ij} - \tilde{P}_k^{ij})\}^2 \right].$$

Due to the following equalities:

$$\sum_{j\neq i} \tilde{P}_k^{ij} = \tilde{P}_k$$

$$\sum_{j\neq k} U^{kj} = U - \tilde{P}_k,$$

the right hand side of the last inequality is equal to:

$$\frac{1}{(\tilde{P}_k U)^2} \left[\{\tilde{P}_k(U - \tilde{P}_k)\}^2 + \{\tilde{P}_k(U - \tilde{P}_i - \tilde{P}_k)\}^2 \right] =$$

$$= (1 - \pi_k)^2 + (1 - \pi_i - \pi_k)^2$$

$$= 2(1 - \pi_k)^2 - 2(1 - \pi_k)\pi_i + \pi^2.$$

In the case that $i = k$, since $\tilde{P}_k^{kj} = 0$, we can write:

$$\sum_{j=k} \left(\frac{\tilde{P}_k^{kj}}{\tilde{P}_k} - \frac{U^{kj}}{U} \right)^2 \leq \frac{1}{U^2} \left(\sum_{k=j} U^{kj} \right)^2 = (1 - \pi_k)^2.$$

With the above results we can finally write:

$$^v\langle \pi_k' \rangle \leq \sigma^2 \pi_k^2 \left[\sum_{i=1}^{N} \{2(1 - \pi_k)^2 - 2(1 - \pi_k)\pi_i + \pi_i^2\} + (1 - \pi_k)^2 \right] + \mathcal{O}(\sigma^3) \leq$$

$$\leq \sigma^2 \pi_k^2 \left[(2N - 1)(1 - \pi_k)^2 - 2(1 - \pi_k) \sum_{i\neq k} \pi_i + \left(\sum_{i\neq k} \pi_i \right)^2 \right] + \mathcal{O}(\sigma^3) \leq$$

$$\leq 2(N - 1)\sigma^2 \pi_k^2 (1 - \pi_k)^2 + \mathcal{O}(\sigma^3)$$

which is a complete proof of the main part of the above theorem.

THEOREM 1.6. *If two vectors given by* $\{p'_{ij} ; j = 1, 2, \dots , N\}, i = 1, 2, \dots , N$ *are mutually independent random vectors and if the conditions (1.59) are satisfied for all* p_{ij}, *then*

$$^\nu \langle \pi'_k \rangle \le 2(2N - 3)\pi_k^2 (1 - \pi_k)^2 \sigma^2 + \mathcal{O}(\sigma^3).$$

d) *Upper bounds of the variances of basic quantities in Markov chains.*

LEMMA 1.6. *If* p'_{ij}, $i \ne j$ *are mutually independent random variables which satisfy* (1.59) *the upper bounds of their variances of basic quantities in an absorbing Markov chain are :*

(1) $\quad ^\nu \langle \langle n'_j \rangle_i \rangle \le (2s - 1)\sigma^2 \left\{ \left(\langle n_j \rangle_i \right)^2 \right\} + \mathcal{O}(\sigma^3),$

(2) $\quad ^\nu \langle \langle t'_j \rangle_i \rangle \le (2s - 1)\sigma^2 \left\{ \left(\langle t \rangle_i \right)^2 \right\} + \mathcal{O}(\sigma^3),$

(3) $\quad ^\nu \langle \langle m'_j \rangle_i \rangle \le (2s - 1)\sigma^2 \left\{ \left(\langle m \rangle_i \right)^2 \right\} + \mathcal{O}(\sigma^3),$

(4) $\quad ^\nu \langle b'_{ij} \rangle \le 2s\, \sigma^2 \left\{ b_{ij} \right\}^2 + \mathcal{O}(\sigma^3),$

(5) $\quad ^\nu \langle h'_{ij} \rangle \le (2s - 1)\sigma^2 \left\{ h_{ij} \right\}^2 + \mathcal{O}(\sigma^3); \ i \ne j,$

(6) $\quad ^\nu \langle ^\nu \langle n'_j \rangle_i \rangle \le (4s - 1)\sigma^2 \left\{ ^\nu \langle n \rangle_i \right\}^2 + \mathcal{O}(\sigma^3),$

(7) $\quad ^\nu \langle h'_{ii} \rangle \le (2s - 1)\sigma^2 \left\{ h_{ii} \right\}^2 + \mathcal{O}(\sigma^3).$

LEMMA 1.7. *If* p'_{ij}, $i \ne j$ *are mutually independent random variables which satisfy* (1.59) *the upper bounds of ther variances of basic quantities in an recurrent Markov chain are :*

(1) $\quad ^\nu \langle \pi'_k \rangle \le 2(N - 1)\, \sigma^2 \left\{ \pi_k \right\}^2 + \mathcal{O}(\sigma^3),$

(2) $\quad ^\nu \langle z'_{kk} \rangle \le 8(N - 1)\, \sigma^2 \left\{ z_{kk} \right\}^2 + \mathcal{O}(\sigma^3),$

(3) $\quad ^\nu \langle m'_{kk} \rangle \le 2(N - 1)\, \sigma^2 \left\{ m_{kk} \right\}^2 + \mathcal{O}(\sigma^3),$

(4) $^v\langle m'_{jk}\rangle \leq (2N-3)\,\sigma^2\,\{m_{jk}\}^2 + \mathcal{O}(\sigma^3);\ j \neq k,$

(5) $^v\langle y'_{kk}\rangle \leq 8(N-1)\,\sigma^2\,\{y_{kk}\}^2 + \mathcal{O}(\sigma^3).$

1.2   Algebraic and analytical methods in the study of Markovian systems

In this paragraph we shall introduce basic definitions on matrix eigenvalues and eigenvectors, Perron-Frobenius theorems and geometrical transform (z-transform) as well as the exponential transform (Laplace transform).

1.2.1. EIGENVALUES AND EIGENVECTORS

In the investigation of a linear operation $A = [a_{ij}]$, an important role is played by the vectors x for which we can have [5], [14], [72]:

$$Ax = \lambda x, \quad x \neq 0.$$

These vectors are called characteristic vectors and the scalars λ corresponding to them are called characteristic values of the matrix A.

To compute the characteristic values and characteristic vectors of a matrix A we must consider an arbitrary basis $z_1, z_2, \dots , z_n$ in $\mathbb{R}$. Let us suppose that $x = \sum_{i=1}^{N} x_i\, z_i$, and $A = [a_{ij}];\ i, j, = 1, 2, \dots , n$ is a matrix in the basis $z_1, z_2, \dots , z_n$. From the above equation we can write:

$$a_{11}x_1 + a_{12}x_2 + \dots + a_{1n}x_n = \lambda x_1$$

$$a_{21}x_1 + a_{22}x_2 + \dots + a_{2n}x_n = \lambda x_2$$

$$\cdots\cdots\cdots\cdots\cdots\cdots\cdots\cdots$$

$$a_{n1}x_1 + a_{n2}x_2 + \dots + a_{nn}x_n = \lambda x_n$$

which will be used in the following form:

$$(a_{11} - \lambda)x_1 + a_{12}x_2 + \dots + a_{1n}x_n = 0$$

$$a_{21}x_1 + (a_{22} - \lambda)x_2 + \dots + a_{2n}x_n = 0$$

$$\cdots\cdots\cdots\cdots\cdots\cdots\cdots\cdots$$

$$a_{n1}x_1 + a_{n2}x_2 + \dots + (a_{nn} - \lambda)x_n = 0.$$

The eigenvector must not be a zero vector, at least one of the x_i's, $i = 1, 2, \ldots ,$ n, must be different from zero. To have a non-zero solution the determinant of the above system must be zero, so that:

$$\begin{vmatrix} a_{11} - \lambda & a_{12} & \cdot & a_{1n} \\ a_{21} & a_{22} - \lambda & \cdot & a_{2n} \\ \cdot & \cdot & \cdot & \cdot \\ a_{n1} & a_{n2} & \cdot & a_{nn} - \lambda \end{vmatrix} = 0.$$

The above equation is known as the characteristic equation of matrix A and every characteristic value λ of A is a root of this equation.

THEOREM 1.7. *For a linear operator A, the characteristic vectors are always linearly independent.*

Proof. For the case when $Ax_i = \lambda_i x_i$ $(x_i \neq 0; \ \lambda_i \neq \lambda_k$ for $i \neq k, \ i, k = 1, 2, \ldots ,$ $m)$, we can consider that $\sum_{i=1}^{m} y_i x_i = 0$. By applying the operator A to both sides of the previous relation we have:

$$\sum_{i=1}^{m} y_i \lambda_i x_i = 0.$$

Following simple algebraic operations, we can write:

$$\sum_{i=2}^{m} y_i (\lambda_i - \lambda_1) x_i = 0.$$

If we apply the operator of the form $(A - \lambda_i I)$ $i = 1, 2, \ldots , m-1$ where $I = [\delta_{ij}]$ — δ_{ij} is the Kronecker symbol — , we obtain the following equation:

$$y_m (\lambda_m - \lambda_{m-1}) (\lambda_m - \lambda_{m-2}) \ldots (\lambda_m - \lambda_i) x_m = 0$$

so that $y_m = 0$, so that, according to Gantmacher [72], $y_1 = y_2 = \ldots = y_m = 0$ (i.e. there is no linear dependence among the vectors x_i, $i = 1, 2, \ldots , m,$) which proves the above theorem.

1.2.2. STOCHASTIC MATRICES

Let us consider a system with Markov behaviour which can be in any of a possible N states $S_1, S_2, \ldots, S_N$ and a sequence of times instants $t_0, t_1, t_2, \ldots$. Transition probabilities $p_{ij}, i, j = 1, 2, \ldots, N$ associated with a Markov process, (see §1.1), constitute the matrix of transition probabilities $P = [p_{ij}]$ with the condition that:

$$p_{ij} \geq 0; \quad \sum_{j=1}^{N} p_{ij} = 1; \quad i = 1, 2, \ldots, N . \tag{1.60}$$

A square matrix with all real elements $A = [a_{ij}], i, j = 1, 2, \ldots n$ is *non-negative* $(A > 0)$ or *positive* $(A > 0)$ if all elements if A are non-negative $(a_{ij} \geq 0)$ or positive $(a_{ij} > 0)$.

A square matrix $P = [p_{ij}], i, j = 1, 2, \ldots n$ is called *stochastic* if P is non-negative and if the relation (1.60) holds.

It is evident that, following the definition of stochastic matrices, for the case of homogeneous Markov processes, the P transition matrix, is such a matrix. Considering this class of matrix we can emphasize the following properties (see Gantmacher [5]):

a) a non-negative matrix $P \geq 0$ is stochastic if, and only if, it has the characteristic vector $\mathbf{1} = [1, 1, \ldots, 1]$ for the characteristic value 1; for such a matrix the maximal characteristic value is 1;

b) a non-negative matrix $A \geq 0$ with the maximal positive characteristic value $r > 0$ and with a corresponding positive characteristic vector $Z = (z_1, z_2, \ldots, z_N) = 0$ is similar to the relation

$$A = Z r P Z^{-1},$$

where:

$$Z \quad \begin{bmatrix} z_1 & 0 & . & . & . & 0 \\ 0 & z_2 & 0 & . & . & 0 \\ . & 0 & z_3 & 0 & . & 0 \\ .. & .. & .. & . & .. & .. \\ .. & .. & .. & .. & . & .. \\ 0 & . & . & . & 0 & z_N \end{bmatrix}$$

c) to the characteristic value 1 of a stochastic matrix there correspond only elementary divisors of the first degree;

d) if a matrix A has the form

$$A = \begin{bmatrix} Q_1 & \vdots & 0 \\ \cdots & \cdots & \cdots \\ S & \vdots & Q_2 \end{bmatrix}$$

where Q_1 and Q_2 are square matrices, and if the characteristic value λ_0 of A, is also a characteristic value of Q_1, but not of Q_2,

$$|Q_1 - \lambda_0 I| = 0; \quad |Q_2 - \lambda_0 I| \neq 0,$$

then the elementary divisors of A and Q_1 corresponding to the characteristic value λ_0 are the same.

1.2.3. PERRON-FROBENIUS THEOREM

A square matrix $A = [a_{ij}]$ is reducible if the index set $1, 2, \ldots, n$ can be split into two complementary sets $i_1, i_2, \ldots, i_a;\ k_1, k_2, \ldots, k_b\ (a + b = n)$ such that

$$a_{i_\alpha k_\beta} = 0 \ (\alpha = 1, 2, \ldots, a ; \ \beta = 1, 2, \ldots, b).$$

Otherwise the matrix is called irreducible.

Next a few basic properties of non-negative irreducible matrices will be presented.

THEOREM 1.8. (see Gantmacher [5]). *A positive square matrix* $A = [a_{ij}]$ *always has a real and positive charateristic value* λ *which is a simple root of the characteristic equation and exceeds the moduli of all the other characteristic values. To this maximal characteristic value* λ *there corresponds a characteristic vector* $Z = (z_1, z_2, \ldots, z_n)$ *of A with* $z_i \geq 0$, i - $1, 2, \ldots, n$.

Next another theorem which investigates the spectral properties of irreducible non-negative matrices will be given.

THEOREM 1.9. (see [5]). *A irreducible non-negative matrix* $A = [a_{ij}]$ *always has a positive charateristic value* λ *that is a simple root of the characteristic equation. The moduli of all the other characteristic values do not exceed* λ. *To this* λ *there corresponds a characteristic vector with positive co-ordinates.*

When the matrix A has r characteristic values $h_0 = \lambda, h_1, \ldots, h_{r-1}$, *then these numbers are all distinct and are roots of the equation :*

$$h^r - \lambda^r = 0.$$

The maximal eigenvalue λ satisfies the main inequality:

$$\min_{1 \leq i \leq N} \frac{(Ax)_i}{x_i} \leq \lambda \leq \max_{1 \leq i \leq N} \frac{(Ax)_i}{x_i}$$

where $x = [x_1, x_2, \ldots, x_N]$, $x \geq 0$, $x \neq 0$.

The matrix A, irreducible and non-negative ($A \geq 0$) can be written in the following form:

$$AZ = \lambda Z; \quad Z \geq 0.$$

1.2.4. THE GEOMETRIC TRANSFORMATION (THE Z-TRANSFORM)

Let $f(t)$ be a discrete function (see Figure 1.3) which can take any real value, positive or negative, for $t \geq 0$. Assume that $f(t) = 0$ for any $t > 0$.

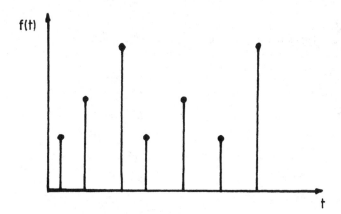

Figure 1.3. Example of a discrete function.

DEFINITION 1.7. The geometric transform is defined by:

$$f^g(z) = f(0) + f(1)z + f(2)z^2 + \dots = \sum_{t=0}^{\infty} f(t)z^t .$$

A discrete function allows a geometric transform,

$$\lim_{t \to \infty} \frac{f(t)}{c^t} = 0,$$

if an arbitrary c exists.

Between the $f(t)$ and the corresponding geometric transformation, the following relation exists:

$$f(t) = \frac{1}{t!} \frac{d^t}{dz^t} f^g(z) \bigg|_{z=0}$$

In Table 1.1(a) a few geometric transformations for the functions utilized in the present work are given.

	Discrete function	Geometric transform
1.	$f(t)$; $t \geq 0$	$f^g(z) = \sum_{t=0}^{\infty} f(t)z^t$
2.	$\delta(t) = \begin{cases} 1 \text{ for } t = 0 \\ 0 \text{ for } t > 0 \end{cases}$	1
3.	$u(t) = 1$	$(1-z)^{-1}$
4.	$af(t)$, where a is a constant	$a f^g(z)$
5.	$\sum_i f_i(t)$	$\sum_i f_i^g(t)$
6.	$\sum_{m=0}^{t} f_1(m) f_2(t-m)$ — convolution	$f_1^g(z) f_2^g(z)$
7.	$f(f(t))$	$z \; d/dx \, f^g(z)$
8.	$a^t(f(t)$	$f^g(tz)$
9.	a^t	$(1-az)^{-1}$
10.	ta^t	$az(1-az)^{-2}$
11.	t	$z(1-z)^{-t}$
12.	$(t+1) f(t+1)$	$d/dz \, f^g(z)$
13.	$(t+1) a^t$	$(1-az)^{-2}$
14.	$t+1$	$(1-z)^{-2}$

15. $M(t)$ $(t \geq 0)$ — matricial function $M^g(z) = \sum_{t=0}^{\infty} M(t)z^t$

16. A^t $\qquad\qquad\qquad\qquad\qquad$ $[1 - Az]^{-1}$

17. $t A^t$ $\qquad\qquad\qquad\qquad\qquad$ $z [1 - Az]^{-1} A [1 - Az]^{-1}$

Table 1.1(a).

1.2.5. EXPONENTIAL TRANSFORMATION (LAPLACE TRANSFORM)

The exponential transformation for a continuous function $f(\cdot)$ is defined by:

$$f^e(s) = \int_0^\infty f(t) e^{-st} dt.$$

In Table 1.1(b) a few exponential transforms for functions utilized in the present work are given.

	Continuous function	Exponential transform
1.	$f(t)$	$\int_0^\infty f(t) e^{-st} dt$
2.	$\delta(t)$	1
3.	$u(t) = 1$ – step-function	1
4.	$f(t - \tau)$ $(\tau > 0)$	$e^{-s} f^{e\,\tau}(s)$
5.	$f(t + \tau)$ $(\tau < 0)$	$e^{-s\tau} [f^e(s) - \int_0^\tau e^{-st} f(t) dt]$
6.	e^{-at}	$(s + a)^{-1}$
7.	$t e^{-at}$	$1/(s + a)^2$
8.	$1/2 (t^2 e^{-at})$	$1/(s + a)^3$
9.	$M(t)$ — matrix function	$\int_0^\infty e^{-st} M(t) dt$
10.	$t M(t)$	$-\dfrac{d}{ds} M^e(s)$
11.	e^{At}	$[sI - A]^{-1}$

Table 1.1(b).

1.3 Transient and recurrent processes

1.3.1. TRANSIENT PROCESSES

A finite Markov chain cannot have all non-recurrent (transient) states; it follows that at least one state is recurrent. If the chain does not have a non-recurrent state (e.g. transient), it is called *recurrent*. A recurrent chain whose states form a single class is called *irreducible* or *ergodic*.

   If the chain *has* recurrent states, it is called *absorbant*. An absorbant chain with only one aperiodic recurrent class is called *non-decomposible*.

   In Markov processes which contain transient states (called transient Markov processes), time and the number of transitions necessary to enter a recurrent state from any of the transient states can be defined and are random variables. When only one recurrent state exists, then this is an absorbant one. The transition probability matrix for a transient process has the form:

$$P = \begin{bmatrix} 0.85 & 0.15 & 0 \\ 0 & 0.75 & 0.25 \\ 0 & 0 & 1.0 \end{bmatrix}$$

A graphical representation is given in Figure 1.4.

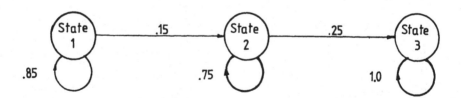

Figure 1.4. Transient process.

   In *transient processes,* transitions among the states of the systm are taking place until the system ends up in an absorbing state, if this exists. To analyse the behaviour of such processes we can use cumulative probability functions. If $p(t)$ is the probability that the system has a stable behaviour at transition t, the cumulative distribution $\phi(t)$ has the form:

$$\phi(t) = \sum_{s=0}^{t} p(s) \ ; \ t = 0, 1, 2, \dots$$

The corresponding geometric transform is:

$$\phi^g(z) = (1 - z)^{-1} p^g(z)$$

and, respectively:

$$p^g(z) = (1 - z) \phi^g(z).$$

Within transient processes, of practical interest, are the k-th order ($k = 1, 2, \dots$) moments of the distribution of random variables.

For the case of transient processes, we suppose that:

$$p^g(z) = \sum_{t=0}^{\infty} p(t) z^t \text{ and } p^g(1) = \sum_{t=0}^{\infty} p(t) = 1$$

then, by successive differentiation, we have:

$$\frac{d}{dz} p^g(z) = \sum_{t=0}^{\infty} t p(t) z^{t-1},$$

$$\frac{d}{dz} p^g(z) = \sum_{t=0}^{\infty} t(t-1) p(t) z^{t-2},$$

.
.
.

For $z = 1$, the above relation becomes:

$$\frac{d}{dz} p^g(z) \Big|_{z=1} = p^g(1) = \sum_{t=0}^{\infty} t p(t) = \bar{v}$$

$$\frac{d^2}{dz^2} p^g(z) \Big|_{z=1} = p^{g''}(1) = \sum_{t=0}^{\infty} (t^2 - t) p(t) = \overline{v^2} - \bar{v}.$$

Variance $\check{v}$, is:

$$\check{v} = \overline{v^2} - (\bar{v})^2 = p^{g''}(1) - p^{g'}(1) - [p^{g'}(1)]^2.$$

For Markov processes, we can identify virtual and real transitions [9], [17].

In the case when the system makes a transition from a state i in the same state i, in practice, the system remains in the same state; this transition is called a *virtual transition*. *Real* transitions are non-virtual transitions.

After a real transition takes place, a transition of the system from one state to the other is happening. Markov processes with real transitions have all the elements from the main diagonal in P matrices equal to zero.

Therefore:

$$p_{ij}^{r} = \begin{cases} \dfrac{p_{ij}}{1 - p_{ij}} \; ; & j \neq i \; ; \\ \\ 0 & j = i \, , \end{cases}$$

where the index r is for real transition processes but $P^r = [p_{ij}^r]$.

The limiting state probability vector associated with this is:

$$\pi^{r} = [\pi_{i}^{r} \; ; \; i = 1, 2, \dots , N].$$

The elements of the vector π^r and π satisfy, respectively, the equations:

$$\sum_{i=1}^{N} \pi_{i}^{r} p_{ij}^{r} = \pi_{j}^{r} \; ; \quad \sum_{i=1}^{N} \pi_{i} p_{ij} = \pi_{j} \, .$$

Taking into account the relation which defines p_{ij}^{r} for real transient processes, we have:

$$\sum_{i=1}^{N} \pi_{i}^{r} \frac{p_{ij}}{1 - p_{ii}} = \pi_{j}^{r} \, ,$$

or, finally:

$$\sum_{i=j}^{N} \pi_{i}^{r} \frac{p_{ij}}{1 - p_{ii}} p_{ij} = \frac{\pi_{j}^{r}}{1 - p_{ij}} \, .$$

Every π_i must be proportional to the quantity $\pi_{i}^{r} / (1 - p_{ii})$ and, therefore:

$$\pi_i = k \, \pi_i^r \frac{1}{1 - p_{ii}},$$

where:

$$k = \frac{1}{\sum_{i=1}^{N} \pi_i^r \, \overline{\tau}_i},$$

or, finally:

$$\pi_i = \frac{\pi_i^r \, \overline{\tau}_i}{\sum_{i=1}^{N} \pi_i^r \, \overline{\tau}_i} \; ; \quad i = 1, 2, \ldots, N$$

where $\overline{\tau}_i$ is the mean time spent by the process in state i.

1.3.2. THE STUDY OF RECURRENT STATE OCCUPANCY IN MARKOV PROCESSES

A class of states is, by definition, that class of states which communicate among themselves, with the property that as soon as the process enters any of them it cannot leave this class of states. Any Markov process has at least one class of states.

The transient states are associated with a class. It is possible for the process to enter this class from a transient state. A transient state must be associated with at least one class [47].

If we exclude the transient states from the definition of a Markov chain, then all states from one class must have a non-zero probability of being occupied, after a large number of transitions have taken place. As has already been mentioned, these states are called recurrent states. The recurrent states for Markovian processes have important statistical properties. The corresponding computational relations will be given next.

We define the quantity $x_{ij}(t)$ equal to one, if the system which started from state i arrived in state j at time t, and equal to zero in any other situation:

$$x_{ij}(t) = \begin{cases} 1, & \text{if } S(t) = j \\ 0, & \text{otherwise} \end{cases},$$

conditioned that $S(0) = i$.

LEMMA 1.8. *If* $v_{ij}(t)$ *is a.random variable which indicates how often a state j is occupied up to the moment t under the condition that, at the moment zero, the system occupied state i, then :*

a) $v_{ij}(t) \sum_{m=0}^{t} x_{ij}(m)$;

b) *mean value for* $v_{ij}(t)$ *is:*

$$\bar{v}_{ij}(t) = \sum_{m=0}^{t} \bar{x}_{ij}(m) = \sum_{m=0}^{t} \phi_{ij}(m) ;$$

c) *second order moments for* $v_{ij}(t)$ *are:*

$$\overline{v_{ij}^2}(t) = 2 \sum_{m=0}^{t} \phi_{ij}(m) \, \bar{v}_{jj}(t-m) - \bar{v}_{ij}(t), \quad t \geq 0 ;$$

d) *the variance for* $v_{ij}(t)$ *of state occupancy is:*

$$\overset{\vee}{v}_{ij}(t) = \overline{v_{ij}^2}(t) - \left(\bar{v}_{ij}(t)\right)^2 ;$$

e) *if* $^*\bar{v}_{ij}(t) = v_{ij}(t) - \delta_{ij}$, $t \geq 0$;

$$^*\bar{v}_{ij}(t) = \bar{v}_{ij}(t) - \delta_{ij} = \sum_{m=1}^{t} \phi_{ij}(m), \quad t \geq 1.$$

$$^*\overline{v_{ij}^2}(k) = \overline{v_{ij}^2}(t) - 2\,\delta_{ij}\,\bar{v}_{ij}(t) + \delta_{ij}$$

$$= 2 \sum_{m=0}^{t} \phi_{ij}(m) \, \bar{v}_{ij}(t-m) - \bar{v}_{ij}(t) - 2\delta_{ij}\,\bar{v}_{ij}(t) + \delta_{ij}$$

$$= 2 \sum_{m=1}^{t} \phi_{ij}(m) \, ^*\bar{v}_{ij}(t-m) + ^*\bar{v}_{ij}(t) ; \quad t \geq 1.$$

1.4 Markovian populations

1.4.1. VECTORIAL PROCESSES WITH A MARKOVIAN STRUCTURE

Many practical cases exist when complex systems built up from a large number of independent units behave in such a way that special interest is paid to the evaluation of the common time trajectory of the system when a large number of Markovian trajectories overlap [79], [91], [94], [99], 101].

Next, we shall analyze systems made up of a large number of units which operate independently, the behaviour of each unit being decided by a Markov process. The basic characteristic of such a process is a vector which represents the population of each state at any moment in time. This kind of process is called a vector Markov process. Let $f_i(m)$ be the number of units introduced at moment m in state i of the Markov process which is characterized by the transition probability matrix P (see Figure 1.5).

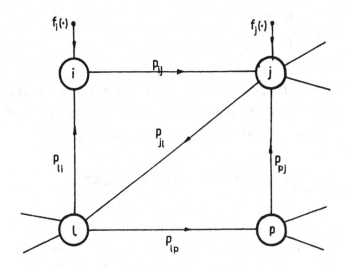

Figure 1.5. Vector Markov process.

Next, we shall assume that the input functions in a vector Markov process are known.

LEMMA 1.9. *The probability at time n, that k units will be in state j, if at the moment $m \leq n$, $f_i(m)$ units are introduced into state i, is denoted by $h_{ij}(k \mid m, n)$ and is given by the following relation :*

$$h_{ij}(k \mid m, n) \;=\; \binom{f_i(m)}{k} \left[\phi_{ij}(n-m) \right]^k \left[1 - \phi_{ij}(n-m) \right]^{f_i(m)}$$

$$0 \leq k \leq f_i(m),$$

where $\phi_{ij}(n-m)$ is the transition probability after $(n-m)$ steps (Howard [9]).

$$\phi_{ij}(n) = \sum_{m=1}^{n-1} f_{ij}(m)\,\phi_{jj}(n-m) + f_{ij}(n)\,; \quad n \geq 2$$

and:

$$f_{ij}(n) = p_{ij}\,\delta(n-1) + \sum_{\substack{k=1 \\ k \neq j}}^{N} p_{ik} f_{kj}(n-1)\,; \quad n \geq 1$$

$$\delta(n-1) = \begin{cases} 1 & \text{for } n = 1 \\ 0 & \text{for } n > 1. \end{cases}$$

COROLLARY 1.3. a) *The probability $h_{ij}(k \mid m)$ that there will be k units in node j at time n for a given input time pattern $f_i(\,\cdot\,)$ is given by the value at k of the convolution of each $h_{ij}(\,\cdot\mid m, n)$ related to m ($m \in [0, n]$), such that :*

$$h_{ij}(k \mid n) = \left[\underset{m=0}{\overset{n}{\times}} h_{ij}(\,\cdot\mid m, n) \right] (k)\,; \quad 0 \leq k \leq \sum_{m=0}^{n} f_i(m)$$

for the case when state i remains the only entrance state in the process ;

   b) *If a finite number of input states N of type i exist in the process, the probability $h_{ij}(k \mid n)$ that k units were in state j at time n, is given by the k-th convolution* [1]:

$$h_{ij}(k \mid n) = \left[\underset{i=1}{\overset{N}{\times}} \underset{m=0}{\overset{n}{\times}} h_{ij}(\,\cdot\mid m, n) \right] (k)\,; \quad 0 \leq k \leq \sum_{i=1}^{N} \sum_{m=0}^{n} f_i(m)\,.$$

Let $r(n) = [r_1(n), r_2(n), \dots , r_N(n)]$ be a vector, where $r_j(n)$ is the number of units in state j at time n of the system. The random variable $r_j(n)$ has the mean and variance:

$$\overline{r}_j(n) = \sum_{i=1}^{N} \sum_{m=0}^{n} f_i(m)\,\phi_{ij}(n-m)\,;$$

$$j = 1, 2, \dots , N\,; \quad n = 0, 1, 2, \dots$$

---

[1] If $f_1(m)$ and $f_2(t-m)$ are two discrete functions, then the convolution of functions $f_1(\cdot)$ and $f_2(\cdot)$ is

$$\sum_{n=0}^{t} f_1(m)\,f_2(t-m) \text{ with the geometric transform } f_1^g(z)\,f_2^g(z).$$

and, respectively [1]:

$$r_j^v(n) = \sum_{i=1}^{N} \sum_{m=0}^{n} f_i(m)\, \phi_{ij}(n-m)\, [1 - \phi_{ij}(n-m)]\,,$$

$$j = 1, 2, \dots, N\,;\quad n = 0, 1, 2, \dots$$

There are many practical situations when the entrance in any of the states i of a vector Markov process is, in turn, governed by a probability distribution, so that all inputs in the process can form a set of random variables. Let $f(\,\cdot\,)$ be the input function in all the nodes of the vector Markov process any moments considered. We assume the function is already known.

COROLLARY 1.5. *The mean and variance of the random variable $r_j(n)$, known as the $f(\,\cdot\,)$ function, are given by the relations* :

$$\langle r_j(n)\,|\,f(\,\cdot\,)\rangle = \bar{r}_j(n) = \sum_{i=1}^{N} \sum_{m=0}^{n} f_i(m)\, \phi_{ij}(n-m)$$

and respectively:

$$^v\langle r_j(n)\,|\,f(\,\cdot\,)\rangle = r_j(n)^2\,|\,f(\,\cdot\,)\rangle - \langle r_j(n)\,|\,f(\,\cdot\,)\rangle^2.$$

If the quantity $f(\,\cdot\,)$ is represented by the repartition $\Pr\{f(\,\cdot\,)\}$, this being the same for all the input functions, then we have

COROLLARY 1.6. *The mean and variance for $r_j(n)$ are given by* :

(a) $$\langle r_j(n)\rangle = \sum_{i=1}^{N} \sum_{m=0}^{n} f_i(m)\, \phi_{ij}(n-m)\,,$$

where $f_i(m)$ represents the mean number of units introduced in state i at moment m ;

(b) $$^v\langle r_j(n)\rangle = \langle (r_j(n))^2 \rangle - \langle r_j(n)\rangle^2\,,$$

where

---

[1]
   The variance is given by the relation:

$$r_j^v(n) = \sum_{i=1}^{N} \sum_{m=0}^{n} f_{ij}(m)\, v_{ij}(n-m),\ \text{for all } j\,;\ n \ge 0,$$

where: $v_{ij}(n) = \phi_{ij}(n)[1 - \phi_{ij}(n)]$; for all i ; $n \ge 0$.

$$\langle (r_j(n))^2 \rangle = \sum_{f(\cdot)} \langle (r_j(n)^2 \,|\, f(\,\cdot\,) \rangle \Pr \{f(\,\cdot\,)\}$$

$$= \sum_{i=1}^{N} \sum_{m=0}^{n} f_i(m) \, \phi_{ij}(n-m) \left(1 - \phi_{ij}(n-m)\right) +$$

$$+ \sum_{i=1}^{N} \sum_{k=1}^{N} \sum_{m=0}^{n} \sum_{r=0}^{n} f_i(m) f_k(r) \, \phi_{ij}(n-m) \, \phi_{kj}(n-r)$$

and:

$$\overline{f_i(m) f_k(r)} = (1 - \delta_{ik} \, \delta_{mr}) f_i(m) f_k(r) + \delta_{ik} \, \delta_{mr} \left(\overline{f_i(m)} \right)^2 .$$

COROLLARY 1.7. *The mean and variance of the quantity* $^c r_j(n)$ *defined as the total number of units which occupied node j in the* $(0, n)$ *interval* $\left(^c r_j(n) = \sum_{l=0}^{n} r_j(l) \right)$ *are given by the relations :*

$$\langle ^c r_j(n) \rangle = \sum_{i=0}^{N} \sum_{m=0}^{n} f_i(m) \, \overline{\eta}_{ij}(n-m),$$

$$^v \langle ^c r_j(n) \rangle = \sum_{i=1}^{N} \sum_{m=0}^{n} f_i(m) \, \overline{\eta}_{ij}^v(n-m),$$

where

$$\overline{\eta}_{ij}(n) = \sum_{m=0}^{n} \phi_{ij}(m) \quad \text{and}$$

$$\eta_{ij}^v(n) = \overline{\eta}_{ij}^2(n) - \left(\eta_{ij}(n) \right)^2 .$$

1.4.2. GENERAL BRANCHING PROCESSES

In real situations the size of a population can change as a consequence of phenomena which appear within the given population. These phenomena can be interpreted as Markovian processes. Generation zero induces the birth of other generations by means of direct descendents. This is the case with the nuclear chain reaction in a nuclear power reactor, or the cascade failures in a complex technical system which includes specific equipment as well as computerized automation and control sub-systems etc. We shall call this type of process, where a state generates new states, branching processes. They involve the notions of *generations* and *successors*; the size of a generation is an important indicator in the study of general branching processes.

If n indicates the index for a generation ($n = 1, 2, ...$) and k the number of direct descendents in generation n, produced by a member of the $(n - 1)$-th generation, then we can write:

$$p_n(k) = \Pr\{k_n = k\}.$$

If m_n is the size of the n-th generation, then the associated probability distribution function is (see Figure 1.6):

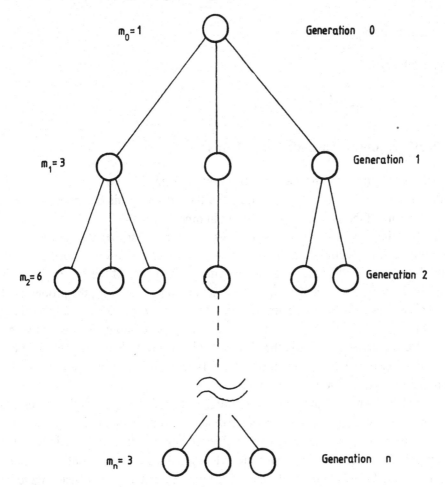

Figure 1.6.   Branching process.

$$q_n(m) = \Pr\{m_n = m\}$$

and, further, we can write:

$$q_{n|n-1}(s \mid r) = \Pr\{m_n = s \mid m_{n-1} = r\},$$

$$q_{n|n-1}(\cdot \mid r) = \mathop{\textstyle\bigtimes}_{r} p_n(\cdot),$$

where the parameters were defined above.

The set of random variables $\{m_n\}$ is called a *general branching process*.

COROLLARY 1.8. *The mean and variance for a general branching process are given by the following relations:*

$$\langle m_n \rangle = k_n m_{n-1},$$

and, respectively:

$$^n\langle m_n \rangle = {}^v\langle m_{n-1} \rangle k_n^2 + \langle m_{n-1} \rangle k_n^v .$$

1.5 Partially observable Markov chains

Generally speaking, for the case of Markov or semi-Markov processes, it is considered that there is an observer external to the system. This observer is able to identify the state of the system at any moment in time.

In practice, we meet many situations when an observer has only partial information on the system states. In many practical cases the observer does not have the possibility of directly observing the dynamic performance of the system. This is the case with the technological equipment of nuclear power stations, the problems of medical diagnoses and clinical treatment and political systems which are only partially observable. A Markov process becomes partially observable by removing the observer from the system; in turn, the observer will have to gather information on the system using external communication channels (e.g. distinct observations on the operational perfomance of the system).

A well known example, in the literature, of a partially observable process, is that determined by the state of illness of a patient (see Smallwood et al. [19], Sondik [20]). In Figure 1.7, an arbitrarily chosen disease is presented. It is modelled by a Markov process which contains four states. It is assumed that three states, "initial". "advanced" and "well" are internal states of the patient, and they are considered to be unobservable. It is further assumed that imperfect information on the state of the patient is obtained by performing certain tests. For each test, we have two outcomes. The probability that outcome θ occurs if the patient is in state i and test k is administered gives the relationship between the internal states and the outcomes for a partially observable Markov process [32], [38].

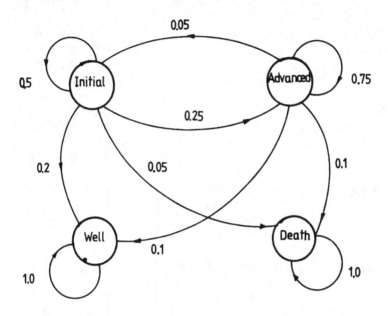

Figure 1.7. States of illness of a patient with an arbitrarily chosen disease.

In matrix form, we have a representation given by R^k (see Table 1.2). For each test, one outcome is probable if the patient is in the advanced state. In Table 1.2 we can also identify the matrices of transition probabilities in different alternatives k, $k = 1, 2, 3, 4, 5$. Under distinct therapies, we have changes in the state transition probabilities from alternative 1 (i.e. no treatment). If the patient is in the initial state of the disease and no treatment alternative is applied, the probability is 0.33 that the patient will die. If the patient is in an advanced state, the corresponding initial probability for this state will be 0.47 [20].

Very often it is not possible to solve a medical diagnosis problem by considering the system (a patient) as a completely observable one. Tests performed to obtain additional infomation concerning the state of the patient give the Markov process the character of a *partially observable process*.

Alternative k	P^k				R^k	
1. No treatment	0.5	0.25	0.2	0.05	1.0	1.0
	0.05	0.75	0.1	0.1	1.0	0
	0	0	1.0	0	1.0	0
	0	0	0	1.0	0	1.0
2. Test I	P^1				0.5	0.5	0
					0.1	0.9	0
					0.6	0.4	0
					0	0	1.0
3. Test II	P^1				0.8	0.2	0
					0.2	0.8	0
					0.9	0.1	0
					0	0	1.0
4. Therapy I	0.1	0.1	0.8	0	1.0	0
	0	0.6	0.4	0	1.0	0
	0	0	1.0	0	1.0	0
	0	0	0	1.0	0	1.0
5. Therapy II	0.1	0.65	0.2	0.05	1.0	0
	0.05	0.2	0.7	0.05	1.0	0
	0	0	1.0	0	1.0	0
	0	0	0	1.0	0	1.0

Table 1.2. Representation of R^k in matrix form.

Partially observable Markov processes are characterized by the core process and the observable process. These two random processes are later coupled and we obtain a Markov process defined by the probability transition matrix, by the inspection matrix, and the inspection and operating costs of the system considered.

1.5.1. THE CORE PROCESS

A Markov process itself represents the core of a partially observable process. Next, as was already mentioned, we shall consider two discrete Markov processes with a finite number of states given by 1, 2, ... , N. The states occupied by the system at time t are expressed by $S(t)$. The probability transition matrix is defined by $P = [p_{ij}]$, where:

$$p_{ij} = \Pr\{S(t+1) = j \mid S(t) = i, \ S(t-1) = l, \ldots, \mathcal{E}(t)\}$$

$$= \Pr\{S(t+1) = j \mid S(t) = i, \ \mathcal{E}(t)\} \tag{1.61}$$

$$i = 1, 2, \ldots N; \quad j = 1, 2, \ldots, N,$$

where $\mathcal{E}(t)$ represents all past information or experience through time t. Relation (1.61) indicates that the system will be in state j at time $(t+1)$, conditional on that it had been in state i at time t. In Figure 1.8 an evolutionary graph for a Markov process with $N = 4$ is presented.

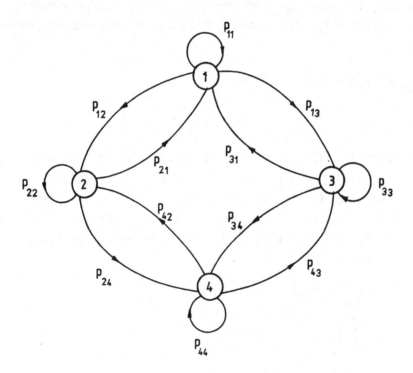

Figure 1.8. Evolutionary graph for a Markov process with $N = 4$.

As was already discussed, all information concerning the past evolution of a Markov process is considered irrelevant to the analysis (Markov processes are considered as memory less processes) and it is possible to compute its future evolution by taking the power of the matrix P. The observations of the process there were not taken into consideration.

When we deal with a semi-Markov process (see §1.3.3), as well as the P matrix, we have to consider storing the probability matrix $H(m) = [h_{ij}(m)]$ where $i, j = 1, 2, \ldots , N$ and $h_{ij}(m) = \Pr\{\tau_{ij} \geq m\}$, where τ_{ij} is an integer positive number indicating the waiting time of the system in state i before making a real transition to state j.

1.5.2. THE OBSERVATION PROCESS

Next, we shall introduce a general model for the observation process which by particularization can degenerate into a completely observable system. We make the assumption that when an external observer examines the process at time t he sees one, and only one, of a finite set of states (e.g. outputs). We consider that there are a finite number, M, of such outputs labelled $1, 2, \ldots , M$. If the output at time t is $Z(t)$, then $Z(t)$ is related to the unknown state of the Markov process $S(t)$ by the probability quantity:

$$r_{j\theta} = \Pr\{Z(t) = \theta \mid S(t) = j\},$$

$$j = 1, 2, \ldots , N ; \quad \theta = 1, 2, \ldots , M.$$

Then the $(N \times M)$ matrix $R = [r_{j\theta}]$ completely describes the observation (output) process. If $N = M$ the corresponding Markov process is completely observable and then $R = I$, where I is the identiy matrix. When $M = 1$, the associated Markov process is completely unobservable and $r_{j1} = 1$ for the core process. The P and R are stochastic matrices and we can write, respectively:

$$p_{ij} \geq 0; \sum_{j=1}^{N} p_{ij} = 1 ; i = 1, 2, \ldots , N ;$$

$$r_{j\theta} \geq 0; \sum_{\theta=1}^{M} r_{j\theta} = 1 ; j = 1, 2, \ldots , N .$$

In Figure 1.9 a graphical representation of a partially observable Markov process is given; the states and outputs (observations) are emphasized. Probabilities which are contained in the R-matrix are also pertinent to partially observable semi-Markov processes.

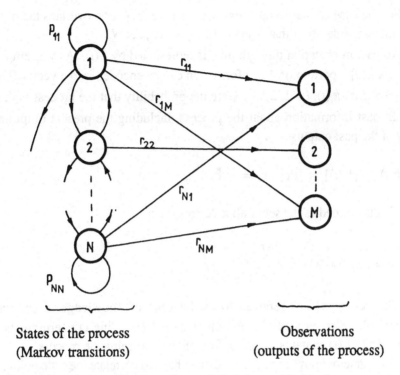

States of the process   Observations
(Markov transitions)   (outputs of the process)

Figure 1.9. Graphical representation of a partially observable Markov process.

1.5.3. THE STATE OF KNOWLEDGE AND ITS DYNAMICS

For the case of completely observable Markov and semi-Markov processes, the state
in which it can be at a given moment can be immediately accessible to an external
observer for any moment t. This can be easily understood by examining Figure
1.10a (Markov processs) and Figure 1.10b (semi-Markov processes).

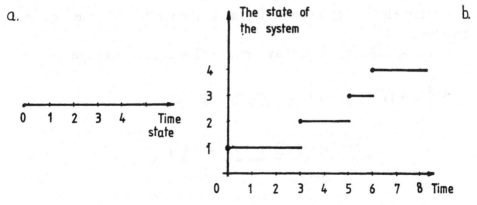

Figure 1.10 (a and b). (a) Markov process; (b) semi-Markov process.

For the case of Markov processes, their time evolution overlaps the transition, taking into consideration that $h_{ij}(1) = 1$; $i, j = 1, 2, \ldots, N$.

All past information through time is represented by $\mathcal{E}(t)$. As was emphasized by Sondik [20] a portion of this information can be encoded into a vector form, $\pi(t)$ whose elements $\pi_i(t)$, $i = 1, 2, \ldots, N$ are the probability that the process is in state i given all past information about the process, including the present output and the history of the past outputs:

$$\pi(t) = [\pi_i(t)] = [\Pr\{S(t) = i \mid \mathcal{E}(t)\}].$$

If Π is defined as the set of all π vectors, then:

$$\Pi = \{\pi_i ; \ \pi_i \geq 0 ; \quad \sum_{i=1}^{N} \pi_i = 1\} .$$

The vector $\pi(t)$ is referred to as the state of knowledge of the partially observable Markov process. It is only a partial encoding concerning the above process, while $\mathcal{E}(t)$ contains all past information concerning the process.

The state of knowledge dynamics will have to correlate the state of knowledge at time t, given by $\pi(t)$ with the corresponding vector at time $(t + 1)$ when we have to deal with a partially observable Markov process. For this it is necessary to explicitly deduce the observations given at $Z(\cdot)$; let us suppose that $\Pi(t)$ represents the state of knowledge on the process after event $Z(t)$ has been observed. Next, it will be necessary to evaluate the dynamic correlation between $S(t)$, $\mathcal{E}(t)$, and $\pi(t)$, taking into account that:

$$\mathcal{E}(t + 1) = [\mathcal{E}(t), Z(t + 1)] ;$$

the information in $\mathcal{E}(t + 1)$ is then encoded as the vector $\pi(t + 1)$ of state occupancy probabilities.

The proof is given by using Bayes' rule (see Sondik [20]), such that:

$$\Pr\{S(t + 1) = j \mid \mathcal{E}(t + 1) = [\mathcal{E}(t), Z(t + 1) = \theta]\} =$$

$$= \frac{\Pr\{S(t + 1) = j, Z(t + 1) = \theta \mid \mathcal{E}(t)\}}{\Pr\{Z(t + 1) = \theta \mid \mathcal{E}(t)\}} =$$

$$= \Pr\{Z(t+1) = \theta \mid \mathcal{E}(t)\}^{-1} \left[\sum_{i=1}^{N} \Pr\{S(t) = i, S(t+1) = j, \right.$$

$$\left. Z(tI+1) = \theta \mid \mathcal{E}(t) \right] = \Pr\{Z(t+1) = \theta \mid \mathcal{E}(t)\}^{-1} \cdot$$

$$\cdot \sum_{i=1}^{N} \Pr\{S(t) = i \mid \mathcal{E}(t)\} \Pr\{S(t+1) = j \mid S(t) = i, \mathcal{E}(t)\} \cdot$$

$$\cdot \Pr\{Z(t+1) = \theta \mid S(t+1) = j, S(t) = i, \mathcal{E}(t)\} =$$

$$= \Pr\{Z(t+1) = \theta \mid \mathcal{E}(t)\}^{-1} \sum_{i=1}^{N} \pi_i(t) \, p_{ij} \, r_{j\theta} \cong$$

$$\cong T_i(\pi \mid \theta).$$

In a vector form we obtain the probability state vector at moment $(t+1)$ conditioned by the observation θ and the previous state of knowledge given by π, so that:

$$T(\pi \mid \theta) = [T_i(\pi \mid \theta); \quad i = 1, 2, \dots, N].$$

By definition, the probability that the next output $Z(t+1)$ will be θ given the current state of knowledge $\mathcal{E}(t)$ is given by the relation:

$$\Pr\{Z(t+1) = \theta \mid \mathcal{E}(t)\} = \sum_{i=1}^{N} \sum_{j=1}^{N} \pi_i(t) \, p_{ij} \, r_{j\theta},$$

and, for the sake of simplicity, we shall use the notation:

$$\{\theta \mid \pi\} = \Pr\{Z(t+1) = \theta \mid \mathcal{E}(t)\} = \sum_{i=1}^{N} \sum_{j=1}^{N} \pi_i(t) \, p_{ij} \, r_{j\theta}.$$

Functions $T(\pi \mid \theta)$ and $\{\theta \mid \pi\}$ will generally be written in terms of vectors and matrices, as follows:

$$\{\theta \mid \pi\} = \pi \, PR_\theta \, \mathbf{1},$$

and

$$T(\pi \mid \theta) = \frac{\pi \, PR_\theta}{\{\theta \mid \pi\}}$$

where $R_\theta = \text{diag } [r_{i\theta}]$, $\mathbf{1} = [1, 1, \ldots , 1]^T$, $\pi = \lim_{t \to \infty} \pi(t)$ and the superscript T denotes transpose. These functions define the dynamics of the state of knowledge of the partially observable Markov process.

If the only source of information from the process is a sequence of outputs, and the state of knowledge $\pi(t)$ is computed after each output is received, then:

$$\Pr\{\pi(t+1) \in A \mid \mathcal{E}(t)\} = \Pr\{\pi(t+1) \in A \mid \pi(t), \pi(t+1), \ldots , \pi(0)\}$$

$$= \Pr\{\pi(t+1) \in A \mid \pi(t)\}$$

$$= \sum_{\theta : T(\pi \mid \theta) \in A} \{\theta \mid \pi\} ,$$

where A is an arbitrary set in Π.

The sequence $[\pi(t), t \geq 0]$ is a Markov process. Dynamics for such a process when $N = 3$ and $M = 2$ is represented in Figure 1.11.

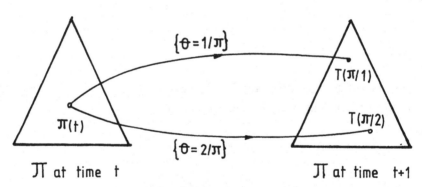

Figure 1.11. Dynamics for a Markov process.

If we are dealing with a semi-Markov process and we can associate it to a finite observation space with cardinal M, then the state dynamics are given by [32], [70]:

$$T_j(\pi \mid \theta, n, k) = \frac{\sum_{i=1}^{N} \pi_i f_{ij}^k(n) r_{j\theta}^k}{\sum_{i=1}^{N} \sum_{j=1}^{N} \sum_{\theta=1}^{M} \pi_i f_{ij}^k(n) r_{j\theta}^k} ,$$

where $f_{ij}(n)$ represents the first passage time for a semi-Markov process with a discrete structure (see Howard [9]) and is given by:

$$f_{ij}(n) = \sum_{\substack{r=1 \\ r\neq j}}^{N} \sum_{m=0}^{n} p_{ir} h_{ir}(m) f_{rj}(n-m) + p_{ij} h_{ij}(n)$$

$$f_{ij}(0) = \begin{cases} 1, & \text{if } i \neq j \\ 0 & \text{otherwise,} \end{cases}$$

$$i, j = 1, 2, \dots, N.$$

The above relation can be written in vector form as:

$$T(\pi \mid \theta, n, k) = [T_j (\pi \mid \theta, n, k); \quad j = 1, 2, \dots, N]$$

$$= \pi F^k(n) R_\theta^k / \{\theta \mid \pi, n, k\}$$

where the denominator (a scalar) implies a matrix multiplication of the form $(1 \times N)(N \times N)(N \times M)(M \times 1)$ and is given by the relation:

$$\{\theta, n, k\} = \pi F^k(n) R_\theta^k \mathbf{1}.$$

In a matrix form, the first time probabilities for a core — semi-Markov process, can be written as:

$$F(n) = \sum_{m=0}^{n} (P \ \Box \ H(m)) F(n-m) + (P \ \Box \ H(n))$$

for $n \geq 0$, and $\Box$ is the symbol for congruent multiplication of two matrices.

It is obvious that the case of a partially observable semi-Markov process can degenerate into a partially observable Markov process when $H(m) = I$, and then $F(n) = P^k$. In the above relation k represents a decision strategy for a decision maker [70].

1.5.4. EXAMPLES

EXAMPLE 1. Diagnosis and treatment of the respiratory system.

Like all physiological processes, the respiratory system exhibits complex dynamic behaviour. In simplistic terms, the system can be regarded as a gas exchange plant in which blood acts as a carrier of gases from the lungs to the tissues and vice versa. In the lungs, blood becomes oxygenated and this extra oxygen is conveyed to the tissues which, in exchange, give up CO_2 to the blood. The blood is then returned to the lungs for further oxygenation. The process of releasing this CO_2 to the atmosphere and the take-in of fresh oxygen is achieved by breathing. A pictorial representation is given in Figure 1.12 (see [69], [70]).

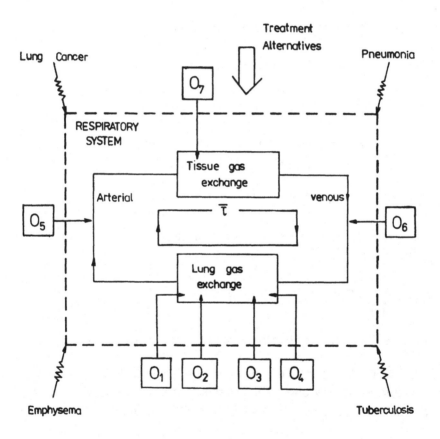

O_i – an observable output O_5 – arterial gas content
O_1 – rate of breathing O_6 – venous gas content
O_2 – tidal volume O_7 – tissue function
O_3 – heart rate
$\bar{\tau}$ – blood circulation time in respiratory system

Figure 1.12. Schematic representation of the respiratory system.

In carrying out the diagnosis, the clinician aims at ensuring that the outcome of his decision is satisfactory both to the patient and to himself. During this process, three stages have to be considered, (i) taking measurements, (ii) reaching a conclusion as to the state of the process and (iii) taking a decision regarding any course of action (e.g. do nothing, apply medication, change enviornment).

Let us consider the respiratory system with $N = 3$ states (ill, satisfactory and well) and $M = 4$ observable outputs (low peak ventilation, i.e. less than normal but greater than $100 \ l \ min^{-1}$, and very low peak ventilation, i.e. less than $100 \ l \ min^{-1}$ etc.). Let us consider that the number of treatment alternatives is $k = 1, 2$. Alternative 1 ($k = 1$) is to apply medication and alternative 2 is to "do nothing". The following transitions represent the state transition probabilities:

Decision alternative 1

$$p^1_{ij} = \begin{bmatrix} 0.1 & 0.1 & 0.8 \\ 0.2 & 0.5 & 0.3 \\ 0.7 & 0.1 & 0.2 \end{bmatrix}$$

Decision alternative 2

$$p^2_{ij} = \begin{bmatrix} 0.1 & 0.8 & 0.1 \\ 0.7 & 0.1 & 0.2 \\ 0.1 & 0.9 & 0 \end{bmatrix}$$

where state 1 = ill, 2 = satisfactory and 3 = well.

The initial state vector is $\pi = [0.3 \quad 0.2 \quad 0.5]$.

Random inspection (clinical observations) probabilities are:

$k = 1$

$$r^1_{j\theta} = \begin{bmatrix} 0.42 & 0.35 & 0.21 & 0.18 \\ 0.06 & 0.05 & 0.63 & 0.54 \\ 0.24 & 0.20 & 0.42 & 0.36 \end{bmatrix}$$

$k = 2$

$$r^2_{j\theta} = \begin{bmatrix} 0.12 & 0.10 & 0.56 & 0.16 \\ 0.24 & 0.20 & 0.42 & 0.12 \\ 0.18 & 0.15 & 0.49 & 0.14 \end{bmatrix}$$

The dynamics of the new state can be calculated using computational relations from this chapter. The results for the two decision alternatives are shown in Tables 1.3 and 1.4.

Decision alternative $k = 1$

$T_i\left(\pi\mid\theta,1\right)$	State j		
	1	2	3
$\theta = 1$	0.622	0.038	0.340
$\theta = 2$	0.622	0.039	0.339
$\theta = 3$	0.239	0.307	0.454
$\theta = 4$	0.239	0.307	0.454

Table 1.3. Decision alternative $k = 1$.

Decision alternative $k = 2$

$T_i\left(\pi\mid\theta,2\right)$	State j		
	1	2	3
$\theta = 1$	0.126	0.814	0.060
$\theta = 2$	0.126	0.814	0.060
$\theta = 3$	0.270	0.654	0.076
$\theta = 4$	0.270	0.654	0.076

Table 1.4. Decision alternative $k = 2$.

EXAMPLE 2. Availability prediction function for partially observable technical systems with semi-Markov behaviour.

A partially observable semi-Markov process is important for modelling a nuclear reactor. As an example, we can choose a BWR system where no direct information about the state of the reactor or sub-systems is available unless the block reactor turbine is stopped. A suitable waiting time is necessary before inspection and maintenance procedures (see Figure 1.13).

We can build up more powerful models for realiabilty computation by defining a set of independent observations using information given by instruments in the control room , as well as the previous experience of operators on the reactor dynamics.

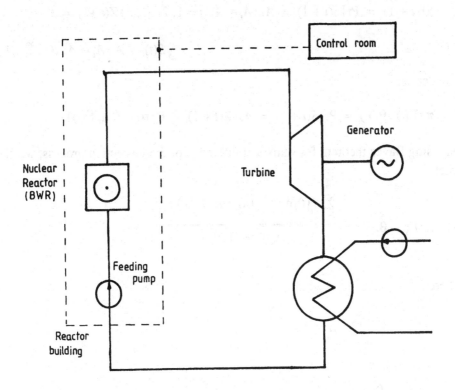

Figure 1.13. Diagrammatic representation of a nuclear reactor.

The system dynamics are described by a finite number N of operational states; the probability transition matrix is $P = [p_{ij}]$; the waiting time probability matrix is $H(m) = [h_{ij}(m)]$ and the random inspection matrix is $R = [r_{j\theta}]$. From considerations concerning the system reliability computation, the state space of the system is partitioned as:

- completely operational states $A(A_i \in A; \ i = 1, 2, \dots, r)$,
- partially operational states $F(F_j \in F; \ j = 1, 2, \dots, m)$,
- non-operational states $C(C_k \in C, \ k = 1, 2, \dots, s)$,

such that $r + m + s = N$ and the state space of the system is given by $S = A \cup F \cup C$.

Let us suppose that from a previous inspection, we can have *a priori* information on the system states distribution:

$$\Gamma(t) = [\gamma_1(t), \gamma_2(t), \dots , \gamma_N(t)].$$

Availability for a partially observable system for the $(t+1)$-th transition is given by the relation:

$$\mathcal{D}^j(t+1) = \Pr\{S(t+1) = A_j,\ A_j \in A,\ j = 1, 2, \dots , r \mid Z(t+1) = \theta,$$

$$S(t) = A_i,\ A_i \in A,\ \tau(t),\ \mathcal{E}(t)\}.$$

If we define:

$$\pi_j(\Gamma(t) \mid \theta, n) = \Pr\{S(t+1) = A_j \mid Z(t+1) = \theta,\ \tau(t) = n,\ \mathcal{E}(t)\},$$

then using Bayes theorem for statistical inference of the inspection process, we can write:

$$\pi_j(\Gamma(t) \mid \theta, n) = \frac{\displaystyle\sum_{i=1}^{N} \gamma_i(t)[\delta_{ij}\ ^{>}w_i(n) + p_{ij}\, h_{ij}(n)\, r_{j\theta}]}{\psi(\theta, n \mid \Gamma(t))}$$

where

$$^{>}w_i(n) = \sum_{m=n+1}^{\infty} p_{ij}\, h_{ij}(m)$$

$$\delta_{ij} = \begin{cases} 0, & \text{if } i \neq j \\ 1, & \text{if } i = j, \end{cases}$$

$$\psi(\theta, n \mid \Gamma(t)) = \sum_{i=1}^{N}\sum_{j=1}^{N} \gamma_i(t)[\delta_{ij}\ ^{>}w_i(n) + p_{ij}\, h_{ij}(n)\, r_{j\theta}].$$

Short term system availability, defined by the above relation, is computed with the following relation:

$$\mathcal{D}^j(t+1) = \pi_j(\Gamma(t) \mid \theta, n, \Delta(t));$$

$$j = 0, 1, \dots , r;\quad \theta = 1, 2, \dots , M,$$

and $\Delta(t)$ represents a maintenance decision on the system at transition t.

In matrix form, the availability is written as:

$$D(t + 1) = \pi(\Gamma(t) \mid \theta, n, \Gamma(t))$$

where:

$$\pi(\cdot \mid \cdot) = \frac{\Gamma(t)[^>W(n) + (P \; \square \; H(n)) \, R_\theta]}{\psi(\theta, n \mid \Gamma(t))}$$

and $^>W(n)$ is a diagonal matrix of waiting times.

Later a computer prográm for computing availability of partially observable technical systems, using the previous formalism, is given (see Figure 1.22).

In the same content, let us now suppose that we have to give an answer concerning the reliability of a car gear box.

After a detailed analysis on the operation dynamics of the gear box we can identify five states. A general description of them is as follows:

State 1: The device operates properly.
State 2: Oil filling opening of the gear box is not accessible.
State 3: Failed gear tooth in the gear sub-system.
State 4: Excessive heat in the gear box due to inadequate heat transfer.
State 5: The level of oil in the gearbox is below the prescribed limit.

Pertinent observations are as follows:

Observation 1: Abnormal noise at the gear box or at the engine.
Observation 2: High temperature in the lubricating system, indicated ashboard of the car.

The probability transition matrix is given by:

$$P = \begin{bmatrix} 0.83 & 0.08 & 0.05 & 0.02 & 0.02 \\ 0.74 & 0.10 & 0.03 & 0.11 & 0.02 \\ 0.18 & 0.15 & 0.45 & 0.15 & 0.07 \\ 0.19 & 0.27 & 0.02 & 0.40 & 0.12 \\ 0.21 & 0.16 & 0.07 & 0.19 & 0.43 \end{bmatrix}$$

and the waiting time probabilities in each state are given by the matrix:

$$H(\cdot) = \begin{bmatrix} 0.57 & 0.25 & 0.11 & 0.05 & 0.02 \\ 0.53 & 0.22 & 0.13 & 0.07 & 0.05 \\ 0.52 & 0.10 & 0.20 & 0.08 & 0.10 \\ 0.45 & 0.11 & 0.15 & 0.20 & 0.09 \\ 0.42 & 0.13 & 0.21 & 0.11 & 0.13 \end{bmatrix}$$

The stochasitic matrix for the system is:

$$R = \begin{bmatrix} 0.5 & 0.5 \\ 0.06 & 0.94 \\ 0.92 & 0.08 \\ 0.06 & 0.94 \\ 0.13 & 0.87 \end{bmatrix}$$

The state probability vector is:

$$\Gamma(\cdot) = [0.85 \quad 0.04 \quad 0.06 \quad 0.04 \quad 0.01].$$

The block diagram for the system reliability computation is presented in Figure 1.14.

If we consider State 1 as a perfect operational state, then short term reliability of the system under the above conditions is equal to 0.85.

Next, we shall consider as dominant states of the system only those denoted by 1 to 4. The system reliability is computed this time for the situation when the system is perfectly observable. Then the stochastic inspection matrix is the unity matrix. The probability transition matrix is of the form:

$$P = \begin{bmatrix} 0.91 & 0.04 & 0.03 & 0.02 \\ 0.80 & 0.09 & 0.07 & 0.04 \\ 0.15 & 0.18 & 0.54 & 0.13 \\ 0.30 & 0.25 & 0.07 & 0.38 \end{bmatrix}$$

and the waiting time probabilities in different states is given in Table 1.5.

In Figure 1.15, the evolution of the system reliability index is presented by considering State 1 as a completely operational state.

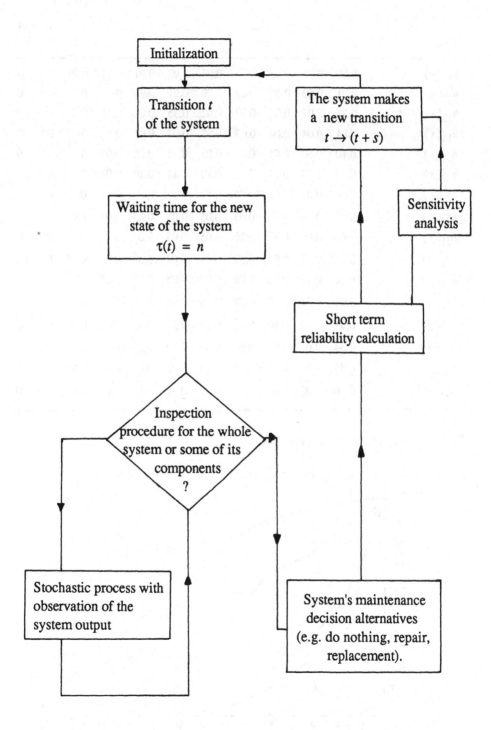

Figure 1.14.   Block diagram for system reliability computation.

State \ Time n	1	2	3	4	5	6	7	8	9	10	11
$h_{11}(\cdot)$	0.80	0.05	0.04	0.03	0.03	0.02	0.01	0.01	0.01	0	0
$h_{12}(\cdot)$	0.01	0.03	0.05	0.07	0.09	0.21	0.54	0	0	0	0
$h_{13}(\cdot)$	0.15	0.07	0.02	0.02	0.02	0.35	0.2	0.15	0.2	0	0
$h_{14}(\cdot)$	0.84	0.04	0.04	0.03	0.02	0.02	0.01	0	0	0	0
$h_{21}(\cdot)$	0.10	0.60	0.05	0.05	0.05	0.05	0.05	0.05	0	0	0
$h_{22}(\cdot)$	0.07	0.07	0.12	0.15	0.20	0.27	0.10	0.02	0	0	0
$h_{23}(\cdot)$	0.04	0.07	0.13	0.76	0	0	0	0	0	0	0
$h_{24}(\cdot)$	0.50	0.10	0.10	0.05	0.05	0.05	0.05	0.05	0.05	0	0
$h_{31}(\cdot)$	0.40	0.05	0.30	0.10	0.05	0.05	0.05	0	0	0	0
$h_{32}(\cdot)$	0.15	0.07	0.02	0.02	0.02	0.02	0.35	0.20	0.15	0	0
$h_{33}(\cdot)$	0.10	0.60	0.05	0.05	0.05	0.05	0.05	0.05	0	0	0
$h_{34}(\cdot)$	0.01	0.02	0.02	0.02	0.05	0.88	0	0	0	0	0
$h_{41}(\cdot)$	0.07	0.07	0.10	0.15	0.22	0.27	0.10	0.02	0	0	0
$h_{42}(\cdot)$	0.84	0.04	0.05	0.02	0.02	0.02	0.01	0	0	0	0
$h_{43}(\cdot)$	0.80	0.05	0.04	0.03	0.03	0.02	0.01	0.01	0.01	0	0
$h_{44}(\cdot)$	0.76	0.02	0.04	0.03	0.02	0.02	0.03	0.04	0.03	0.01	0

Table 1.5. Waiting time probabilities.

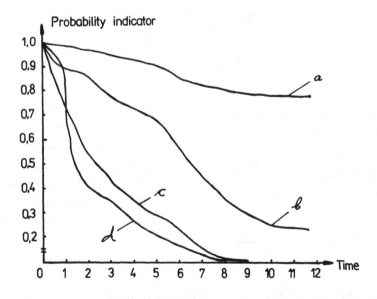

Figure 1.15. Evolution of system reliability index.

1.6 Rewards and discounting

1.6.1. REWARDS FOR SEQUENTIAL DECISION PROCESSES

Markovian decision processes are sequential by their nature. They imply transformations in the state space of the analyzed system.

To a given transformation k we can attach a value $r^k(s)$ which indicates the reward (gain or loss) from the system if an alternative k is applied, where the system state space is described by the state vector N.

The reward sequence is represented by $r_1(s), r_2(s), ... , r_n(s)$; the state vector changes at a given transformation. The terminal reward function is given by $r_0(s)$ which indicates the extra reward which can be obtained if the system has the state vector s, after the system accepts a set of pertinent transformations.

The practical solution for problems with rewards is by using the method of dynamic programming (see Chapter 2).

1.6.2. REWARDS IN DECISION PROCESSES WITH MARKOV STRUCTURE.

For semi-Markov processes, if the system was in state j, then $y_{ij}(\varepsilon)$ is defined as a reward we can obtain at time ε after the system entered state i. The process yields the fixed amount $c_{ij}(\tau)$ when the transition from state i to state j has been taking place at time τ. In this way, the system could enter state i ; a transition could take place to state j after a waiting time τ in state i, so the total reward has the form:

$$c_{ij}(\tau) + \sum_{\varepsilon=0}^{\tau-1} y_{ij}(\varepsilon) .$$

(1.61)

A special reward feature is that for which the following relations occur:

$$y_{ij}(\varepsilon) = y_{ij} ,$$

$$c_{ij}(\tau) = c_{ij} .$$

(1.62)

For a simplification of semi-Markov decisison models we can consider that the reward y_{ij} does not depend on the following state which is selected by the process, so that $y_{ij} = y_i$.

Semi-Markov decision processes generate the total expected reward when the system finds itself in state i and t time units remain. This is computed with the following relation:

$$v_i(t) = \,^>\!w_i(t)[v_i(0) + y_i(t)] \ + \ \sum_{j=1}^{N} p_{ij} \int_0^t h_{ij}(\tau)[c_{ij} + y_i\,\tau + v_j(t-\tau)]\,d\tau$$

$$(1.63)$$

where $i = 1, 2, \ldots , N$ and $t \geq 0$.

Semi-Markov decision processes could degenerate into appropriate Markov processes if $H(m) = I$ and $y_i = 0$, for all i, and in this case the corresponding expected value for the reward is given by the relation:

$$v_i(t) = \sum_{j=1}^{N} p_{ij}(c_{ij} + v_j(t-1)); \quad i = 1, 2, \ldots , N ; \ t \geq 0.$$

In Chapter 4 a wider presentation for the reward structure and optimization models for policy evaluation in sequential Markov and semi-Markov decision processes with complete information on states is presented [72], [74], [75].

1.6.3. MARKOVIAN DECISION PROCESSES WITH AND WITHOUT DISCOUNTING.

In a large number of practical situations the payments should be rewarded after a long time period (e.g. in the case of investment situations). In this context, we can use the concept of *discounting* of future monetary units. This concept implies that a monetary unit at time t into the future has a present value given by $e^{-\alpha t}$, where $\alpha \geq 0$, represents the discount rate. In Figure 1.16, a schematic representation for the value of a monetary unit (m.u.) into the future is outlined in the case where the observer finds himself at moment $t = 0$. In Figure 1.17 a diagrammatic representation of the case when he finds himself at time t, and looks retrospectively, is given.

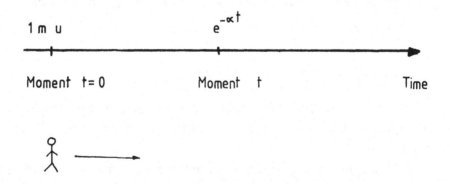

Figure 1.16. Schematic representation for the value of a monetary unit.

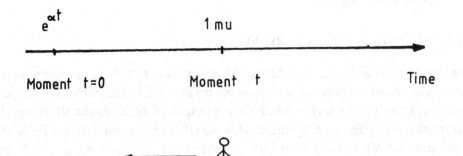

Figure 1.17. Diagrammatic representation of retrospective view of value of monetary unit.

We define the discount factor β by $\beta = e^{-\alpha}$; the present value is given by β^t.

If $g(t)$ is the reward at time t, the present value of this reward is given by the relation:

$$\sum_{t=0}^{\infty} g(t)\, \beta^t \,.\tag{1.67}$$

However, this represents a geometric transformation for the reward value with the condition $z = \beta$, and therefore:

$$\sum_{t=0}^{\infty} g(t)\, \beta^t \ = \ g^g(\beta).\tag{1.68}$$

If we pay a fixed value σ at each moment $t = 0, 1, 2, \ldots$, the present value for this sum, using the discount value, becomes:

$$\sigma(1 + \beta + \beta^2 + \ldots) \ = \ \sigma(1 - \beta)^{-1}.$$

The payment σ will have the present value given by:

$$\sigma(1 - \beta)^{-1} \ = \ g^g(\beta),$$

and then, $\sigma = (1 - \beta)\, g^g(\beta)$.

Markov decision proecesses for which $\beta = 1$ are considered to be processes without discounting.

1.7 Models and applications

1.7.1. REAL SYSTEMS WITH A MARKOVIAN STRUCTURE

The management and systems engineeering practice, the modelling of engineereing and economic and social systems, emphasized the utility of models with a Markovian structure, as well as the way in whcih the application of these models offers useful information for systems management. Howard (1971) presented different fields of application for Markov and semi-Markov models. In the following, a few of the possible fields of application of Markov decision models will be presented, with a description of states and transition processes, as well as associated optimization problems, in order to introduce the reader to the problematique of the present book [9].

In inventory control theory, the system state is characterized by the number of available elements at a given moment in stock. Its transition is determined by the renewal or depletion activiety of the stock. The reward structure of the process is given by a general profit, by selling the existing elements in stock and the optimal inventory policy [56], [59], [60], [67].

1.7.2. MODEL FORMULATION AND PRACTICAL RESULTS

Next, a few practical examples taken from the literatire, or developed by the author, will be presented which emphasize the applied prospective of Markov and semi-Markov models with total or partial information. The examples were taken from a variety of concrete situations; they refer to hospital planning, the reliability of technical systems and, also, to the Markovian interpretation of PERT graphs [12], [53], [55], [58], [63].

1.7.2.1. *A semi-Markov model for hospital planning.* The dynamics of patient movement in a hospital plays an important role in planning the occupation of different areas within it. Kao (1973) [12] has used, in this respect, a semi-Markov population model determined by two important components:

a) a stochastic process which approximates, in a corresponding manner, the movement of different patients within the hospital departments;

b) a second stochastic process which describes the way the patients arrive in the system.

The model allows the inclusion of a finite number of patients as inputs into the system as well as a finite number of states of health in which they find themselves for a given period of time. In this way, we deal with a semi-Markov population model. The results obtained by applying the model refer to the determination of the expected number of patients who can find themselves in every zone of the system (e.g. the hospital). The model has direct applicatiblity in long term planning and the optimal allocation of human resources as well as the financial aspects for the system.

If the vector $\pi(n) = [\pi_i(n), \ i = 1, 2, \ldots , N]$ represents the patient state of health (e.g. $\pi_i(n)$ indicates that the patient will be in state i at time n), by extension $\pi(0)$ is the corresponding state vector for the patient at time zero. In view of determining the vector $\pi(n)$ we shall introduce the concept of interval transition probability $\phi_{ij}(n)$, which indicates that the patient will be in state j at time (e.g. day) n, given that at time zero he was in state i. The health states in which one patient can be found are $1, 2, \ldots , (N + 1)$. The state $(N + 1)$ emphasizes the fact that the patient leaves the system at a given time in the future (the state of departure from the system). The quantity $\phi_{ij}(n)$ is computed with the relation:

$$\phi_{ij}(n) = \delta_{ij}\,^{>}w_i(n) + \sum_{k=1}^{N+1} p_{ik} \sum_{m=1}^{n} h_{kj}(m)\, \phi_{ij}(n-m) \tag{1.66}$$

$$i, j = 1, 2, \ldots , N + 1; \ \ n = 1, 2, \ldots$$

$$\phi_{ij}(0) = \begin{cases} 1, & \text{if } i = j \\[2mm] 0, & \text{otherwise} \end{cases}$$

p_{ij} represents transition probabilities:

$$\left(\sum_{j=1}^{N+1} p_{ij} = 1; \ \text{for all } i \,; \ \ p_{ij} \geq 0, \ \text{for all } i \,; \ \text{for all } j \right),$$

and $h(\ \cdot\ )$ represents the holding time probabilities in the system. In the above relation $^{>}w_i(n)$ is computed with the following relation:

$$^{>}w_i(n) = \sum_{k=1}^{N+1} \sum_{m=n+1}^{\infty} p_{ik}\, h_{ik}(m) \tag{1.67}$$

$$i, j = 1, 2, \ldots , N + 1; \ \ n > 1.$$

Knowing all these probabilities, we can compute $\pi_j(n)$, the health state probability of the patient:

$$\pi_j(n) = \sum_{i=1}^{N+1} \pi_i(0)\, \phi_{ij}(n); \ \ j = 1, 2, \ldots , N+1; \ \ n \geq 1. \tag{1.68}$$

The probability distribution for the waiting duration is computed with the relation:

$$
D(n) = \begin{cases} \pi_{N+1}(0), & \text{of } n = 0 \\ \sum_{i=1}^{N+1} \pi_i(0) f_{i,N+1}(n), & \text{if } n \geq 1, \end{cases} \tag{1.69}
$$

where $f_{ij}(\cdot)$ represents the first passage probability from state i to state j :

$$
f_{ij}(n) = \sum_{\substack{k=1 \\ k \neq j}}^{N+1} \sum_{m=1}^{n} p_{ik} h_{ik}(m) f_{ij}(n-m) + p_{ij} h_{ij}(n) \tag{1.70}
$$

$$
i, j = 1, 2, \dots, N+1,
$$

$$
f_{ij}(0) = 0, \quad \text{for all } i ; \text{ for all } j .
$$

The state of health prediction for a patient can be made by determining the value of $\xi_{ik}(s \mid r)$ which gives the probability of when the patient enters state i after s days, given that he was in state k for r days.

However, we can write the following relation:

$$
\xi_{ik}(s \mid r) = \frac{p_{ki} h_{ki}(s+r)}{>w_k(r)} ; \quad \text{for all } i, \text{ for all } k . \tag{1.71}
$$

Next we shall define the event $E_k(r)$ which indicates the fact that the patient was in state k for r time units from the moment he entered that state. The value of $\Psi_i(t \mid E_k(r))$ indicates the probability that after t days the system (the patient) will be found in state i, and therefore:

$$
\Psi_i(t \mid E_k(r)) = \sum_{j=1}^{N+1} \sum_{s=1}^{t} \xi_{kj}(s \mid r) \phi_{ji}(t-s) + \delta_{ki} \sum_{j=1}^{N+1} \sum_{r=t+1}^{\infty} \xi_{kj}(s \mid r) \tag{1.72}
$$

$$
i, k = 1, 2, \dots, N ; \ t \geq 1,
$$

where δ_{ki} is the Kronecker symbol.

It is a well known fact that a patient with a large number of illnesses could arrive in a hospital. These illnesses are assessed in different departments within the hospital.

Each patient can be modelled by a semi-Markov process, completely characterized by the values of $\pi(\cdot)$, $P = [p_{ij}]$ and $H(\cdot) = [h_{ij}(m)]$. With an increased degree of generalization, we can consider the case when the patients can be grouped in two types of symptoms or categories of illnesses. The value α_g^c indicates the patient arrival probability from group g, classified by a c-type connection ($g = 1, 2, \dots , G$; $c = 1, 2, \dots , C$). A schematic representation of the analyzed semi-Markov model is given in Figure 1.18.

Practical advantages for applying this model come from the fact that it offers statistical evaluation concerning the mean number of possible patients to be found in each of the system states, as well as giving an image of the way in which a patient moves in the different hospital departments.

Let n_j denote the number of patients in a state j: inputs of patients in the system are considered to have an arbitrary distribution. Next we shall compute the mean and variance for the number of patients in each state, considering that the system was observed for a long time. If μ_g and σ_g^2 are the mean and variance, respectively, for the group arrival distribution, and for the connection c, the mean and variance are μ^c and $(\sigma^2)^c$, respectively, then:

$$\mu^c = \sum_{g=1}^{G} \alpha_g^c \mu_g$$

$$(\sigma^2)^c = \sum_{g=1}^{G} (\sigma_g^c)^2 \sigma_g^2 .$$

(1.73)

For Markov populations the following result is known (see Staff and Vagholken (1971) as referred to by Kao [12]):

Using the notation $Z^c(E_k(r) \mid s) = \Pr\{E_k(r) \mid Z_c, Y_g, X_s\}$, we can write (see Kao (1973) [12]):

$$Z^c(E_k(r) \mid s) = \sum_{i=1}^{N+1} \pi_i^c(0)\, e_{ik}^c(s - r)\,^> w_k^c(r),$$

where:

$$e_{ik}^c(n) = \delta_{ik}\, \delta(n) + \sum_{j=1}^{N+1} \sum_{m=1}^{n} p_{ij}^c\, h_{ij}^c(m)\, e_{jk}^c(n - m);$$

$$\delta(n) = \begin{cases} 1, & \text{if } n = 0, \\ 0, & \text{otherwise.} \end{cases}$$

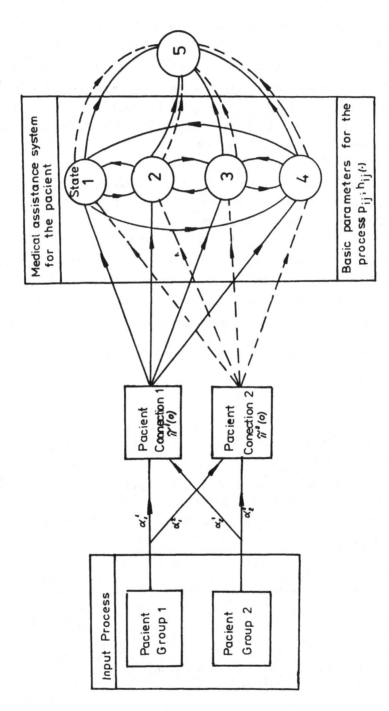

Figure 1.18. Schematic representation of the semi-Markov model.

Next, if $\gamma_g(c) \mid E_k(r), s) = \Pr\{Z_c \mid E_k(r), Y_g, X_s\}$ we can have the computational probability that a patient indexed by K_{gskr} follows the c-connection such that:

$$\langle n_i \rangle = \sum_{c=1}^{C} \langle n_i^c \rangle = \sum_{c=1}^{C} \mu^c \sum_{n=1}^{\infty} \pi_i^c(n)$$

$$^v\langle n_i \rangle = \sum_{c=1}^{C} {}^v\langle n_i^c \rangle = \sum_{p=1}^{C} \left\{ \left[\mu^c \sum_{n=1}^{\infty} \pi_i^c(n) \right] + \left[(\sigma^2)^c - \mu^c \right] \sum_{n=1}^{C} \pi_i^c(n) \right]^2 \right\}$$

(1.75)

$$^{cov}\langle n_i , n_j \rangle = \sum_{c=1}^{C} {}^{cov}\langle n_i^c , n_j^c \rangle = \sum_{c=1}^{C} \left\{ \left[(\sigma^2)^c - \mu^c \right] \left(\sum_{n=1}^{\infty} \pi_i^c(n) \, \pi_j^c(n) \right) \right\} \qquad (1.76)$$

$$i \neq j$$

where π_i^c is the probability a patient will follow a given c-connection, if he were found to be in state i after n days and n_i^c represents the number of patients which follow connection c in state i.

In the case where L represents the total number of patients in the system, then:

$$\langle L \rangle = \sum_{i=1}^{N} \langle n_i \rangle$$

$$^v\langle L \rangle = \sum_{i=1}^{N} {}^v\langle n_i \rangle + 2 \sum_{i=1}^{N} \sum_{j=i=1}^{N} {}^{cov}\langle n_i , n_j \rangle .$$

We shall introduce, next, the information vector $\mathfrak{I}$, whose components L_{gskr} indicate the number of the group g of patients who were in state k for r days, and spent s days within the system:

$$\mathfrak{I} = (K_{1111}, \dots , K_{gskr}, \dots , K_{GTNM_N})$$

where T (in days) represents the planning horizon for the system and:

$$M_i = \max \{M_{ij}/p_{ij} > 0\}, \quad i = 1, 2, \dots , N,$$

represents the maximum number of days for which the patient will find himself in state i.

If X_s is the event which indicates the patient was in the system for s days, Y_g is the event which denotes the patient belongs to group g and Z_c is the event a patient will follow the c-connection and then:

$$\Pr\{Z_c \mid E_k(r), Y_g, X_s\} =$$

$$= \frac{\Pr\{E_k(r) \mid Z_c, Y_g, X_s\} \Pr\{X_s \mid Z_c, Y_g\} \Pr\{Z_c \mid Y_g\}}{\sum\limits_{u=1}^{P} \Pr\{E_k(r) \mid Z_u, Y_g, X_s\} \Pr\{X_s \mid Z_u, Y_g\} \Pr\{Z_u \mid Y_g\}}$$

$$\gamma_g(c \mid E_k(r), s) = \frac{Z^c(E_k(r) \mid s) {}^{>}f_{N+1}^{c}(s) \alpha_g^c}{\sum\limits_{u=1}^{C} Z^u(E_k(r) \mid s) {}^{>}f_{N+1}^{u}(s) \alpha_g^u} \tag{1.77}$$

where:

$$^{>}f_{N+1}^{c}(s) = \sum\limits_{i=1}^{N+1} \sum\limits_{m=s+1}^{\infty} \pi_i^c(0) f_{i, N+1}(m), \tag{1.78}$$

and $f_{ij}(\cdot)$ represents the first passage time probability. The probability $\gamma_g(\cdot \mid \cdot)$ is an *a posteriori* value; it indicates the fact that a patient who is described by characteristics corresponding to vector $\Im$ will follow the connection c.

If $P_i(t)$ represents the number of patients who were found for t days in state i conditional on that the present state of the system is fully characterized by the informational vector $\Im$, then the mean and dispersion for this random value can be obtained by the following relations:

$$\langle P_i(t) \rangle = \sum\limits_{g=1}^{G} \sum\limits_{s=1}^{S} \sum\limits_{k=1}^{N} \sum\limits_{r=1}^{M_k} \left\{ K_{gskr} \, \psi_i^c \left(t \mid E_k(r) \right) \gamma_g(c \mid E_k(r), s) \right\} \tag{1.79}$$

$$i = 1, 2, \dots, N; \quad t = 1, 2, \dots, T$$

and

$$^{v}\langle P_i(t) \rangle = \sum\limits_{g=1}^{G} \sum\limits_{s=1}^{S} \sum\limits_{k=1}^{N} \sum\limits_{r=1}^{M_k} K_{gskr} \left\{ \sum\limits_{c=1}^{P} \psi_i^c \left(t \mid E_k(r) \right) \gamma_g(c \mid E_k(r), s) \right\} \bullet$$

$$\bullet \left\{ 1 - \sum\limits_{c=1}^{C} \psi_i^c \left(t \mid E_k(r) \right) \gamma_g(c \mid E_k(r), s) \right\}; \tag{1.80}$$

$$i = 1, 2, \dots, N; \quad t = 1, 2, \dots, T$$

where $\psi_i^c(t \mid E_k(r)$ is given by (1.72) for a given c.

If $\mu_g(t)$ and $\sigma^2(t)$ are the mean and dispersion, respectively, correspoding to the number of groups g who arrive within the system at a given moment t, $(t = 1, 2, \dots, T)$ and if μ_t^c and $(\sigma^2)^c$ induce the connection c, then:

$$\mu_t^c = \sum_{g=1}^{G} \alpha_g^c \, \mu_g(t) \qquad (1.81)$$

$$(\sigma^2)_t^c = \sum_{g=1}^{G} (\alpha_g^c)^2 \, \sigma_g^2(t). \qquad (1.82)$$

If $D_i(t)$ represents an event which indicates the number of new patients who follow connection c and who will be in state i for t days, the mean and variance for this random value are given, in accordance with those given previously, by the relations:

$$\langle D_i(t) \rangle = \sum_{c=1}^{C} \sum_{m=1}^{t} \mu_m^c \, \pi_i^c(t-m) \qquad (1.83)$$

$$i = 1, 2, \dots, N \; ; \; t = 1, 2, \dots, T,$$

$$^v\langle D_i(t) \rangle = \sum_{c=1}^{C} \left\{ \sum_{m=1}^{t} \mu_m^c \, \pi_i(t-m) \, [1 - \pi_i(t-m)] + \sum_{m=1}^{t} (\sigma^2)_m^c \, [\pi_i(t-m)]^2 \right\}$$

$$\qquad (1.84)$$

$$i = 1, 2, \dots, N \; ; \; t = 1, 2, \dots, T.$$

As was emphasized, previously, at a given moment t, in state i, we can find old and new patients in a total amount $A_i(t)$; the mean and variance for this new stochastic value are given by the relations:

$$\langle A_i(t) \rangle = \langle P_i(t) \rangle + \langle D_i(t) \rangle \qquad (1.85)$$

$$^v\langle A_i(t) \rangle = \,^v\langle P_i(t) \rangle + \,^v\langle D_i(t) \rangle \qquad (1.86)$$

EXAMPLE (Kao [12]). Let us consider patients with a coronary disease. We are interested in making a study concerning the forecasting of their state of health for a given period of time. Possible states in which a patient could be found are defined by a set of particular parameters (e.g. medication, quantity of oxygen administered, etc.). The states are, respectively, (1) acute, (2) primary activity, (3) secondary activity, and (4) normal activity. The grouping of patients is made on age criteria; group 1 refers to patients under 65 years of age, and group 2 for patients who are over this age.

By using realistic data, we can write:

$$\alpha_1^1 = 0.8540 \text{ and } \alpha_2^1 = 0.1460$$

$$\alpha_1^2 = 0.6919 \text{ and } \alpha_2^2 = 0.3081.$$

For any connection c - 1, 2 the intial state vector as well as the corresponding transition probabilties were chosen with the following values:

$$\pi^1(0) = [0.7736; \ 0.1887; \ 0.0377; \ 0.0; \ 0.0];$$

$$\pi^2(0) = [0.8214; \ 0.0357; \ 0.0; \ 0.0; \ 0.1429];$$

$$P^1 = \begin{array}{c} i \backslash \\ \\ 1 \\ 2 \\ 3 \\ 4 \\ 5 \end{array} \begin{array}{ccccc} 1 & 2 & 3 & 4 & 5 \\ \left[\begin{array}{ccccc} 0 & 1 & 0 & 0 & 0 \\ 0.0189 & 0 & 0.8113 & 0.0755 & 0.0943 \\ 0 & 0 & 0 & 0.8444 & 0.1556 \\ 0.0238 & 0 & 0 & 0 & 0.9762 \\ 0 & 0 & 0 & 0 & 1 \end{array}\right] \end{array}$$

$$P^2 = \begin{array}{c} i \backslash \\ \\ 1 \\ 2 \\ 3 \\ 4 \\ 5 \end{array} \begin{array}{ccccc} 1 & 2 & 3 & 4 & 5 \\ \left[\begin{array}{ccccc} 0 & 0.2917 & 0 & 0 & 0.7083 \\ 0.1250 & 0 & 0.3750 & 0 & 0.5 \\ 0 & 0 & 0 & 0 & 1 \\ 0 & 0 & 0 & 0 & 1 \\ 0 & 0 & 0 & 0 & 1 \end{array}\right] \end{array}$$

Waiting time distributions for the semi-Markov process, when $p_{ij} \geq 0$ are given in Table 1.6.

m	$h_{12}(m)$	$h_{15}(m)$	$h_{21}(m)$ $h_{23}(m)$ $h_{24}(m)$	$h_{25}(m)$	$h_{34}(m)$	$h_{35}(m)$	$h_{41}(m)$ $h_{45}(m)$
1	0.12	0.1176	0.0385	0	0	0	0
2	0.08	0.1176	0.1346	0.1111	0.0789	0.1	0.0476
3	0.14	0.0588	0.0962	0.3333	0.1579	0.1	0.0476
4	0.12	0.0588	0.0577	0	0.2632	0.1	0.1190
5	0.08	0.1765	0.0577	0.2222	0.1053	0.1	0.1190
6	0.1	0	0.1731	0.1111	0.0789	0.2	0.0476
7	0.12	0.0588	0.0769	0	0.0789	0	0.0952
8	0.04	0.0588	0.0385	0.1111	0.0263	0.1	0.0714
9	0.08	0.0588	0.0769	0.1111	0.1053	0.1	0.1667
10	0.06	0	0.0769	0	0	0	0.0714
11	0	0.1176	0.0577	0	0.0263	0.2	0.0476
12	0.04	0	0.0385	0	0.0263	0	0.0476
13	0	0.0588	0	0	0	0	0.0476
14	0	0	0.0385	0	0	0	0
15	0.02	0	0	0	0	0	0
16	0	0	0.0191	0	0.0526	0	0
17	0	0	0	0	0	0	0.0238
18	0	0	0	0	0	0	0
19	0	0	0	0	0	0	0
20	0	0.0588	0	0	0	0	0
21	0	0	0	0	0	0	0
22	0	0	0	0	0	0	0
23	0	0.0588	0	0	0	0	0
24	0	0	0	0	0	0	0
25	0	0	0	0	0	0	0
26	0	0	0	0	0	0	0
27	0	0	0	0	0	0	0.0479
28	0	0	0.0192	0	0	0	0

Table 1.6. Waiting time parameters for the semi-Markov process when $p_{ij} \geq 0$.

Considering the input distribution (arrivals/day) for the two age groups characterized by the parameters $\mu_1 = 0.8$, $\sigma_1^2 = 2$, $\mu_2 = 1.2$, $\sigma_2^2 = 1.4$, by using the relations given above, we can write:

$$\mu^1 = 1.5135; \quad (\sigma^2)^1 = 2.1288$$

$$\mu^2 = 0.4865; \quad (\sigma^2)^2 = 0.1758$$

The mean and variance for the number of patients in each state of the system are presented in Table 1.7. The covariance of the different parameters are given in the same table.

State i	$\langle n_i \rangle$	$^v\langle n_i \rangle$
1	7.9	8.2
2	10.7	12.1
3	7.9	8.5
4	8.9	9.7

$$^{cov}\langle n_i, n_j \rangle$$

$$^{cov}\langle n_1, n_2 \rangle = 0.7 \quad ^{cov}\langle n_1, n_3 \rangle = 0.2 \quad ^{cov}\langle n_1, n_4 \rangle = 0.1$$

$$^{cov}\langle n_2, n_3 \rangle = 0.7 \quad ^{cov}\langle n_2, n_4 \rangle = 0.5 \quad ^{cov}\langle n_3, n_4 \rangle = 0.6$$

Final Result:

$$\langle A_i \rangle = 35.5 \qquad ^v\langle A_i \rangle = 44.1$$

Table 1.7. Mean, variance and covariance for the number of patients in each state of the system.

In the case where we would like to make a short term forecast for the system to determine the *a posteriori* probability, conditioned by connection 1, we can obtain, by applying the corresponding relations, the results given in Table 1.8.

Group of patients g	Days s	Present state of patient k	Days r	K_{gskr}	$\gamma_g(1|E_k(r), s)$
1	5	1	5	1	0.9102
2	3	1	3	3	0.7468
2	10	1	2	1	0.7598
1	10	2	3	1	0.9819
1	12	2	7	1	0.9873
2	2	2	1	1	0.9083
2	10	2	4	4	0.9528
2	14	2	6	1	0.9748
1	9	3	2	1	0.9914
1	11	3	4	1	0.9912
2	18	3	3	1	0.9857
2	14	3	6	1	0.9829
1	16	4	2	2	1.0000
1	18	4	3	1	1.0000
2	29	4	4	1	1.0000
2	19	4	6	1	1.0000

Table 1.8. Results for short term forecast.

If the duration of a planning period is a week and if the number of new patients who could arrive in the system coming from group g in day t of the week follows a distribution with the mean $\mu_g(t)$ and variance $\sigma_g^2(t)$, then from the basic data shown in Table 1.9, we can have the results given in Table 1.10.

Arrival Distribution for the Patients in the System

t	Day of the Week	Group 1		Group 2	
		$\mu_1(t)$	$\sigma_1^2(t)$	$\mu_2(t)$	$s_2(t)$
1	Monday	1.1	2.0	1.4	1.6
2	Tuesday	0.8	1.8	0.9	1.7
3	Wednesday	0.8	1.8	0.8	1.9
4	Thursday	0.8	1.9	0.9	1.4
5	Friday	1.0	2.0	1.5	1.8
6	Saturday	0.7	1.2	1.2	1.4
7	Sunday	0.6	1.0	1.1	1.5

Table 1.9. Basic data of arrival distribution of patients in the system.

Computational results are diagrammatially represented in Figure 1.19.

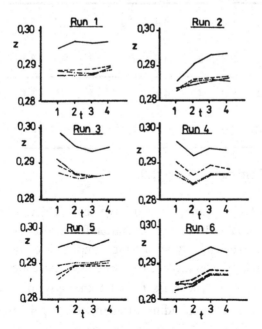

Figure 1.19. Computational results (see text).

t	Parameter	$P_i(t)$	$D_i(t)$	$A_i(t)$	$P_i(t)$	$D_i(t)$	$A_i(t)$	$P_i(t)$	$D_i(t)$	$A_i(t)$	$P_i(t)$	$D_i(t)$	$A_i(t)$
1	$\langle\cdot\rangle$	4.2	2.0	6.2	7.8	0.4	8.2	3.8	0.1	3.9	5.3	0.0	5.3
	$^v\langle\cdot\rangle$	0.7	1.9	2.6	1.5	0.4	1.9	1.3	0.1	1.4	1.2	0.0	1.2
2	$\langle\cdot\rangle$	3.5	3.0	6.5	6.8	0.8	7.6	4.4	0.1	4.5	5.3	0.0	5.3
	$^v\langle\cdot\rangle$	1.1	3.4	4.5	2.7	0.9	3.5	2.4	0.1	2.5	1.9	0.0	1.9
3	$\langle\cdot\rangle$	2.9	3.9	6.8	6.2	1.4	1.7	4.7	0.2	4.9	5.2	0.0	5.2
	$^v\langle\cdot\rangle$	1.3	4.6	5.9	3.1	1.5	4.6	2.9	0.2	3.1	2.4	0.0	2.4
4	$\langle\cdot\rangle$	2.1	4.8	6.9	5.9	1.9	7.8	4.9	0.4	5.3	5.1	0.0	5.1
	$^v\langle\cdot\rangle$	1.3	5.6	6.9	3.2	2.1	5.3	3.0	0.4	3.4	2.7	0.0	2.7
5	$\langle\cdot\rangle$	1.7	6.1	7.8	5.2	2.7	7.9	4.9	0.5	5.4	5.1	0.1	5.2
	$^v\langle\cdot\rangle$	1.2	6.7	7.9	3.0	3.0	6.0	3.1	0.5	3.6	3.0	0.1	3.1
6	$\langle\cdot\rangle$	1.3	6.7	8.0	4.6	3.4	8.0	4.9	0.7	5.6	5.2	0.1	5.3
	$^v\langle\cdot\rangle$	1.0	7.0	8.0	2.7	3.9	6.6	3.1	0.7	3.8	3.2	0.1	3.3
7	$\langle\cdot\rangle$	0.9	7.1	8.0	4.1	4.1	8.2	4.8	1.0	5.8	5.1	0.2	5.3
	$^v\langle\cdot\rangle$	0.8	7.2	8.0	2.4	4.7	7.1	3.1	1.0	4.1	3.3	0.2	3.5

Table 1.10. Results from data in Table 1.9.

1.7.2.2. *System reliability.* Complex and large scale technical systems become extremely difficult to model. The state of operation for a system can be divided into three classes: the system capable of performing complete operation (A), undergoing repair (F), and incapable of operation (C). The class A states are fully operational; A_i, ($i = 1, 2, \ldots, r$); $A_i \in A$. The class F is for the system undergoing repair; F_j ($j = 1, 2, \ldots, s$); $F_j \in F$ corresponding to intermediate performance levels for the technical system. The class C is for non-operational states; C_k, ($k = 1, 2, \ldots, c$);

$C_k \in C$. The state space is the union $S = A \cup F \cup C$. The reliability function is given by the probability that the system will be in a fully operational state:

$$\mathcal{R}(n) = \Pr\{S(n) \in A\},$$

at time n.

Consider a completely observable technical system with semi-Markov behaviour. At each transition of the system we can identify the system state. Generally, we may consider that two kinds of states exist for a system; the operational states $Q_k \in Q$ ($k = 1, 2, \ldots, L$) where $Q = A \cup F$ and the trapping state, $d \in C$.

The probability of going from state $k \in Q$ to class C in t time periods is defined as an exit probability and:

$$\gamma_k(n) = \Pr\{S(n + u) \in C \mid S(u) = k\}; \quad k = 1, 2, \ldots, N.$$

Figure 1.20 shows an evolution diagram for such a behaviour. As was mentioned earlier, a semi-Markov process used in the study of systems operational behaviour is given by the probability rate transition p_{ij} for the system to make a transition from state $i, i \in Q \cup C$, to state $j, j = Q \cup C$ and by the waiting time probabilities $h_{ij}(m) = \Pr\{\tau_{ij} = m\}$, $(m \geq 0)$, for which $h_{ij}(0) = 0$.

Using basic concepts from the theory of semi-Markov processes, the probability exit functon satisfies the relation:

$$\gamma_k(n) = \sum_{m=0}^{n} \sum_{l \notin C} p_{kl} \, h_{kl}(m) \, \gamma_l(n - m) + \sum_{m=0}^{n} \sum_{l \in C} p_{kl} \, h_{kl}(m);$$

$$k = 1, 2, \ldots, N \, ; \quad n \geq 0.$$

For the case when the system accepts only one trapping state d, the above equation becomes:

$$\gamma_k(n) = \sum_{l=1}^{N} p_{kl} \sum_{m=0}^{n} h_{kl}(m) \, \gamma_l(n - m),$$

$$k = 1, 2, \ldots, N \, ; \quad n \geq 0;$$

$$\gamma_k(0) = 0.$$

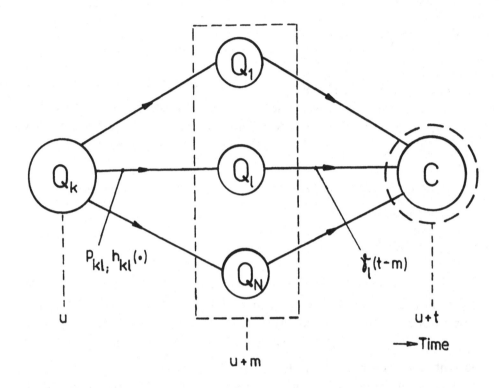

Figure 1.20. Evolution diagram for exit probability.

In matrix form, the exit probability of the system from an operational state Q is $\Gamma(n)$ i.e. a diagonal matrix with the dimensions $(N \times N)$ and $\gamma_k(n)$ as matrix components such that:

$$\Gamma(n) = \sum_{m=0}^{n} (P \square H(m))\, \Gamma(n-m); \quad n = 0, 1, 2, \dots .$$

Consider a complex technical system described only by operational kinds of states A and F. States in class F are lexicographically ordered – the system experiences states $F_1, F_2, \dots, F_l, \dots, F_s$ in the increasing order of the index l, such that $F_j \succ F_l$. The system is degraded (see Figure 1.21); for reliability prediction we use the concept of "evolution diagram" given by Girault.

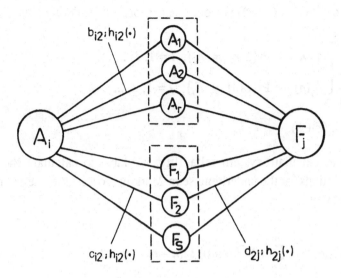

Figure 1.21. Diagram of a degraded system (see text).

The partition matrix is:

$$P = \begin{bmatrix} B & \vdots & C \\ \cdots\cdots & \vdots & \cdots\cdots \\ 0 & \vdots & D \end{bmatrix}$$

where D has the special form (1.87) including information about the lexicographical order of states in class F. The elements of P are:

$$p_{ij}(p_{ij} > 0, \ \sum_{j=1}^{r+2} p_{ij} = 1. \ i = 1, 2, \dots , (r+s))$$

i.e. the probability that the system moves from state i to state j in the next transition. The holding time function is $H(m)$.

$$D = \begin{bmatrix} 0 & d_{12} & . & . & d_{1r} \\ . & 0 & . & . & d_{2r} \\ . & . & 0 & . & . \\ . & . & . & . & . \\ . & . & . & . & 1 \end{bmatrix} \qquad (1.87)$$

Let $\alpha(n)$ be the $(1 \times (r + s))$ vector of state probabilities defined by:

$$\alpha_k(n) = \begin{cases} \alpha_i(n) = \Pr\{S(n) = A_i\}; \ i = 1, 2, \ldots, r \\ \\ \alpha_j(n) = \Pr\{S(n) = F_j\}; \ j = 1, 2, \ldots, s, \end{cases} \tag{1.88}$$

for $k = 1, 2, \ldots, (r + s)$.

In order to calculate system reliability, we have to calculate the interval transition probabilities for the underlying semi-Markov process; the interval transition probability matrix is:

$$\Omega(n) = {}^{>}W(n) + \sum_{m=0}^{n} (P \, \square \, H(m)) \, \Omega(n - m), \quad n \geq 0 \tag{1.89}$$

where ${}^{>}W(t)$ is the diagonal matrix of the complementary cumulative probability distribution for the waiting time $\left({}^{>}w_i(t) = \sum_{m=t+1}^{\infty} w_i(m)\right)$, and $w_i(\cdot)$ is the mass probability function for τ_i (waiting time in any of the states $i \in S$), such that:

$$w_i(\cdot) = \sum_{j=1}^{N} p_{ij} h_{ij}(\cdot).$$

For a semi-Markov process $\alpha(n) = \alpha(0) \, \Omega(n)$, $n \geq 0$, where $\alpha(0)$ is the vector of the initial (time $t = 0$) state probabilities. The system reliability is given by:

$$R(n) = \sum_{i : i \in A_i \in A} \alpha_i(n) \, ; \ n \geq 0. \tag{1.90}$$

In Figure 1.22 a computer program for system reliability evaluation is given.

The computer program gives the following results in the caluclation framework of systems engineering performance:
— reliability figures for completely observable sytems under semi-Markov behaviour;
— reliability figures for a multi-functional system, partially observable under semi-Markov behaviour.

```
0008                     MASTER ADRIANMAINT
0009          C
0010          C THIS PROGRAM IS TO CALCULATE THE RELIABILITY FIGURES FOR
0011          C COMPLETELY OBSERVABLE SYSTEMS UNDER SEMI / MARKOV
0012          C BEHAVIOUR
0013          C ••••••••••••••••••••••••••••••••••••••••••••••••••••••••••••••••••••••••••
0014          C THIS PROGRAM IS TO CALCULATE THE INTERVAL TRANSITION
0015          C PROBABILITIES AND FIRST PASSAGE TIME PROBABILITIES
0016          C THE LAST PART OF THE PROGRAM CALCULATES THE STATE
0017          C DURATION
0018          C ••••••••••••••••••••••••••••••••••••••••••••••••••••••••••••••••••••••••••
0019          C FOR DETAILS SEE HOWARD VOL 2
0020          C ••••••••••••••••••••••••••••••••••••••••••••••••••••••••••••••••••••••••••
0021          C
0022          C MODIFIED
0023          C ••••••••••••••••••••••••••••••••••••••••••••••••••••••••••••••••••••••••••
0024                     DIMENSION P(4,4), H(4,4,12), FI(4,4,12), DELTA(4,4 ),
0025                1WMIC(4,12), F(4,4,12)
0026                     DIMENSION SUM1(4,4), DIF(4,4), PI2(4), HI(4,4)
0027                     DIMENSION D1(4,12), D2(4,12), DELTA1(12), DELTA2(12)
0028                     READ (1,1001) L, NMIC
0029                     READ (1,1000) KMIC
0030                     READ (1,1000) ((P(I,J), J = 1,L), I = 1,L)
0031                     READ (1, 999) (((H(I, J, M), M = 1, NMIC), J = 1, L), I = 1, L)
0032                     WRITE (2, 400) ((P(I, J), J = 1, L), I = 1, L)
0033                     WRITE (2, 401) (((H(I, J, M), M = 1, NMIC), J = 1, L), I = 1, L)
0034          C
0035          C ••••••••••••••••••••••••••••••••••••••••••••••••••••••••••••••••••••••••••
0036          C CALCULATE COMPLEMENTARY PROB DISTRIBUTION
0037          C ••••••••••••••••••••••••••••••••••••••••••••••••••••••••••••••••••••••••••
0038          C
0039                     DO 800   N = 1, NMIC
0040                     DO 801   I = 1,L
0041                     PI1 = 0.0
0042                     DO 802   J = 1, L
0043                     SUM = 0.0
0044                     DO 803   M = 1, N
0045                803  SUM = SUM + H(I, J, M)
0046                     SUM1(I, J) = SUM
0047                     DIF(I, J) = 1.0 – SUM1(I, J)
```

Figure 1.22. Program for system reliability evaluation (p.1),

```
0048              802   PI1 = PI1 + P(I, J) • DIF(I, J)

0049                    PI2(I) = PI1
0050                    WMIC(I, N) = PI2(I)
0051              801   CONTINUE
0052              800   CONTINUE
0053                    WRITE (2, 333)
0054                    WRITE (2, 202) ((WMIC(I, N), N = 1, 12), I = 1, 4)
0055                    WRITE (2, 1007)
0056                    DO 724  I = 1, L
0057                    DO 725  J = 1, L
0058                    IF (I .EQ. J) GO TO 600
0059                    GO TO 601
0060              600   FI(I, J, 1) = 1.
0061                    GO TO 725
0062              601   FI(I, J, 1) = 0.
0063              725   CONTINUE
0064                    WRITE (2, 1003)  (FI(I, J, 1), J = 1, L)
0065              724   CONTINUE
0066                    WRITE (2, 7325)
0067       C
0068                    DO 100  N = 2, NMIC
0069                    WRITE (2, 666)  N
0070                    DO 101  I = 1, L
0071                    DO 1101  J = 1, L
0072                    IF (I .EQ. J) GO TO 20
0073                    DELTA(I, J) = 0.
0074                    GO TO 22
0075               20   DELTA(I, J) = 1.
0076       C
0077       C ••••••••••••••••••••••••••••••••••••••••••••••••••••••••••••••••••••••••••••••••••
0078       C CALCULATE PROBABILITY FI(I, J, N)
0079       C ••••••••••••••••••••••••••••••••••••••••••••••••••••••••••••••••••••••••••••••••••
0080       C
0081               22   PI = 0.0
0082                    DO 3  K = 1, L
0083                    HA = 0.0
0084                    DO 2  M = 1, N − 1
0085                2   HA = HA + H(I, K, M) • FI(K, J, N − M)
0086                    HI(I, K) = HA
0087                3   PI = PI + P(I, K) • HI(I, K)
```

Figure 1.22. Program for system reliability evaluation (p.2),

```
0088                    FI(I, J, N) = DELTA(I, J) • WMIC(I, N – 1) + PI
0089           1101 CONTINUE
0090                WRITE (2, 1003)   (FI(I, J, N), J = 1, L)
0091           101  CONTINUE
0092           100  CONTINUE
0093        C  ••••••••••••••••••••••••••••••••••••••••••••••••••••••••••••••••••••••••••••••
0094        C  CALCULATE THE FIRST PASSAGE TIME
0095        C  ••••••••••••••••••••••••••••••••••••••••••••••••••••••••••••••••••••••••••••••
0096                WRITE (2, 640)
0097                DO 300   N = 1, NMIC
0098                WRITE (2, 666)  N
0099                IF (N .EQ. 1) FI(I, J, N) = 0.
0100                DO 301   I = 1, L
0101                DO 302   J = 1, L
0102                S1 = 0.0
0103                DO 303   K = 1, L
0104                IF (K .EQ. J) GO TO 303
0105                DO 304   M = 1, N
0106                S1 = S1 = P(I,K) • H(I, K, M) • F(K, J, N – M + 1)
0107           304  CONTINUE
0108           303  CONTINUE
0109                F(I, J, N) = S1 + P(I, J) • H(I, J, N)
0110           302  CONTINUE
0111                WRITE (2, 1003)   (F(I, J, N), J = 1, L)
0112           301  CONTINUE
0113           300  CONTINUE
0114        C
0115        C  ••••••••••••••••••••••••••••••••••••••••••••••••••••••••••••••••••••••••••••••
0116        C  THIS SECTION IS TO CALCULATE THE DURATION OF A STATE
0117        C  WHEN EITHER THE NUMBER OF TRANSITION OR AMOUNT OF TIME
0118        C  REQUIRED FOR THE FIRST REAL TRANSITION ARE AVAILABLE
0119        C  ••••••••••••••••••••••••••••••••••••••••••••••••••••••••••••••••••••••••••••••
0120        C
0121        C  THIS IS THE MAIN CORE OF DURATION CALCULATION FOR
0122        C  COMPLEX SYSTEMS
0123        C  ••••••••••••••••••••••••••••••••••••••••••••••••••••••••••••••••••••••••••••••
0124        C
0125        C  ••••••••••••••••••••••••••••••••••••••••••••••••••••••••••••••••••••••••••••••
0126        C  THIS PART OF THE PROGRAM CALCULATES DURATION OF THE
```

Figure 1.22. Program for system reliability evaluation (p.3),

```
0127            C  PROCESS IN TRANSITIONS   ••••••••••••••••••••••••••••••••••••••••••••••
0128            C
0129                       DO 200  I = 1, L
0130               200   D1(I, 1)  = 1.
0131                       DO 201  I = 1, L
0132                       DO 222  K = 2, KMIC
0133                       IF  (K .EQ. 2)   DELTA1(K) = 1.
0134                       DELTA1(K) = 0.
0135                       D1(I, K) = P(I, I) • D1(I, K – 1) + DELTA1(K) • (1. – P(I, I))
0136               222   CONTINUE
0137               201   CONTINUE
0138                       WRITE  (2, 7731)
0139                       WRITE  (2, 202)  ((D1(I, K),  K = 1, KMIC),  I = 1, L)
0140            C
0141            C  ••••••••••••••••••••••••••••••••••••••••••••••••••••••••••••••••••••••••••
0142            C  THIS PART OF THE PROGRAM CALCULATES DURATION OF THE
0143            C  PROCESS IN TIME
0144                       DO 203  I = 1, L
0145               203   D2(I, 1) = 1.
0146                       DO 204  I = 1, L
0147                       DO 205  N = 2, NMIC
0148                       IF  (N .EQ. 2)  DELTA2(N) = 1.
0149                       DELTA2(N) = 0.
0150                       D2(I,N) = P(I, I) • D2(I, N – 1) + DELTA1(N) • (1. – P(I, I))
0151               205   CONTINUE
0152               204   CONTINUE
0153                       WRITE  (2, 7732)
0154                       WRITE (2, 202)  ((D2(I, N),  N = 1, NMIC),  I = 1, L)
0155            C
0156            C  ••••••••••••••••••••••••••••••••••••••••••••••••••••••••••••••••••••••••••
0157               202   FORMAT ((1H, 12(F7.5, 2X)))
0158               333   FORMAT (20X, 15H  VALUES OF WMIC, / )
0159               400   FORMAT ((1H, 4(F4.2, 3X)))
0160               401   FORMAT ((1H, 12(F4.2, 3X)))
0161               640   FORMAT ( / / / , 26H PROB OF FIRST PASSAGE TIME, / / )
0162               666   FORMAT (10X, 5H STEP, I2, / )
0163               999   FORMAT (12 F 4.2)
0164              1000 FORMAT (4 F 4.2)
0165              1001 FORMAT (2I2)
0166              1003 FORMAT (1H0, 4(F6.4, 4X))
```

Figure 1.22. Program for system reliability evaluation (p.4),

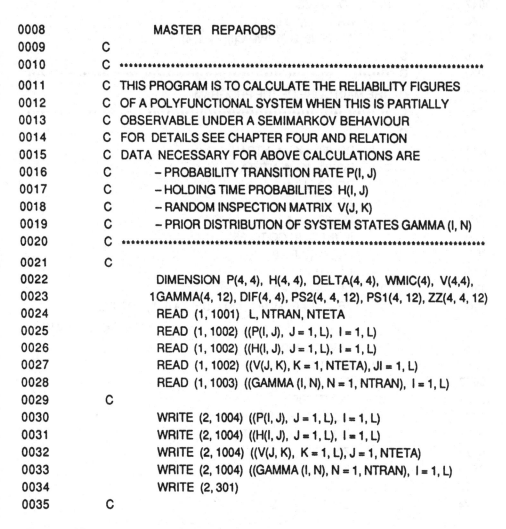

```
0167                1007 FORMAT (1H0, 10X, 7H STEP 1, / )
0168                7325 FORMAT ( / / , 44H  RELIABILITY FIGURES COMPLETELY
                                                    OBSERVABLE, / / )
0169                7731 FORMAT (10X, 32H PROCESS DURATION IN TRANSITION, / / )
0170                7732 FORMAT (10X, 25H PROCESS DURATION IN TIME, / / )
0171         C  ••••••••••••••••••••••••••••••••••••••••••••••••••••••••••••••••••••
0172         C
0173                STOP
0174                END


END OF SEGMENT, LENGTH 1051, NAME ADRIANMAINT


0008                    MASTER   REPAROBS
0009         C
0010         C  ••••••••••••••••••••••••••••••••••••••••••••••••••••••••••••••••••••
0011         C THIS PROGRAM IS TO CALCULATE THE RELIABILITY FIGURES
0012         C OF A POLYFUNCTIONAL SYSTEM WHEN THIS IS PARTIALLY
0013         C OBSERVABLE UNDER A SEMIMARKOV BEHAVIOUR
0014         C FOR  DETAILS SEE CHAPTER FOUR AND RELATION
0015         C DATA  NECESSARY FOR ABOVE CALCULATIONS ARE
0016         C        – PROBABILITY TRANSITION RATE P(I, J)
0017         C        – HOLDING TIME PROBABILITIES H(I, J)
0018         C        – RANDOM INSPECTION MATRIX V(J, K)
0019         C        – PRIOR DISTRIBUTION OF SYSTEM STATES GAMMA (I, N)
0020         C  ••••••••••••••••••••••••••••••••••••••••••••••••••••••••••••••••••••
0021         C
0022                DIMENSION  P(4, 4),  H(4, 4),  DELTA(4, 4),  WMIC(4),  V(4,4),
0023                1GAMMA(4, 12), DIF(4, 4), PS2(4, 4, 12), PS1(4, 12), ZZ(4, 4, 12)
0024                READ (1, 1001)  L, NTRAN, NTETA
0025                READ (1, 1002)  ((P(I, J),  J = 1, L),  I = 1, L)
0026                READ (1, 1002)  ((H(I, J),  J = 1, L),  I = 1, L)
0027                READ (1, 1002)  ((V(J, K),  K = 1, NTETA), JI = 1, L)
0028                READ (1, 1003)  ((GAMMA (I, N), N = 1, NTRAN),  I = 1, L)
0029         C
0030                WRITE (2, 1004)  ((P(I, J),  J = 1, L),  I = 1, L)
0031                WRITE (2, 1004)  ((H(I, J),  J = 1, L),  I = 1, L)
0032                WRITE (2, 1004)  ((V(J, K),   K = 1, L), J = 1, NTETA)
0033                WRITE (2, 1004)  ((GAMMA (I, N), N = 1, NTRAN),  I = 1, L)
0034                WRITE (2, 301)
0035         C
```

Figure 1.22. Program for system reliability evaluation (p.5),

```
0036        C  ••••••••••••••••••••••••••••••••••••••••••••••••••••••••••••••••••••••••
0037        C  CALCULATE THE PROBABILITY MASS FUNCTION FOR TAU(I), THE
0038        C  WAITING TIME, W(I).
0039        C  ••••••••••••••••••••••••••••••••••••••••••••••••••••••••••••••••••••••••
0040        C
0041               DO 801  I = 1, L
0042               PI1 = 0,
0043               DO 802  J = 1, L
0044               DIF(I, J) = 1.0 – H(I, J)
0045        802    PI1 = PI1 + P(I, J) • DIF(I, J)

0046               WMIC(I) = PI1
0047        801    CONTINUE
0048               WRITE  (2, 333)
0049               WRITE  (2, 202)  (WMIC(I), I = 1, 4)
0050        C
0051        C  ••••••••••••••••••••••••••••••••••••••••••••••••••••••••••••••••••••••••
0052        C  THE MAIN CORE OF THE PROGRAM
0053        C  ••••••••••••••••••••••••••••••••••••••••••••••••••••••••••••••••••••••••
0054        C
0055               DO 30  KK = 1, NTRAN
0056               DO 31  K = 1, NTETA
0057               DO 32  J = 1, L
0058               ZB = 0.
0059               DO 33  I = 1, L
0060               IF (I .EQ. J)  GO TO 35
0061               DELTA (I, J) = 0.
0062               GO TO 33
0063        35     DELTA (I, J) = 1.
0064        33     ZB = ZB + GAMMA (I, KK) • (DELTA (I, J)) • WMIC(I) +
                                             P(I, J) • H(I, J) • V(J, K)

0065               PS2(J, K, KK) = ZB
0066        32     CONTINUE
0067        31     CONTINUE
0068        30     CONTINUE
0069               DO 21  KK = 1, NTRAN
0070               DO 22  K = 1, NTETA
0071               ZA = O.
0072               DO 23 J = 1, L
0073               Z = 0.
0074               DO 24  I = 1, L
```

Figure 1.22. Program for system reliability evaluation (p.6),

```
0075                    IF (I .EQ. J) GO TO 25
0076                    DELTA (I, J ) = 0.
0077                    GO TO 24
0078            25      DELTA (I, J) = 1.
0079            24      Z = Z + GAMMA (I, KK) * (DELTA (I, J)) * WMIC(I) +
                                                  P(I, J) * H(I, J) * V(J, K))
0080                    PS(J, K, KK) = Z
0081            23      ZA = ZA + PS(J, K, KK)
0082                    PS1(K, KK) = ZA
0083            22      CONTINUE
0084            21      CONTINUE
0085                    DO 41   KK = 1, NTRAN
0086                    DO 42   K = 1, NTETA
0087                    DO 43   J = 1, L
0088                    ZZ(J, K, KK) = PS2(J, K, KK) / PS1(K, KK)
0089            43      CONTINUE
0090            42      CONTINUE
0091            41      CONTINUE
0092                    DO 700   KK = 1, NTRAN
0093                    DO 703   K = 1, NTETA
0094                    WRITE  (2, 702) KK, K
0095                    WRITE  (2, 1004) (ZZ(J, K, KK), J = 1, L)
0096            703     CONTINUE
0097            700     CONTINUE
0098     C
0099     C  ••••••••••••••••••••••••••••••••••••••••••••••••••••••••••••••••••
0100            202     FORMAT (1H, 4(F7.5, 2X))
0101            301     FORMAT (1H, 10X, 53H RELIABILITY FIGURES FOR PARTIALL
0102                    1Y OBSERVABLE SYSTEMS, / / )
0103            333     FORMAT (20X, 15H VALUES OF WMIC, / )
0104            702     FORMAT (10X, 8H STEP NO, I2, 11H OUTPUT NO , I2, / / )
0105            1001    FORMAT (3I2)
0106            1002    FORMAT (4F4.2)
0107            1003    FORMAT ( 4 F5.3 )
0108            1004    FORMAT ( / / / ,1H , 4(F4.2, 3X) )
0109     C  ••••••••••••••••••••••••••••••••••••••••••••••••••••••••••••••••••
0110     C
0111                    STOP
0112                    END
```

END OF SEGMENT, LENGTH 724, NAME REPAROBS

Figure 1.22. Program for system reliability evaluation (p.7).

NUMERICAL EXAMPLE. Consider a completely observable system with three
fully operational states and two states undergoing repair. For computation purposes
the probability transition matrix is:

$$
\begin{array}{cccccc}
 & 1 & 2 & 3 & 4 & 5 \\
\end{array}
$$

$$
P = \begin{array}{c} 1 \\ 2 \\ 3 \\ 4 \\ 5 \end{array}
\left[
\begin{array}{ccccc}
0.8431 & 0.0092 & 0.0310 & 0.0932 & 0.0235 \\
0.1835 & 0.6192 & 0.0927 & 0.0946 & 0.0100 \\
0.1545 & 0.1920 & 0.4920 & 0.0932 & 0.0683 \\
0.0 & 0.0 & 0.0 & 0.0 & 1.0 \\
0.0 & 0.0 & 0.0 & 0.0 & 1.0
\end{array}
\right]
$$

The holding-time mass functions for the system under consideration are given
in Table 1.11 for $p_{ij} \geq 0$.

The results of a computer simulation for the reliability function $R(n)$ with three
different initial state vectors $\alpha(0)$ are given in Figure 1.23.

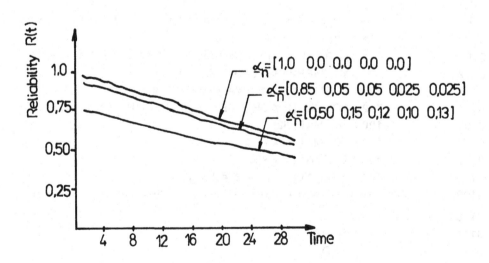

Figure 1.23. Results of computer simulation.

Day m	$h_{11}(m)$ $h_{12}(m)$	$h_{13}(m)$ $h_{14}(m)$ $h_{15}(m)$	$h_{21}(m)$ $h_{22}(m)$	$h_{23}(m)$ $h_{24}(m)$	$h_{25}(m)$ $h_{31}(m)$	$h_{32}(m)$ $h_{33}(m)$	$h_{34}(m)$ $h_{35}(m)$
1	0.0342	0.0418	0.0031	0.1821	0.1514	0.1824	0.1613
2	0.0421	0.0523	0.0049	0.1120	0.0924	0.0925	0.0949
3	0.1525	0.1828	0.1124	0.0930	0.1134	0.0947	0.2144
4	0.1145	0.1516	0.1344	0.0750	0.0824	0.1229	0.1312
5	0.2612	0.0743	0.2184	0.0820	0.0944	0.0043	0.0074
6	0.0321	0.1204	0.0039	0.1520	0.1825	0.2144	0.1085
7	0.1140	0.0985	0.1185	0.1347	0.0320	0.1320	0
8	0	0.0003	0.0040	0.0025	0.1024	0.0987	0.0943
9	0.2164	0	0.0325	0	0.0003	0	0.1291
10	0.0032	0	0	0.0444	0	0.0072	0.0520
11	0.0200	0.0643	0	0.0932	0	0	0
12	0	0	0	0	0.0954	0	0
13	0	0	0.0002	0.0092	0.0040	0.0369	0
14	0.0098	0.0749	0	0	0	0	0.0069
15	0	0.1241	0	0.0005	0	0	0
16	0	0.0143	0	0	0.0009	0	0
17	0	0	0	0	0	0.0100	0
18	0	0	0.0004	0	0.0421	0	0
19	0	0	0.1524	0	0	0	0
20	0	0.0004	0.0985	0	0	0	0
21	0	0	0	0	0	0	0
22	0	0	0.1085	0	0	0	0
23	0	0	0	0.0097	0	0.0040	0
24	0	0	0	0	0	0	0
25	0	0	0	0	0	0	0
26	0	0	0	0	0.0032	0	0
27	0	0	0	0	0	0	0
28	0	0	0	0	0	0	0
29	0	0	0	0	0	0	0
30	0	0	0.0079	0.0097	0.0032	0	0

Table 1.11.   Holding-time mass functions.

1.7.2.3. *A Markovian Interpretation for PERT networks.* As is well known from graph theory, the *PERT* method is a technique which describes the networks of inter-related activities which must be undertaken in the course of accomplishing some specified objective(s).

In a PERT network (a probabilistic version of the Critical Path Method) we can emphasize the START and END activities as well as the set of the n independent activities within the graph (see Figure 1.24).

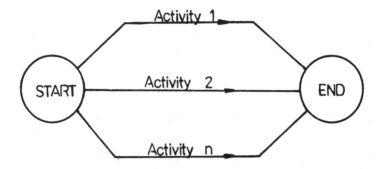

Figure 1.24. Diagrammatic representation of PERT network.

A particular case of such a network is when $n = 2$; then both activities can be performed simultaneously and which must both be completed in order to finish the project. In the case where activity one takes a longer period of time to perform than activity two, then activity one is said to be on the critical path since its completion establishes the completion time of the given project.

As is well known in the PERT method the periods of time required to perform the activities are considered as random variables with a Beta probability density function. As was emphasized by Holmes [55] "with the assumption of random times, the concept of critical path becomes somewhat fuzzy, since one is no longer certain which activity takes the longer time".

By extension of the above problem, Holmes takes a different approach; he considers that the PERT probability distribution problem lies in the field of Markov chains.

In Figure 1.25 a simple two-state Markov chain is shown.

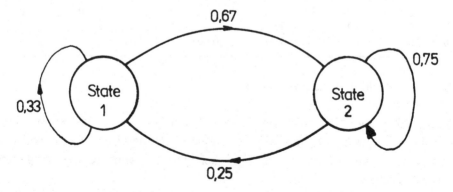

Figure 1.25. Simple two-state Markov chain.

In the case that $p_{22} = 1$, state 2 is defined as a *trapping state*. In the case of a Markov chain with an absorbing state, the question of the steady state of the system is not of practical interest. The problem is to compute the time needed to arrive in such a state. If we consider the example represented in Figure 1.26 for the situation when the system starts in state 1, with probability 0.67, it will be in state 1 for exactly one time period before making a transition to state 2. The probability of the system being in state 1 for two time periods before making a transition to state 2 is (0.33) (0.75). Proceeding in a similar manner, the probability distribution of the times to START at 1 and END at 2 can be shown to be a geometric probability function.

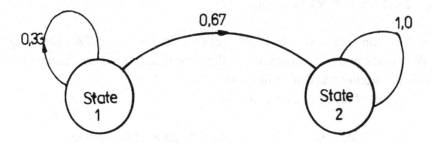

Figure 1.26. Example of simple two-state Markov chain.

When a system is described by three states (see Figure 1.27), the probability distribution of the times required to START at 1 and END at 3 is the convolution of the times to get from 1 to 2 and from 2 to 3.

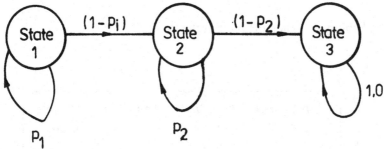

Figure 1.27. Three-state Markov chain.

In the case where $p_1 = p_2$, the distribution is the convolution of the two geometrics with identical values of their parameters and is referred to as a Pascal or a negative binomial.

Next, taking into consideration the above assumptions, we can show the way to a direct connection between Markov processes and PERT networks.

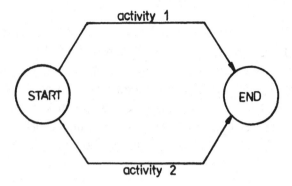

Figure 1.28. Simple PERT network.

Let us consider a simple PERT network (see Figure 1.28); the simplest equivalent Markov chain is obtained if the time distributions are assumed to be geometric probability distribution functions.

The states of the system are defined as:

State number	Possible states of the system
1	1-start and 2-start
2	1-start and 2-end
3	1-end and 2-start
4	1-end and 2-end

The probability transition matrix is of the form:

$$
P = \begin{array}{c} \\ 1 \\ 2 \\ 3 \\ 4 \end{array}
\begin{array}{cccc}
1 & 2 & 3 & 4 \\
\left[\begin{array}{cccc}
(1-p_1)(1-p_2) & (1-p_1)p_2 & p_1(1-p_2) & p_1 p_2 \\
0 & 1-p_1 & 0 & p_1 \\
0 & 0 & 1-p_2 & p_2 \\
0 & 0 & 0 & 1
\end{array}\right]
\end{array}
$$

Let us consider a new state X which interacts with states 1 and 2 as represented in a PERT network (see Figure 1.29).

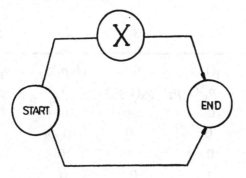

Figure 1.29. PERT network with new state X.

The corresponding system states are of the form:

State number	Possible states of the system
1	1-start and 2-start
2	1-X and 2-start
3	1-end and 2-start
4	1-start and 2-end
5	1-X and 2-end
6	1-end and 2-end

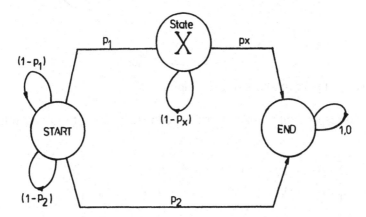

Figure 1.30. Transition diagram for network with new X state.

The transition diagram is given in Figure 1.30 and the probability transition matrix is of the form:

	1	2	3	4	5	6
1	$(1-p_1)(1-p_2)$	$p_1(1-p_2)$	0	$(1-p_1)p_2$	p_1p_2	0
2	0	$(1-p_x)(1-p_2)$	$p_x(1-p_2)$	0	$(1-p_x)p_2$	p_xp_2
3	0	0	$1-p_2$	0	0	p_2
4	0	0	0	$1-p_1$	p_1	0
5	0	0	0	0	$1-p_x$	p_x
6	0	0	0	0	0	1

$P =$

The solution of this system will give the probability distribution of times required to complete the project.

NUMERICAL EXAMPLE. Assume the PERT network given in Figure 1.29 for which we have the information on the following probabilities: $p_1 = 0;9$, $p_2 = 0.2$, $p_x = 0.7$ (Figure 1.31).

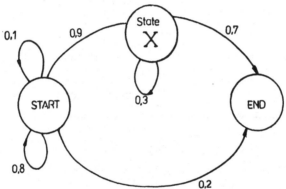

Figure 1.31. Diagram for numerical example.

The corresponding probability transition matrix, using the above relation is as follows:

$$P = \begin{bmatrix} 0.08 & 0.72 & 0 & 0.02 & 0.18 & 0 \\ 0 & 0.24 & 0.56 & 0 & 0.06 & 0.14 \\ 0 & 0 & 0.8 & 0 & 0 & 0.2 \\ 0 & 0 & 0 & 0.1 & 0.9 & 0 \\ 0 & 0 & 0 & 0.3 & 0.7 & \\ 0 & 0 & 0 & 0 & 0 & 1 \end{bmatrix}$$

If we consider the starting vector $\pi = [1, 0, 0, 0, 0, 0]$ then by the multiplication πP we have the following product vector $\pi = [0.08, 0.72, 0, 0.02, 0.18, 0]$ which indicates the probability of being in states 1 - 6 one time period hence. Since $\pi_6 = 0$, we know that the system cannot reach the END state in 1 time period. Next, computational results will be given in Table 1.12.

Number of Transitions N	Probability of reaching END state after N transitions (last element in vector π)	Probability that END state is reached in exactly N time periods
0	1	2
0	0	-
1	0	0
2	0.2268	0.2268
3	0.4304	0.2036
4	0.5691	0.1387
5	0.6650	0.0959
6	0.7534	0.0704
7	0.7895	0.0541
8	0.8320	0.0425
9	0.8657	0.0337
10	0.8926	0.0269
11	0.9141	0.0215
12	0.9313	0.0172
13	0.9450	0.0137
14	0.9560	0.0110
15	0.9648	0.0088
16	0.9719	0.0071
17	0.9775	0.0056
18	0.9820	0.0045
19	0.9856	0.0036
20	0.9885	0.0029
21	0.9908	0.0023
22	0.9926	0.0018
23	0.9941	0.0015
24	0.9953	0.0012
25	0.9962	0.0009
26	0.9969	0.0007
27	0.9976	0.0007
28	0.9981	0.0005
29	0.9984	0.0003
30	0.9988	0.0004
31	0.9990	0.0002
33	0.9992	0.0002
32	0.9994	0.0002

Table 1.12. Computational results.

The numerical values from the above table indicate the probability of reaching END state after n transitions as well as the probability that this state is reached in exactly N time periods.

1.8   Dynamic decision models for clinical diagnosis

A unified approach to clinical decision making will be investigated in Chapter 5.

The complex process of medical diagnosis has traditionally relied on the experience and judgement of the clinician. With the increased application of systems techniques to medical problems in general, it is timely to consider their application to diagnosis and treatment situations as, inherently, clinicians are using some form of pattern recognition techniques, cause-effect relationships and decision rules in the diagnosis-treatment process.

Decision theory and dynamic probabilistic models are well established and several decision models have been applied to medical problems. Equally pattern recognition techniques have proved useful in medical diagnosis and classification where large quanitities of information have to be processed. Unlike the real clinical situation, however, there have been few attempts at combining these techniques in an overall diagnostic decision model.

In an attempt to remedy this deficiency, partially observable Markov decision processes (Markov or semi-Markov) are here combined with cause-effect models as a probabilistic representation of the diagnostic process. Pattern recognition techniques are used in a first stage of disease classification within which the patient states can be identified. The methodology is given for combining the patient state of health, the clinician's state of knowledge of the cause-effect representation from the observation space (measurements), feature selection, using pattern recognition techniques, and finally, the treatment decisions with which to restore the patient to a more desirable state of health. A cost functional for the decision process has then to be optimized according to some pre-assigned objective function (social return from the patient state of health or treatment cost for the patient), when the process has an infinite time horizon.

1.8.1. PATTERN RECOGNITION

The cause-effect model provides a mapping from a set of input measurements to a diagnosis/primary event. In principle, therefore this alone is sufficient for medical diagnosis. In practice the number of possible measurements may be of the order of 20,000 and the number of possible diagnoses of the order of 6,000. A single cause-effect model to achieve such a mapping would clearly be unwieldy and probabily impossible to design.

Similarly a pattern recognition system provides an adequate mapping from a set of input measurements to the appropriate pattern class (diagnosis/primary event). Again the complexity of the medical diagnosis problem renders a one-step pattern recognition solution unrealistic.

Both cause-effect models and pattern recognition, however, have advantages worth retaining. With the former, problem knowledge is readily incorporated into the decision structure. How to introduce such problem knowledge into pattern recognition is, currently, not well understood.

Pattern techniques, however, are ideally suited to extracting information from the large clinical data base. The system operates on the raw measurements together with any features derived from those measurements with the idea pf producing a broad classification (cluster) into disease classes (e.g. respiratory disease, circulatory disease, thyroid disease, etc.). The classification obtained is transfered to the cause-effect model, together with the raw measurements and derived features, for a more detailed classification. This will comprise identification of the primary cause and associated recommended treatment decision.

In this way the information to be handled by the pattern recognition system is reduced to a feasible level and the demands made upon the cause-effect model are similarly eased. Effectively, the pattern recognition system could drive the cause-effect model to an appropriate area of uncertainty. This could be realised, for example, by having a multiplicity of cause-effect sub-models, one for each specific disease class assigned by the pattern recognition system.

The problem of the possible number of measurements which may be taken can be eased in several ways. First, the clinician can exclude certain measurements after a preliminary examination of the patient. Second, features and complex features can be defined, using problem knowledge, as functions of raw measurements. Whilst not reducing the number of measurements to be taken, it does reduce the number to be handled by the pattern recognition system and the cause-effect model — since these are needed to examine the features and complex features only. Third, sequential pattern recognition techniques can be employed. With these techniques it is possible to decide whether the measurements already taken carry sufficient information for a classification to be made and, if they do not, which measurement should be taken next to optimally increase the information. The optimal choice amongst the remaining measurements can incorporate the relative cost and difficulty of taking each particular one. Similarly, the cause-effect model can incorporate the facility of calling for particular additional measurements to achieve the final diagnosis.

1.8.2. MODEL OPTIMIZATION

The model envisaged in this section is the result of careful analysis of the real world clinical diagnosis and treatment process. The implementation of the theoretical results and methodology given here appears feasible and, with the present computational facilities, compact clinical systems (i.e. with small numbers of patient states and cause-effect models) can be analysed. For example, the development of dynamic decision models for thyroid disease is part of the on-going research programme of some of the authors. Data is available, or being collected, so as to calculate the appropriate probability rate transitions, holding time cumulative distribution functions and inspection probabilitites (see model formulation in Chapter 5).

Large scale diagnostic systems, incorporating the dynamic behaviour of the patient, can reveal the practical limitations of the class of model developed here. For example, large cause-effect models and a high order decision space reveal the inadequacies of existing computational techniques, indicating the need for more O.R. work, developing improved algorithms with which to evaluate the optimal decision strategy, and to use AI and knowledge based systems.

The class of models given here, defining theoretically semi-Markov decision processes when the states are partially observable, is novel. It is potentially applicable to a wide range of diagnostic systems, including those in the field of systems engineering, and could stimulate further theoretical research. The analysis could, however, be taken a stage further, applying risk-sensitive Markov decision models to the diagnosis-treatment process and defining a utility functional over the rewards of the treatment process. The maximizing of the expected utility functions in a diagnosis-treatment process has previously been used by Ginsberg for the case of steady-state system behaviour [70]. This application of using Markov models when the decision maker (i.e. the clinician) is risk sensitive and the process is only partially observable leads to a better understanding over the spectrum of "optimal strategies" when the clinician, in taking a treatment decision, exhibits risk aversion or risk preference behaviour [85], [89], [90], [92], [97].

Advanced expert systems based on knowledge processing techniques using inferential schemes could embody dynamic models as Markovian ones for improving the ability of the model itself to discriminate amone dynamics and internal structures of the diagnosed system.

CHAPTER 2

DYNAMIC AND LINEAR PROGRAMMING

Basic optimization models in decision problems in systems with dynamic probabilistic structures are of the dynamic programming and linear programming type.

Next, basic elements of dynamic programming and linear programming will be presented from the viewpoint of their application in the solution of Markovian decision processes.

2.1. DISCRETE DYNAMIC PROGRAMMING

Dynamic programming is an optimization method which transforms a complex problem into a sequence of simple problems. It is an optimization procedure characterized by a multiple nesting of states and decisions pertinent to the system. This method, at the same time, ensures a general framework for the analysis of a large variety of problems [7], [8], [13], [17], [24], [33].

Within the terminology of dynamic programming for each moment when decisions are taken, we use the notion of *stage* in the decision making process.

The principle of optimality (Bellman) is of basic importance in dynamic programming and has the following formulation:

"Any optimal policy has the property that, regardless of the current state and the applied decision for the system, decisions which are to be taken next represent an optimal policy related to the resultant state from the current decision".

Let us consider a multi-step decision process where the yield/loss for a given step has the form:

$$g_n(d_n, s_n),$$

where d_n is an admissible decision out of D_n and s_n is the state of the process when n steps are left till the end of the process. The set D_n of the decisions at time n depends on the state of the process and then we can write $D_n(d_n)$. We make the assumption that there exist N steps within the process , and n represents the number of stages left till the end of the process.

The next state of the process depends entirely on the current state of the process as well as the current decision taken in that state. The transition function of the system is of the following form:

$$s_{n-1} = f_n(d_n, s_n), \tag{2.1}$$

where $d_n \in D_n$ is a selected decision for the current state and system state. In the simplest case, the uncertainties in the transitions among the system states are not included. In Figure 2.1 the multi-step decision process is represented. The benefits for each step are considered independently.

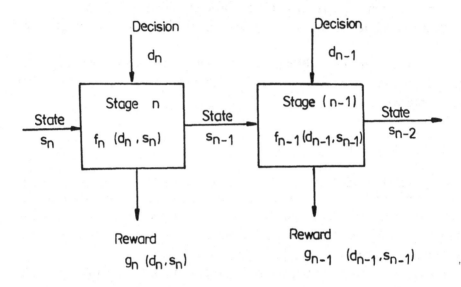

Figure 2.1. The multi-step decision process.

In the case of a general dynamic programming problem, the goal consists in maximizing the sum of the yield functions (or minimizing the cost functions) for all the steps within the decision process.

The constraint consists only in the finite character of the decisions. In each of the system states, a transition in the state space is given by the relation of type (2.1).

When the system is in the state s_n with n steps before the end of the process, the optimization problem consists of choosing decision variables $d_n, d_{n-1}, \dots, d_0$ to solve the following problems:

$$v_n(s_n) = \max \{g_n(d_n, s_n) + g_{n-1}(d_{n-1}, s_{n-1}) + \dots + g_0(d_0, s_0)\} \qquad (2.2)$$

with the following constraints:

$$s_{m-1} = f_m(d_m, s_m); \quad m = 1, 2, \dots, n,$$

$$d_m \in D_m; \quad m = 0, 1, \dots, n.$$

In the above relation $v_n(s_n)$ is called the *objective value function*.
Equation (2.2) can be written successively, so that:

$$v_n(s_n) = \max \{g_n(d_n, s_n) + \max[g_{n-1}(d_{n-1}, s_{n-1}) + \dots + g_0(d_0, s_0)]\} \qquad (2.3)$$

under the constraints:

$$s_{n-1} = f_n(d_n, s_n)$$

$$d_n \in D_n,$$

$$s_{m-1} = f_m(d_m, s_m); \quad m = 0, 1, \dots, n-1,$$

$$d_m \in D_n; \quad m = 0, 1, \dots, n-1.$$

Because the right hand side of equation (2.3) is an optimal value function for the dynamical programming with $(n-1)$ steps remaining till the end of the process, we can write the following relation:

$$v_n(s_n) = \max [g_n(d_n, s_n) + v_{n-1}(s_{n-1})] \qquad (2.4)$$

with the constraints:

$$s_{n-1} = g_n(d_n, s_n)$$

$$d_n \in D_n.$$

To emphasize in a more useful way that we are dealing with an optimization process, d_n, the relation (2.4) can be written so that:

$$v_n(s_n) = \max \{g_n(d_n, s_n) + v_{n-1}[f_n(d_n, s_n)]\} \qquad (2.5)$$

with the constraint $d_n \in D_n$.

The above representation is in full agreement with the principle of optimality given above.

The problem at step zero is not defined in a recursive way and it can be expressed as:

$$v_0(s_0) = \max g_0(d_0, s_0)$$

with the constraint $d_0 \in D_0$.

Next, discounted dynamic programming will be presented.

The discounting factor is β, so that $\beta = 1/(1 + i)$ or, in a more general way, $\beta^n = 1/(1 + i)^n$, where i is the interest rate and n is the year of the discounting.

The future discounted earnings are given by the relation:

$$v_n(s_n) = \max \{g_n(d_n, s_n) + \beta g_{n-1}(d_{n-1}, s_{n-1}) + \beta^2 g_{n-2}(d_{n-2}, s_{n-2}) +$$

$$+ \ldots + \beta^n g_0(d_0, s_0)\}, \tag{2.6}$$

where we must add the following constraints:

$$s_{n-1} = f_n(d_{n, s_n}); \quad d_n \in D_n,$$

$$s_{m-1} = f_m(d_m, s_m); \quad m = 1, 2, \ldots, n - 1,$$

$$d_m \in D_m; \quad m = 0, 1, 2, \ldots, n - 1.$$

Relation (2.6) can also be written with the form:

$$v_n(s_n) = \max [g_n(d_n, s_n) + \beta v_{n-1}(s_{n-1})] \tag{2.7}$$

with the constraints:

$$s_{n-1} = f_n(d_{n, s_n}); \quad d_n \in D_n.$$

The case of Markovian decision models involves the introduction of dynamic programming under uncertainty.

In this case it is assumed that an uncertain event $\tilde{e}_n$ exists, for which the results e_n are not controlled by the decision maker. The functions $g_n(\,.\,)$ are of the form $g_n(d_n, s_n, \tilde{e}_n)$; for the definition of the state s_{n-1} we can use the notation:

$$\tilde{s}_{n-1} = f_n(d_n, s_n, \tilde{e}_n).$$

By introducing the above random variable to the decision process under uncertainty, we can associate a probability measure given by $p_n(e_n | d_n, s_n)$.

In this case we can write:

$$v_n(s_n) = \max E \{(g_n(d_n, s_n, \tilde{e}_n) + v_{n-1}(\tilde{s}_{n-1})\} \tag{2.8}$$

with the following set of constraints:

$$\tilde{s}_{n-1} = f_n(d_n, s_n, \tilde{e}_n); \quad d_n \in D_n,$$

where $E\{ \cdot \}$ indicates the expected value of the quantities in parenthesis. To initiate the recursive computations we can define:

$$v_0(s_0) = \max E \{g_0(d_0, s_0, \tilde{e}_0)\},$$

with $d_0 \in D_0$.

2.2   A linear programming formulation and an algorithm for computation

The model of linear programming is well known in the literature of operations research. Next, we shall present, in a canonical and matrix form, the LP problem and the adequate simplex algorithm. This model (LP) is used as an alternative optimization solution for the decision process in dynamic probabilistic systems.

2.2.1. A GENERAL FORMULATION FOR THE LP PROBLEM AND THE SIMPLEX METHOD

Under the more general form the LP model implies maximization/minimization of the objective function under a given set of constraints which are linear in form.

To solve a model using an LP means finding n variables, $x_1, x_2, \dots, x_n$ to maximize the objective function Z, where:

$$Z = c_1 x_1 + c_2 x_2 + \dots + c_n x_n \tag{2.9}$$

under the following constraints:

$$a_{11}x_1 + a_{12}x_2 + \dots + a_{1n}x_n \leq b_1$$

$$a_{21}x_1 + a_{22}x_2 + \dots + a_{2n}x_n \leq b_2$$

$$\cdots \cdots \cdots \cdots \cdots \cdots$$

$$a_{m1}x_1 + a_{m2}x_2 + \dots + a_{mn}x_n \leq b_m \tag{2.10}$$

and:

$$x_1 \geq 0, \ x_2 \geq 0, \dots, x_n \geq 0 \tag{2.11}$$

where c_j, a_{ij}, b_i are given constraints for the process to be modelled.

The values of decision variables $x_1, x_2, \dots, x_n$ which satisfy, simultaneously, the constraints (2.10) and (2.11) are called admissible solutions for the LP. The admissible solution which also optimizes the objective function Z is called the *optimal admissible solution.*

The model to solve an LP is well known as the *simplex model.*

Concerning the canonical form of an LP problem we can mention the following:

a) all the decision variables are non-negative;

b) all the constraints, except those of non-negativity, will be given under the equality form;

c) left hand coefficients are all non-negative.

Any LP-problem can have an adequate canonical form.

Optimality criteria : We can consider that in a maximization problem each non-basic variable has a non-positive coefficient within the objective function for the canonical form.

Then the basic feasible solution given by the canonical form maximizes the objective function relative to the feasibility region.

The procedure by which a new basic variable is generated is called *pivoting* and represents the main element within the Simplex method.

After pivoting, the modified equations still represent the original problem and have the same feasible solutions, and objective-function values, as for any basic feasible solution.

Improving criteria : We can make the assumption that, in a maximization problem, one of the negative variables has a positive coefficient within the objective function given in a canonical form. For the case when that variable has a positive coefficient within one of the constraints, by pivoting, we can find a new basic feasible solution.

There are a number ways of isolating basic variables within the constraints if this is not directly transparent. The Simplex method introduces new variables known as *artificial variables* !

Because, in this chapter, an LP problem is presented for the case of maximizing an objective function, Step I of the algorithm for solving an LP problem maximizes w as the the negative of the sum of the artificial variables, instead of directly minimizing the sum of these variables.

The canonical form for Step I in an LP is determined by adding the artificial variables to the w equation.

If $w = 0$, then all artificial variables must be zero. If, moreover, each artificial variable is non-basic within this optimal solution, the basic variables can be determined from the original variables.

The artificial variables were added to obtain a canonical form for the LP problem. By maximizing w we obtain:

a) max $w < 0$ and then the problem is not feasible and the optimization process stops;

b) max $w = 0$ and then we obtain a canonical form to initiate an original problem.

Then apply the optimality criteria and the improvement of the original objective function Z, starting from the canonical form.

In reducing an LP problem in a canonical form, it is easier to manage the necessary transformations by using the following sequence

(i)   replace each decision variable without constraint in sign by a   difference between two non-negative variables;

(ii)   replace inequalities by equalities, introducing some additionl variables;

(iii) multiply equations by (-1);

(iv)   add an artificial variable for each equation.

In Figure 2.2 a flow-chart for the Simplex algorithm is given.

In Step II of the algorithm, the problem can be written in the following canonical form:

$$x_1 \qquad +\bar{a}_{1,m+1}x_{m+1} + \ldots + \bar{a}_{1s}x_s + \ldots + \bar{a}_{1n}x_n = \bar{b}_1$$

$$x_2 \qquad +\bar{a}_{2,m+1}x_{m+1} + \ldots + \bar{a}_{2s}x_s + \ldots + \bar{a}_{2n}x_n = \bar{b}_2$$

$$x_r \qquad +\bar{a}_{r,m+1}x_{m+1} + \ldots + \bar{a}_{rs}x_s + \ldots + \bar{a}_{rn}x_n = \bar{b}_r$$

$$x_m \qquad +\bar{a}_{m,m+1}x_{m+1} + \ldots + \bar{a}_{ms}x_s + \ldots + \bar{a}_{mn}x_n = \bar{b}_m$$

$$(-Z) \qquad +\bar{c}_{m+1}x_{m+1} + \ldots + \bar{c}_s x_s + \ldots + \bar{c}_n x_n = \bar{Z}_0$$

$$x_j \geq 0; \ j = 1, 2, \ldots, n .$$

Parameters $\bar{a}_{ij}$, $\bar{b}_i$, $\bar{Z}_0$, $\bar{w}_0$ and $\bar{c}_j$ are known values. these values are even the original data (non-based) of the LP-problem. The assumption was made that $x_1, x_2, \ldots, x_m$ are basic variables. In a canonical form, $\bar{b}_i \geq 0$, $i = 1, 2, \ldots, m$.

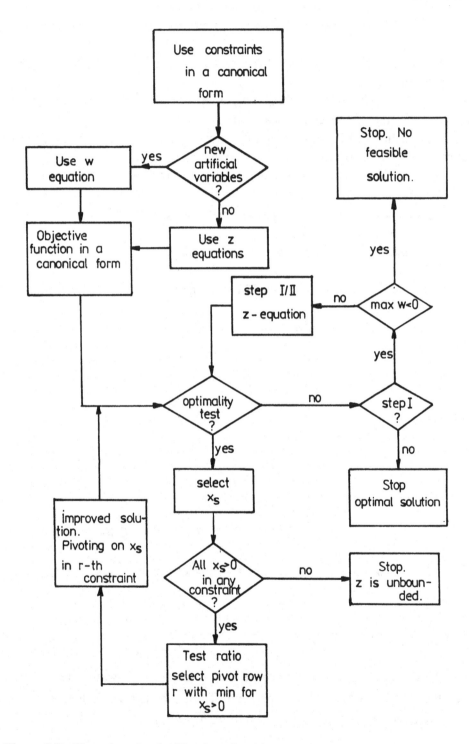

Figure 2.2. Flow chart for the Simplex algorithm.

The Simplex algorithm in a maximization form involves the following procedure:

Step 1: Initially the problem is in a canoncal form and all $\bar{b}_i \geq 0$.
Step 2: If $\bar{c}_j \leq\, = 0$ for $j = 1, 2, \dots , n$ then Stop; the optimal solution is found. If $\bar{c}_j > 0$ exist, then continue the same procedure.
Step 3: Choose a column for which the pivoting operation is performed such that:

$$\bar{c}_s = \max_{j} \{ \bar{c}_j \mid \bar{c}_j \geq 0 \}.$$

Iff $\bar{a}_{is} \leq 0$ for $i = 1, 2, \dots , m$, then Stop. The primal problem is unbounded. If we continue, then $\bar{a}_{is} \leq 0$ for some $i = 1, 2, \dots , m$.
Step 4: Choose row r for pivoting by using the following test:

$$\frac{\bar{b}_r}{\bar{a}_{rs}} = \min_{i} \left\{ \frac{\bar{b}_i}{\bar{a}_{is}} \mid \bar{a}_{is} > 0 \right\}.$$

Step 5: Replace the basic variable in row r with variable s and re-evaluate the canonical form (e.g. the pivot for $\bar{a}_{rs}$ coefficient).
Step 6: Go to Step 2.
The pivot can be appreciated as being performed in the following two stages:
a) normalization for the r-th constraint such that x_s has a (+1) coefficient;
b) eliminate variable x_s by using adequate procedures.

The algorithm and the general procedure for the Simplex method attached to an LP-problem is given in Figure 2.2. It operates in a finite number of iterations.

2.2.2. LINEAR PROGRAMMING - A MATRIX FORMULATION

As was already shown, an LP model is of the following form:

$$\text{Max} \sum_{i=1}^{n} c_i x_i$$

with the set of constraints:

$$x_i \geq 0 \text{ for } i = 1, 2, \dots , n .$$

In matrix form we can write:

$$\text{Max } \underline{c}\ \underline{x}$$

under the constraints:

$$A\underline{x} \le \underline{b}\ ;\ \underline{x} \cdot \ge 0.$$

The LP dual is given by:

$$\text{Minimize } \sum_{i=1}^{m} b_i y_i$$

with the following set of constraints:

$$\sum_{i=1}^{m} a_{ij} y_j \ge c_j\ ;\ j = 1, 2, \ldots, n\ ,$$

$$y_i \ge 0\ ;\ i = 1, 2, \ldots, m.$$

and in matrix form is given by:

$$\text{Minimize } \underline{y}\ \underline{b}$$

under the constraints:

$$\underline{y}\ A \ge \underline{c}$$

$$\underline{y} \ge \underline{0}\ .$$

CHAPTER 3

UTILITY FUNCTIONS AND DECISIONS UNDER RISK

3.1 Informational lotteries and axioms for utility functions

Risk preference problematique is closely related to the design and evaluation of utility functions and, in turn, by the whole complex process of decision-making under risk and uncertainty. In the following chapters of the present work, risk-sensitive Markovian models will be presented; in this respect, it is necessary to give a short representation of basic notions from decision theory which will be used in the formal mechanism of the above mentioned processes.

One of the most difficult choices to which a person, or even an organization, could be exposed refers to the case when they must decide among alternatives whose outcome is uncertain. There is a logical way to structure these processes and, in the end, to offer the decision maker the best solution in respect of the degree of acceptability of risk. In this context,, we shall introduce new terms and notions, such as informational lottery, certain equivalent of a lottery, utility function, risk aversion coefficient, etc. [8].

An uncertain proposition is described by means of a lottery. It is completely determined by costs $v_i, i = 1, 2, \dots , n$ and probability $p_i ; i = 1, 2, \dots , n$ of events occurring in a lottery (see Figure 3.1(a)). The results in a lottery are not necessarily measurable or comparable (expressed in an equivalent unitary measure). The events in a lottery are chosen in such a way that: $\sum_{i=1}^{n} p = 1 ; \ 0 \le p_i \le 1 ; \ i = 1, 2, \dots , n$.

If an informational lottery is described by the discrete random variable v then we can have the representation from Figure 3.1(b). In the case where it is described by a continuous random variable, it is not possible to use the tree representation, but only by using a probability distribution function (see Figure 3.2).

Expected value, $\bar{v}$, of an informational lottery represents the sum of the products between probabilities and values in the lottery, so that $\bar{v} = \sum_{i=1}^{n} p_i v_i$.

Certain equivalent, $\tilde{v}$, of a lottery represents its selling price, instead of playing off the lottery (see Figure 3.3).

Risk premium is defined by the relation $v_p = \bar{v} - \tilde{v}$. When $v_p = 0$, the decision maker is a risk venture person, and if $v_p \ne 0$, then the person is considered as risk sensitive.

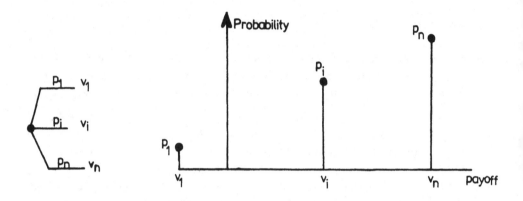

Figure 3.1 (a) and (b). (a) Representation of a lottery; (b) Probability vs payoff for a lottery.

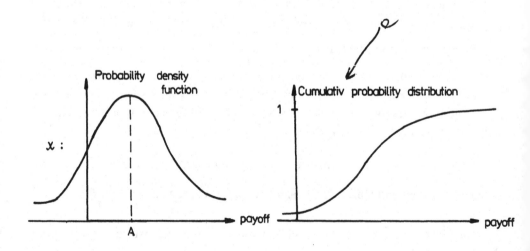

Figure 3.2 (a) and (b). (a) Probability density vs payoff; (b) Cumulative probability vs payoff.

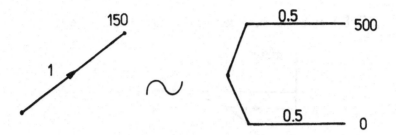

Figure 3.3. Diagram of certain equivalent.

Axioms of the theory of risk preference. These must refer to a set of minimal conditions which a person, or a group of persons, should accept in a general decision process.

Howard (1970) demonstrates that for a person whose preferences satisfy the axioms of orderly continuity, substitutability, and monotony, as well as that of decomposability for the utilities, these preferences can be imbedded in a utility function. A utility function associates numbers for any reward/value in an informational lottery [8].

Property 3.1. A utility function which describes the risk preference of a person has the following properties:

   a) utility u in any lottery $\mathcal{L}$ is the expected lottery of the values v in the lottery:

$$\langle u \mid \mathcal{L}, \mathcal{E} \rangle = \int_v \langle u \mid v, \mathcal{E} \rangle \Pr \{ v \mid \mathcal{L}, \mathcal{E} \} \, ;$$

   b) if a decision maker prefers a lottery $\mathcal{L}_1$ to $\mathcal{L}_2$ ($\mathcal{L}_1 \succ \mathcal{L}_2$) then $\mathcal{L}_1$ has the greater utility:

$$(\langle u \mid \mathcal{L}_1, \mathcal{E} \rangle > \langle u \mid \mathcal{L}_2, \mathcal{E} \rangle).$$

It should be mentioned that, due to some linearity properties for the expected values, (a) any utility function can be multiplied by a positive number, and (b) we can add a constant to any utility value without changing the utility function preferences.

A schematic utility function is shown in Figure 3.4.

Taking the utility concept into consideration, the certain equivalent of a lottery represents the value for a lottery which has the same utility as the expected utility of the lottery and, then we can write:

$$\langle u \mid v = \ {}^{\sim}\langle v \mid \mathcal{L}, \mathcal{E} \rangle, \mathcal{E} \rangle = \langle u \mid \mathcal{L}, \mathcal{E} \rangle,$$

where $^{\sim}\langle v \mid \mathcal{L}, \mathcal{E} \rangle$ represents the certain equivalent for the $\mathcal{L}$ lottery.

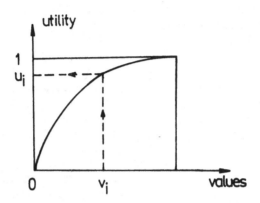

Figure 3.4. Schematic utility function.

Axiom 3.1 (orderability). The decision maker must be able to state his preferences among the prizes of any lottery. That is, if a lottery has prizes, A, B, and C, he must be able to say which he likes best, second best, etc.

Axiom 3.2 (continuity). If the decision maker has expressed the transitive preference A > B > C, then we must be able to construct a lottery with prizes A and C and determine a probability p of winning A such that he is indifferent between receiving B for certain and participating in the lottery (see Figure 3.5). We can observe that for the appropriate p, B is the certain equivalent of the lottery.

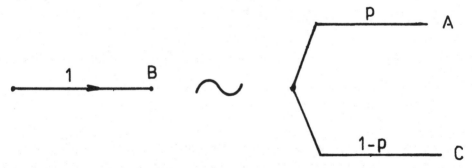

Figure 3.5. Diagram to show continuity.

Because u_i (the utility of any value v_i) can eventually be associated with a probability ($0 \le u_i \le 1$; $i = 1, 2, \dots, n$), we can extend the continuity axiom to the utility case (see Figure 3.6).

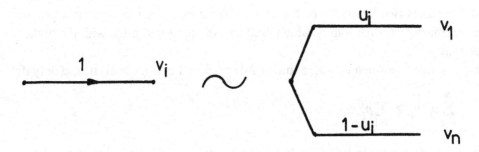

Figure 3.6. Diagram to show utility.

Axiom 3.3 (substitutability). If a decision maker has stated his certain equivalent of a given lottery, then he must be truly indifferent between the lottery and the certain equivalent (the lottery and its certain equivalent must be interchangable without affecting preferences).

In Figure 3.7 a graphical representation for this axiom is given; we can observe, within the lottery given above, only two values, namely v_1 and v_n.

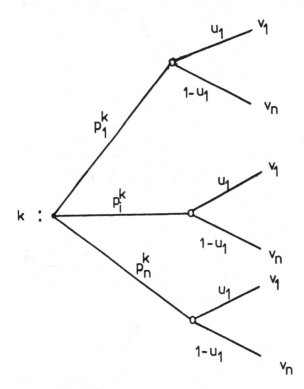

Figure 3.7. Graphical representation for Axiom 3.3.

Axiom 3.4 (monotonicity). If a decision maker has a preference between two prizes and if he faces two lotteries which each have these two prizes as their only prizes, then he must prefer the lottery which produces the preferred prize with the higher probability.

For the case when $v_1 > v_n$, a lottery A is preferred to B (A $\succ$ B) if, and only if:

$$\sum_{i=1}^{n} p_i^A u_i \geq \sum_{i=1}^{n} p_i^B u_i.$$

Axiom 3.5 (decomposability). If a lottery has a complex structure, then it may have other lotteries as prizes (e.g. compound lotteries). In the case of such a complex lottery, when faced with such a proposition, the decision maker will consider only the final prize he might win, and then compute the probability of winning each prize using the laws of probability.

This axiom is presented graphically in Figure 3.8 for the case when p and q are probabilities in the lottery and, for the general case, we can show the application of decomposability to a lottery (see Figure 3.9).

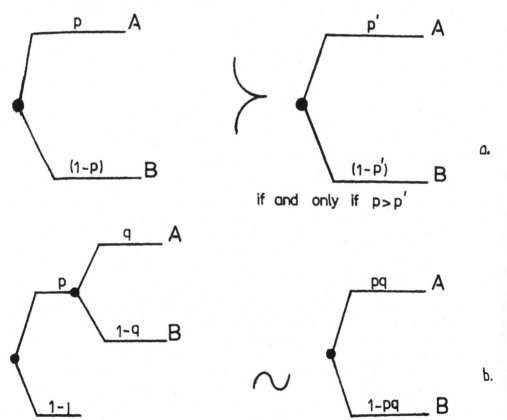

Figure 3.8(a and b). Graphical representation of Axiom 3.5 (decomposability).

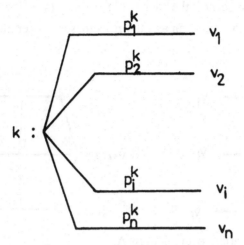

Figure 3.9. Application of decomposability to a lottery.

The equivalent form for a lottery $\mathcal{L}$, considering u_i as the utility of the value v_i, is presented in Figure 3.10.

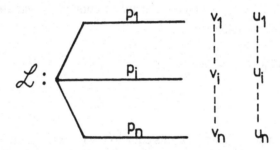

Figure 3.10. Decomposability for a lottery $\mathcal{L}$.

3.2 Exponential utility functions

An exponential utility function is of the form $e^{-\gamma v}$, or, in an alternative presentation, we can write:

$$u(v) = \frac{1 - e^{-\gamma v}}{1 - e^{-\gamma}},$$

where γ is the Pratt coefficient for the risk sensitivity:

$$\gamma = -\frac{u''(v)}{u'(v)}; \quad -\infty < v < \infty.$$

This type of function accepts the "delta" (Δ) property (see Howard (1970)) "an increase of all prizes in a lottery by an amount Δ increases the certain equivalent by Δ" (see Figure 3.11) [8].

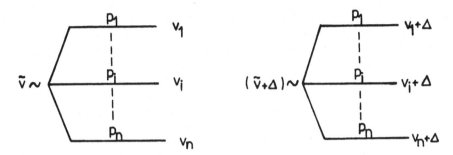

Figure 3.11. Increase of certain equivalent by Δ.

LEMMA 3.1. *The "delta" property of a utility function implies, in turn, that this can be either linear or exponential in its form.*

DEMONSTRATION. Let $g_v(\cdot)$ be the density function of the variable v, describing the lottery. Then we can write:

$$\int dv_0 \, g_v(v_0) \, u(v_0 + \Delta) \; = \; u(\tilde{v} + \Delta)$$

for any Δ.

If we differentiate both sides of this equation twice, we obtain, successively:

$$\int dv_0 \, g_v(v_0) \, u'(v_0 + \Delta) \; = \; u'(\tilde{v} + \Delta)$$

and

$$\int dv_0 \, g_v(v_0) \, u'(v_0 + \Delta) \; = \; u''(\tilde{v} + \Delta).$$

If $\Delta = 0$, we can write the following relation:

$$\frac{u'(v)}{u'(v)} \; = \; \frac{\int dv_0 \, g_v(v_0) \, u''(v_0)}{\int dv_0 \, g_v(v_0) \, u'(v_0)} \; .$$

A variety of probability density functions $g_v(\cdot)$ will produce the same certain equivalent, $\tilde{v}$. It is proved (Pratt (1963), Howard (1970)), that (see [8]):

$$\frac{u''(v)}{u'(v)} = -\gamma; \quad -\infty < v < \infty \tag{3.1}$$

where γ is a constant.

By interpreting the relation (3.1), we can write:

$$\ln u'(v) = -\gamma v + k_0$$

$$u'(v) = k_1 e^{-\gamma v} \tag{3.2}$$

where k_0, k_1 are constants. If $\gamma = 0$, we can write $u'(v) = k_1$; $u(v) = k_1 v + k_2$, which represents a linear utility function. If $\gamma \neq 0$, by integrating relation (3.2), we have $u(v) = k_2 e^{-\gamma v} + k_3$, which is an exponential utility function.

LEMMA 3.4. *The certain equivalent for a utility function is given by the relation*:

$$u(v) = (1 - e^{-\gamma v}) / (1 - e^{-\gamma})$$

which has the "delta" property with $\gamma \neq 0$, can be written as:

$$\tilde{v} = -(\gamma)^{-1} \ln e^{-\gamma v} = -(\gamma)^{-1} \ln g^e_v(\gamma) \tag{3.3}$$

and if $\gamma \to 0$, then $\tilde{v} = \bar{v}$, where $g^e_v(\cdot)$ represents the exponential transform of the density function $g^e_v(\cdot)$.

LEMMA 3.3. *The certain equivalent for a lottery described by the mean m and the standard dispersion σ, is given by*:

$$\tilde{v} = m - \tfrac{1}{2}\gamma\sigma^2.$$

LEMMA 3.4. *The certain equivalent for two successive lotteries with the values v_1 and v_2, respectively, and if $v = v_1 + v_2$, then we can write*:

$$\tilde{v} = -(\gamma)^{-1} \ln e^{\overline{-\gamma(v_1 + v_2)}} \quad \text{for } \gamma \neq 0,$$

$$\tilde{v} = \tilde{v}_1 + \tilde{v}_2 \quad \text{for } \gamma = 0.$$

3.3 Decisions under risk and uncertainty; event trees

In the theory of decision analysis, we must take three phases such as deterministic, probabilistic and informational (Howard (1968)), into consideration [8]. The Markovian decision processes could be included in the probabilistic phase. A graphical representation of such a process is given in Figure 3.12.

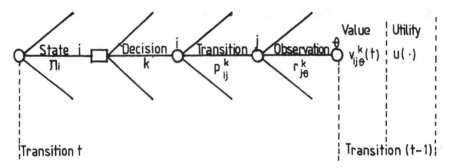

Figure 3.12. Graphical representation of Markov process.

The marginal distribution probability is $\Pr\{S_i \mid \mathcal{E}\}$, $i = 1, 2, \ldots, n$ for each variable within the model (Markov or semi-Markov, completely or partially observable). Because the state variables are given by the vector $s = [s_1, s_2, \ldots, s_n]$ then we can write $\Pr\{s \mid \mathcal{E}\}$. The decision vector d, with its components d_i has the dimension $(1 \times m)$. For any given d, and the associated probability distributions over the states of the process $\Pr\{s \mid \mathcal{E}\}$, we can write $\Pr\{v \mid d, \mathcal{E}\}$, called a *value lottery*. The decision problem deals with defining the optimal vector d_{opt} which produces the most aggregated value lottery.

   If the state of knowledge of the decision maker on the process is incorportaed in $\mathcal{E}$, then the objective function for the process can be written as:

$$d(\mathcal{E}) = \max_{d}^{-1} \langle u \mid d, \mathcal{E} \rangle$$

$$= \max_{d}^{-1} \,\tilde{}\langle v \mid d, \mathcal{E} \rangle.$$

The expected utility and the certain equivalent for the lottery which describes the optimal solution for the decision process are:

$$\langle u \mid \mathcal{E} \rangle = \langle u \mid d = d(\mathcal{E}), \mathcal{E} \rangle$$

and, respectively,

$$\tilde{}\langle v \mid \mathcal{E} \rangle = \tilde{}\langle v \mid d = d(\mathcal{E}), \mathcal{E} \rangle.$$

    The Markovian decision processes can be described using the concept of event trees, so that the decision moments will alternate with the state of the system (see Figure 3.13).

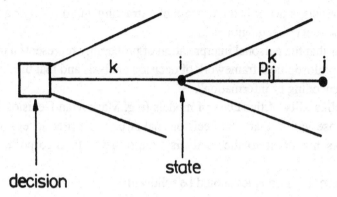

Figure 3.13. Decision moments alternating with the state of the system.

    If we consider a specific case for Markov or semi-Markov partially observable decision processes with a risk-sensitive decision maker, then the corresponding decision structure is given in Figure 3.14.

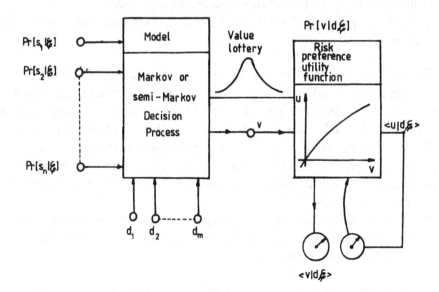

Figure 3.14. Decision model for a specific case of Markovian decision processes.

3.4 Probability encoding

As employed by Spetzler et al. [105], "probability encoding plays an important role in decision analysis, since it is the process of extracting and quantifying individual judgement about uncertain quantities".

It is clear that the personal interpretation of probability represents a main aspect in the general philosophical framework of decision analysis and that the "probability represents an encoding of information".

The applicability of this class of models (e.g. Markovian Decision Processes) leads to the observation that "the decision maker can ... either accept the expert's information as his input to the analysis or modify it to incorporate his own judgements".

Some encoding principles should be followed:

a) the uncertain quantity is important to the decision maker, as determined by a sensitivity analysis;

b) the quantity is defined for the subject as a non-ambiguous state variable;

c) the level of detail required from the encoding process depends on the importance of the quantity and should be determined by sensitivity analysis before the interview;

d) the quantity should be well structured and clearly defined;

e) the quantity must be described by the analysis on a scale that is meaningful to the subject.

The relevance of probability encoding is very well perceived by others:

"People perceive and assess uncertainty in a manner similar to the way they perceive and assess distances. They use intuitive assessment procedures that are based on clues of limited reliability and validity. At the same time the procedures generally produce reasonable answers. For example, an automobile driver is generally able to estimate distance accurately enough to avoid accidents, and a business executive is generally able to evaluate uncertainties well enough to make his enterprise profitable", [105].

As is well known, a variety of encoding methods are based on questions for which the answers can be represented as points on a cumulative distribution function, so we can have the following classification:

a) *P*-method asks questions on the probability scale with the values fixed.

b) *V*-method asks questions on the value scale with the probabilitites fixed.

c) *PV*-methods which ask questions to be answered on both scales jointly.

One important tool used within different encoding methodologies (see Table 3.1) is the probability wheel.

| | Response Mode | | |
| Encoding Method | Direct | Indirect | |
		External Reference Process	Internal Events
Probability (*P*-method)	Cumulative Probability	Wheel	Odds
Probability (*P*-method)	Fractiles	Wheel; fixed probability events	Interval techniques
Probability-Value (*PV*-method)	Parametric description	—	—

Table 3.1. Encoding methodologies.

Next, a detailed explanation of the wheel-technique will be given from Spetzler et al.

"The probability wheel is useful with most subjects. As an external reference process it can be used as a *P* method, or a *V*-method, but the former is the method generally preferred. The probability wheel is a disk with two sectors, one blue and the other red, with a fixed pointer in the centre of the disk. The disk is spun, finally stopping with the pointer either in the blue or in the red sector. A simple adjustment changes the relative size of the two sectors and, thereby, also the probability of the pointer indicating either sector when the disk stops spinning. The subject is asked whether he would prefer to bet either on an event relating to the uncertain quantity, e.g that next year's production will not exceed X units, or on the pointer ending in the red sector. The amount of red in the wheel is then varied until the expert becomes indifferent. When indifference has been obtained, the relative amount of red is assigned as the probability of the event. This is a *P*-method since the event (value) is fixed and the probability is determined through the process".

The interview process involves five distinct steps: motivating, structuring, conditioning, encoding and verifying.

Motivating: introduces the subject to the encoding task and at the same time explores whether any motivational biases might operate (e.g. we must point out that no commitment is inherent in a probability distribution).

Structuring within this stage we must clearly define and structure the uncertain quantity, which is assumed to be of extreme importance to the decision maker.

Conditioning: this step heads off biases that otherwise might surface during the encoding, and conditions the subject to think fundamentally about his judgement.

Encoding: this phase uses the probability wheel as the encoding technique.

Verifying: is the phase when the judgements are tested to see if the subject really believes them.

CHAPTER 4

MARKOVIAN DECISION PROCESSES (SEMI-MARKOV AND MARKOV) WITH COMPLETE INFORMATION (COMPLETELY OBSERVABLE)

4.1 Value iteration algorithm (the finite horizon case)

4.1.1. SEMI-MARKOV DECISION PROCESSES

In Figure 4.1(a) a diagrammatic representation of a semi-Markov decision process is presented; special attention to transition probabilities p^k_{ij}, and waiting time probabilities $h^k_{ij}(m)$ is given. A corresponding decision tree for a completely observable semi-Markov process is presented in Figure 4.1(b).

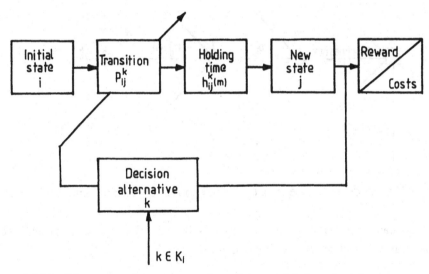

Figure 4.1(a). Diagrammatic representation of a semi-Markov decision process.

The value iteration algorithm is based on the dynamic programming principle; the procedure identifies the strategy which maximizes the total expected gain whether or not the discounting coefficient is taken into account [54], [57], [61], [69].

If we consider $v_i(t, \beta)$ the expected present value of the future reward generated by the process by time t, if it is placed in state i when $t = 0$, and the discounting coefficient is β then:

$$v_i(t, \beta) = \max_k \left\{ \sum_{j=1}^{N} p_{ij}^k \sum_{m=n+1}^{\infty} h_{ij}^k (m) \cdot \left[\sum_{l=0}^{t-1} \beta^l y_{ij}^k (l) + \beta^t v_i(0) \right] + \right.$$

$$+ \sum_{j=1}^{N} p_{ij}^k \sum_{m=1}^{t} h_{ij}^k (m) \left[\sum_{l=0}^{m-1} \beta^l y_{ij}^k (l) + \beta^m b_{ij}^k (m) + \right.$$

$$\left. + \beta^m v_j(t - m, \beta) \right] \right\} ; \quad i = 1, 2, \dots , N ; \quad t \geq 1. \tag{4.1}$$

Decision alternative k, which will finally generate the functional $v_i(\cdot, \cdot)$, $i = 1, 2, \dots , N$, represents the optimal decision. We can define the optimal decision vector given by $d^*(m, \beta)$, $1 \leq m \leq t$.

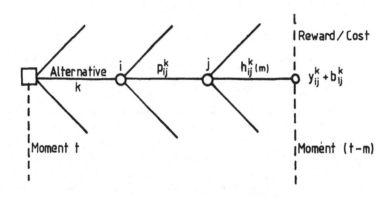

Figure 4.1(b). A decision tree for a completely observable semi-Markov process.

The case without discounting implies $\beta = 1$. We can simplify the above functional if we accept the following notations:

(a) $y_{ij}^k (m, \beta) = \sum_{l=0}^{m-1} \beta^l y_{ij}^k (l)$

(b) $^>y_i^k (t, \beta) = \sum_{j=1}^{N} p_{ij}^k y_{ij}^k (t, \beta) \sum_{m=t+1}^{\infty} h_{ij}^k (m) = \sum_{j=1}^{N} p_{ij}^k y_{ij}^k (t, \beta) {}^>h_{ij}^k (t)$

(c) $r_i^k (t, \beta) = \sum_{j=1}^{N} p_{ij}^k \sum_{m=1}^{t} h_{ij}^k (m) \cdot [y_{ij}^k (m, \beta) + \beta^m b_{ij}^k (m)].$

$$\tag{4.2}$$

The simplified form for the reward functional given by relation (4.1) for a finite horizon semi-Markov process has the following expression:

$$v_i(t, \beta) = \max_k \left\{ {}^{>}y_i{}^k (t, \beta) + \beta^t v_i(0) {}^{>}w_i{}^k (t) + r_i^k (t, \beta) + \right.$$

$$\left. + \sum_{j=1}^{N} p_{ij}^k \sum_{m=1}^{t} h_{ij}^k (m) \beta^m v_j (t - m, \beta) \right\} ;$$

$$i = 1, 2, \dots , N ; \ t \geq 1. \qquad (4.3)$$

A flow chart for the value iteration algorithm is shown in Figure 4.2.

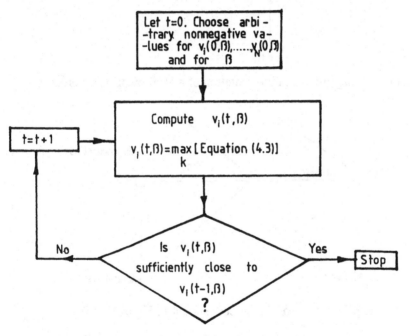

Figure 4.2. Flow chart for the value iteration algorithm.

In relation (4.3), if $\beta = 1$, we obtain the computational algorithm for the finite horizon semi-Markov decision process without discounting [77], [78], [98].

4.1.2. MARKOV DECISION PROCESSES

If we consider e_{ij}^k as the reward generated by the process when it is in state i and a decision k is applied, the process will make the next transition to state j (see Figure 4.3(a)). The corresponding decision tree for such a process is represented in Figure 4.3(b).

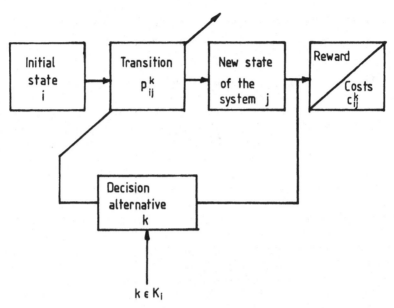

Figure 4.3(a).   Diagrammatic representation of a Markov decision process.

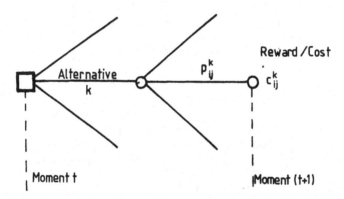

Figure 4.3(b). Decision tree for the process shown in Figure 4.3(a).

Defining $d_i(t)$ as the number of alternatives accessible to the system in state i, at time t, then $d_i(t)$ for all states i at time t, forms a policy (strategy) for the Markov decision process.

Let $v_i(t)$ denote the total expected reward form the process when t time units remain till the process terminates. The computational relation for $v_i(t)$, if we follow an optimal policy in accordance with the Bellman principle, can be written as the following relation:

$$v_i(t+1) = \max_k \sum_{j=1}^{N} p_{ij}^k [c_{ij}^k + \beta v_j(t)] ; \quad t \geq 0. \qquad (4.4)$$

By denoting the immediate expected reward by $q_i^k = \sum_{j=1}^{N} p_{ij}^k c_{ij}^k$, the relation (4.4) can be written in the form:

$$v_i(t+1) = \max_k \left[q_i^k + \beta \sum_{j=1}^{N} p_{ij}^k v_j(t) \right]; \quad t \geq 0. \tag{4.5}$$

In Figure 4.4 a computational flow chart for the value iteration algorithm for the finite horizon Markov decision process is given. For $\beta = 1$, relation (4.5) gives the computational algorithm for the finite horizon Markov decision process, without discounting.

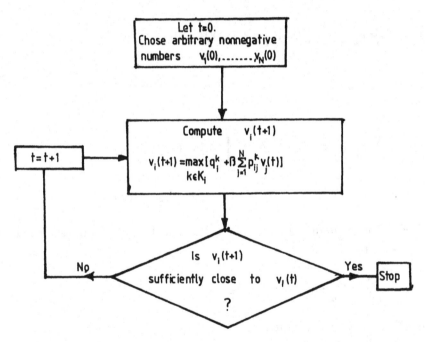

Figure 4.4. Computational flow chart for a finite horizon Markov decision process.

4.2 Policy iteration algorithm (the finite horizon optimization)

4.2.1. SEMI-MARKOV DECISION PROCESSES

The reward structure is super-imposed on the semi-Markov process; the system will yield at a transition, so that a yield as well as a waiting time, will be associated with the transition. If c_{ij} is the corresponding reward for a transition of the system from state i to state j, and y_i is the yield rate of state i at any time during its occupancy, the total expected reward $v_i(t)$ is of the following form:

$$v_i(t) = {}^>w_i(t)\,[v_i(0) + y_i(t)] + \sum_{j=1}^{N} P_{ij} \int_0^\tau d\tau h_{ij}(\tau)\,[c_{ij} + y_i\,\tau + v_j(t-\tau)] ;$$

$$i = 1, 2, \dots , N ; \quad t \geq 0. \tag{4.6}$$

If $t \to \infty$, then ${}^>w_i(t) \to 0$, because here we operate with systems with finite values for the mean waiting times. We can prove, also, that $t\,{}^>w_i(t) \to 0$, and under this assumption, for $t \to \infty$, we can obtain the following reward functional:

$$v_i(t) = \sum_{j=1}^{N} P_{ij} \int_0^\tau d\tau h_{ij}(\tau)\,[c_{ij} + y_i\,\tau + v_j(t-\tau)] ;$$

$$i = 1, 2, \dots , N ; \quad \text{for } t \to \infty, \tag{4.7}$$

or:

$$v_i(t) = \sum_{j=1}^{N} P_{ij}\,c_{ij} + y_i\,\bar{\tau}_i + \sum_{j=1}^{N} P_{ij} \int_0^\tau d\tau h_{ij}(\tau)\,v_j(t-\tau) ;$$

$$i = 1, 2, \dots , N ; \quad \text{for } t \to \infty. \tag{4.8}$$

If we define the immediate value for the expected reward for the semi-Markov decision process by:

$$q_i = \frac{1}{\bar{\tau}_i} \sum_{j=1}^{N} P_{ij}\,c_{ij} + y_i ; \quad i = 1, 2, \dots , N \tag{4.9}$$

then the relation (4.8) can be written as:

$$v_i(t) = q_i\,\bar{\tau}_i + \sum_{j=1}^{N} P_{ij} \int_0^\tau d\tau h_{ij}(\tau)\,v_j(t-\tau) ;$$

$$i = 1, 2, \dots , N ; \quad \text{for } t \to \infty. \tag{4.10}$$

It can be proved that for t sufficiently large, $v_i(t)$ is given by the relation (see Howard (1971)) [9]:

$$v_i(t) = v_i + gt ; \quad i = 1, 2, \dots , N ; \quad t \to \infty \tag{4.11}$$

Relation (4.11) indicates that the expected reward will grow linearly with time if this tends to very large values. The value g is considered as the goal value for the process. Using relation (4.11), the relation (4.10) can be written as:

$$v_i(t) + gt = q_i \bar{\tau}_i + \sum_{j=1}^{N} p_{ij} v_j + gt - g \bar{\tau}_i ,$$

or, finally, we have:

$$v_i(t) + g \bar{\tau}_i = q_i \bar{\tau}_i + \sum_{j=1}^{N} p_{ij} v_j ; \quad i = 1, 2, \dots , N . \qquad (4.12)$$

The above system of equations is homogenous in v_i ; it can be easily considered that one of the v_i 's has zero value ($v_N = 0$). We obtain a compatible system with N equations and N unknowns ($v_1, v_2, \dots , v_{N-1}, g$).

If we multiply (4.12) by π_i, $i = 1, 2, \dots , N$, then we can write:

$$\sum_{i=1}^{N} \pi_i v_i + g \sum_{i=1}^{N} \pi_i \bar{\tau}_i = \sum_{i=1}^{N} \pi_i q_i \bar{\tau}_i + \sum_{j=1}^{N} v_j \sum_{i=1}^{N} \pi_i p_{ij} . \qquad (4.13)$$

Because $\pi_j = \sum_{i=1}^{N} \pi_i p_{ij}$, relation (4.13) can be written with the following form:

$$g \sum_{i=1}^{N} \pi_i \bar{\tau}_i = \sum_{i=1}^{N} \pi_i q_i \bar{\tau}_i \qquad (4.14)$$

or:

$$g = \frac{\sum_{i=1}^{N} \pi_i q_i \bar{\tau}_i}{\sum_{i=1}^{N} \pi_i \bar{\tau}_i} , \qquad (4.15)$$

or, finally:

$$g = \sum_{i=1}^{N} \phi_i q_i ,$$

where:

$$\phi_i = \pi_j \bar{\tau}_j \Big/ \sum_{j=1}^{N} \pi_j \bar{\tau}_j .$$

Next, we shall consider the situation when, the system being in state i, there exist decision alternatives. Corresponding to each alternative k, for any state i, the input parameters for the optimization model are the following:

$$p^k_{ij}, \; h^k_{ij}(\,.\,), \; r^k_{ij}, \; y^k_i \; .$$

The decision alternatives in each state are finite in number but distinct for each state experienced by a semi-Markov process. The optimization model is of the policy iteration type developed by Howard.

Let us assume that there exists the choice of selecting a decision alternative when the system can experience any operational state. The set of decision alternatives for all the system states defines a *policy*. It is necessary to identify the policy which maximizes the objective function, g, the mean reward per unit of time when the system evolves in the state space.

Solving equation (4.12), we can compute g as given by the following relation:

$$g \; = \; q_i + \frac{1}{\overline{\tau}_i} \left[\sum_{j=1}^{N} p_{ij} \, v_j - v_i \right], \; i = 1, 2, \dots , N \; . \tag{4.16}$$

To maximize the value of g it is necessary that the right hand side value of relation (4.16) is as large as possible. This can be realized by the iterative scheme of Howard's algorithm.

We must choose an arbitrary policy to start the iterative process; the system of equations (4.12) must be solved so that we are able to compute the values of g and v_i, $i = 1, 2, \dots , N - 1$, under the condition that $v_N = 0$.

The next step uses the values of the v_i 's computed in the previous stage to improve the critical policy by identifying the alternative in each state i which maximizes the test function:

$$\Gamma \; = \; q^k_i + \frac{1}{\overline{\tau}^k_i} \left[\sum_{j=1}^{N} p^k_{ij} \, v_j - v_i \right] \tag{4.17}$$

When at two successive iterations the same policy remains, then this policy is the optimal one.

If this is not true, the new policy replaces the initial one and the policy iteration scheme is repeated. The computational algorithm does converge in a finite number of iterations; its efficiency depends on the initial start policy used within the model.

For the above model, we can observe that the waiting time probabilities are influenced only by the value of $\overline{\tau}_i$.

In Figure 4.5 a schematic computational representation for the optimization of semi-Markov decision processes is given.

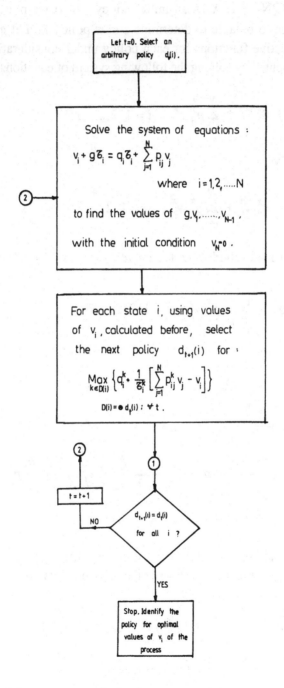

Figure 4.5.   Decision process optimization.

THEOREM 4.1. *The policy iteration algorithm for a semi-Markov decision process without discounting converges in a finite number of iterations.*

DEMONSTRATION: Let X be an initial policy. Next we must use the policy improvement stage to be able to determine a new policy Y. Let g^x and g^y be the values of the objective functions for each policy under consideration. These two values can be computed by solving the following system of equations:

$$v_i^y + g^y \, \tilde{\tau}_i^y = q_i^y \, \tilde{\tau}_i^y + \sum_{j=1}^{N} p_{ij}^y \, v_j^y \; ; \quad i = 1, 2, \ldots, N \tag{4.18}$$

and, respectively:

$$v_i^x + g^x \, \tilde{\tau}_i^x = q_i^x \, \tilde{\tau}_i^x + \sum_{j=1}^{N} p_{ij}^x \, v_j^x \; ; \quad i = 1, 2, \ldots, N \; . \tag{4.19}$$

If relation (4.18) is divided by $\tilde{\tau}_i^y$, and relation (4.19) by $\tilde{\tau}_i^x$ we obtain:

$$\frac{v_i^y}{\tilde{\tau}_i^y} + g^y = q_i^y + \frac{1}{\tilde{\tau}_i^y} \sum_{j=1}^{N} p_{ij}^y \, v_j^y \; ; \quad i = 1, 2, \ldots, N \; . \tag{4.20}$$

$$\frac{v_i^x}{\tilde{\tau}_i^x} + g^x = q_i^x + \frac{1}{\tilde{\tau}_i^x} \sum_{j=1}^{N} p_{ij}^x \, v_j^x \; ; \quad i = 1, 2, \ldots, N \; . \tag{4.21}$$

Subtracting (4.21) from (4.20) we obtain:

$$\frac{v_i^y}{\tilde{\tau}_i^y} - \frac{v_i^x}{\tilde{\tau}_i^x} + g^y - g^x = q_i^y - q_i^x + \frac{1}{\tilde{\tau}_i^y} \sum_{j=1}^{N} p_{ij}^y \, v_j^y - \frac{1}{\tilde{\tau}_i^x} \sum_{j=1}^{N} p_{ij}^x \, v_j^x \; ;$$

$$i = 1, 2, \ldots, N \; . \tag{4.22}$$

Because policy Y has been generated by using the values generated by X, from the policy iteration stage, we can, in turn, write the following relation:

$$q_i^y + \frac{1}{\tilde{\tau}_i^y} [\sum_{j=1}^{N} p_{ij}^y \, v_j^y - v_i^x] \geq q_i^x + \frac{1}{\tilde{\tau}_i^x} [\sum_{j=1}^{N} p_{ij}^x \, v_j^x - v_i^x] \; ;$$

$$i = 1, 2, \ldots, N \; . \tag{4.23}$$

Next, we shall define ε_i given by the relation:

$$\varepsilon_i = q_i^y - q_i^x + \frac{1}{\bar{\tau}_i^y}\left[\sum_{j=1}^{N} p_{ij}^y\, v_j^y - v_i^y\right] - \frac{1}{\bar{\tau}_i^x}\left[\sum_{j=1}^{N} p_{ij}^x\, v_j^x - v_i^x\right] \geq 0, \qquad (4.24)$$

Subtracting (4.24) from (4.22), we can write:

$$\frac{1}{\bar{\tau}_i^y}(v_i^y - v_i^x) + g^y - g^x = \varepsilon_i + \frac{1}{\bar{\tau}_i^y}\sum_{j=1}^{N} p_{ij}^y\,(v_j^y - v_i^x);$$

$$i = 1, 2, \dots, N. \qquad (4.25)$$

Multiplying the above relations by $\bar{\tau}_i^y$ we obtain:

$$v_i^\delta + g^\delta\,\bar{\tau}_i^y = \varepsilon_i\,\bar{\tau}_i^y + \sum_{j=1}^{N} p_{ij}^y\, v_j^\delta;$$

$$i = 1, 2, \dots, N, \qquad (4.26)$$

where, in general, $x^\delta = x^y - x^x$.

Equations (4.26) are similar in form to the policy equations.
It is clear that in this case

$$g^\delta = \sum_{j=1}^{N} \phi_j^y\,\varepsilon_j.$$

Improvement in value for the objective function from policy X to policy Y is given by the sum of increasing in value of the test due to applying the policy improvemenent stage. We can immediately see that $g^\delta \geq 0$.

We can prove that it is possible for an objective function greater than X to exist for policy Y but this cannot be identified by the policy improvement stage from the above algorithm. If such a policy exists, then $g^\delta > 0$. Next, we shall consider that the policy iteration procedure converges with reference to policy X; then for all the states we can write that $\varepsilon_i \leq 0$. Because $\phi_i^B \geq 0$ for all i, then $g^\delta > 0$ and it has been proved that such a situation cannot exist. Policy iteration must converge for a policy with the largest value for the objective function.

4.2.2. MARKOV DECISION PROCESSES

Let us consider a Markov decision process characterized by N states, $P = [p_{ij}]$, the reward matrix $C = [c_{ij}]$. The process is considered to be completely ergodic when all the limit state probabilities π_i are independent of the starting state of the process.

The objective function for such a process is of the following form:

$$g = \sum_{i=1}^{N} \pi_i q_i ,$$

where $q_i = \sum_{j=1}^{N} p_{ij} c_{ij}$; $i = 1, 2, \dots , N$ represents the earning rate of state i.

The policy iteration algorithm for the Markov decision process is determined by two stages, namely: value determination and policy improvement, respectively.

In Figure 4.6 a decision tree for the process investigated above is given.

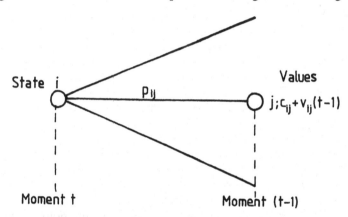

Figure 4.6. A decision tree for a Markov process.

We can write the functional for the lottery value of the system under analysis, when this is in state i and t transitions are left for the process to finish.

The following relation can be written:

$$v_i(t) = q_i + \sum_{j=1}^{N} p_{ij} v_j(t - 1) ; \quad i = 1, 2, \dots , N ; \quad t \geq 1. \tag{4.27}$$

The asymptotic value for a completely ergodic Markov process can be computed with the relation:

$$v_i(t) = tg + v_i ; \quad i = 1, 2, \dots , N \tag{4.28}$$

for $t \to \infty$.

Using (4.27) in (4.28) we obtain:

$$tg + v_i = q_i + \sum_{j=1}^{N} p_{ij} [(t-1)g + v_j] ; \quad i = 1, 2, \dots , N$$

$$tg + v_i = q_i + (t-1)g \sum_{j=1}^{N} p_{ij} + \sum_{j=1}^{N} p_{ij} v_j . \tag{4.29}$$

Using basic definitions for Markov processes (e.g. $\sum_{j=1}^{N} p_{ij} = 1$; for all i), we can write the relation:

$$g + v_i = q_i + \sum_{j=1}^{N} p_{ij} v_j , \quad i = 1, 2, \dots , N . \tag{4.30}$$

In this case, it is necessary to solve a set of N equations to determine $v_1, v_2,$ ... , v_{N-1} and g with the initial condition $v_N = 0$.

In the case of the policy improvement stage, similar to the semi-Markov decision process, it is necessary to introduce the test function with the following form:

$$\Gamma = q_i^k + \sum_{j=1}^{N} p_{ij}^k v_j(t) \tag{4.31}$$

for all states i taken into consideration for the process under investigation.

The policy improvement stage for the Markov decision process is formulated as:

"For each state i, we must identify a decision alternative k which maximizes the test value Γ given by relation (4.31), using the relative values, v_i, computed in the previous stage".

In Figure (4.7) a general flow chart for the computational application of the policy iteration in the case of Markov processes is given.

We can prove, in a similar way to semi-Markov decision processes, that:

THEOREM 4.2. *A policy iteration algorithm applied for a Markov decision process is convergent in a finite number of iterations.*

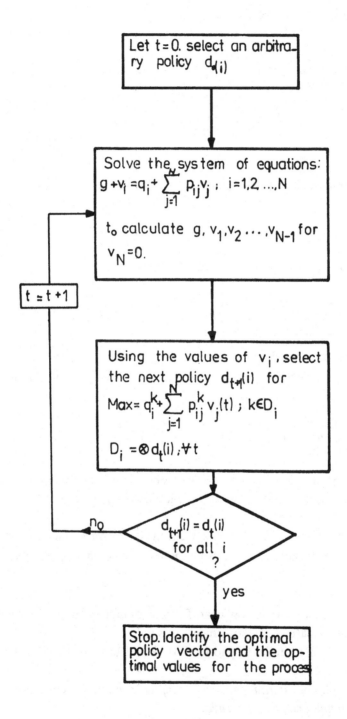

Figure 4.7. General computational flow chart for policy iteration in Markov processes.

4.3 Policy iteration with discounting

4.3.1. SEMI-MARKOV DECISION PROCESSES

In many decision processes concerning investments, or expenses involved in different activities (e.g. the maintenance and repair of technical systems), it is necessary to introduce the discounting coefficient into the model structure. The effect of discounting in the case of Markov decision models leads to a linear increase of the total expected reward for a given policy, $v_i(t)$.

The present value for the expected reward for the system is given by:

$$v_i(t) = {}^> w_i(t) \, \{ e^{-\alpha t} \, v_i(t) + y_i \, (1/\alpha) \, [1 - e^{-\alpha t}] \} +$$

$$+ \sum_{j=1}^{N} p_{ij} \int_0^t d\tau \, h_{ij}(\tau) \, \{ c_{ij} \, e^{-\alpha r} + y_i \, (1/\alpha) \, [1 - e^{-\alpha r}] + e^{-\alpha r} \, v_j(t - r) \};$$

$$i = 1, 2, \ldots, N; \ t \ge 0. \tag{4.32}$$

If v_i represents the present value for $v_i(t)$ when $t \to \infty$, then we can write:

$$v_i = \lim_{t \to \infty} v_i(t)$$

$$= \lim_{t \to \infty} \sum_{j=1}^{N} p_{ij} \int_0^t d\tau \, h_{ij}(\tau) \, \{ c_{ij} \, e^{-\alpha \tau} + y_i \, (1/\alpha) \, [1 - e^{-\alpha \tau}] + e^{-\alpha \tau} \, v_j(t - \tau) \};$$

$$i = 1, 2, \ldots, N .$$

$$v_i = \sum_{j=1}^{N} p_{ij} \, c_{ij} \, h_{ij}^{T}(\alpha) + y_i \, (1/\alpha) \cdot [1 - w_i^{T}(\alpha)] + \sum_{j=1}^{N} p_{ij} \, h_{ij}^{T}(\alpha) \, v_j$$

$$i = 1, 2, \ldots, N . \tag{4.33}$$

If we define:

$$q_i(\alpha) = \sum_{j=1}^{N} p_{ij} \, c_{ij} \, h_{ij}^{T}(\alpha) + y_i \, (1/\alpha) \, [1 - w_i^{T}(\alpha)];$$

$$i = 1, 2, \ldots, N; \ \alpha > 0,$$

as the present value for the expected reward, then we can write:

$$v_i = q_i(\alpha) + \sum_{j=1}^{N} p_{ij} c_{ij} h_{ij}^T (\alpha) v_j \; ; \quad i = 1, 2, \dots, N .$$

In this respect, the above system can be solved for the present value — the starting value — in each state of the system.

In a matrix form, the above system is represented by:

$$v = q(\alpha) + [P \; \square \; H^T (\alpha)] \, v ,$$

or

$$v = [I - P \; \square \; H^T (\alpha)]^{-1} \, q(\alpha) .$$

Because the matrix for the product $P \; \square \; H^T (\alpha)$ has the eigenvalues strictly less than one, we have:

$$[I - P \; \square \; H^T (\alpha)]^{-1} = \sum_{m=0}^{\infty} (P \; \square \; H^T (\alpha))^m = C.$$

Because all the elements for the congruent product $P \; \square \; H^T (\alpha)$ are positive, it follows that the elements of the C matrix are positive. The flow chart for the case with discounting (α) is presented in Figure 4.8a.

For the discrete semi-Markov decision processes, the computational flow chart for the same algorithm — the case with discounting coefficient β — is presented in Figure 4.8b.

The algorithm involves the selection, from the beginning, of an arbitrary policy: Next we must compute the present value for all v_i 's. We can improve the initial policy by identifying in each state an alternative k which maximizes the test functional:

$$\Gamma(\alpha) = q_i^k (\alpha) + \sum_{j=1}^{N} p_{ij}^k h_{ij}^T (\alpha) \, v_j ,$$

using all the values of v_j, computed previously.

THEOREM 4.3. *The policy iteration algorithm for Markov decision processes with discounting does converge in a finite number of iterations.*

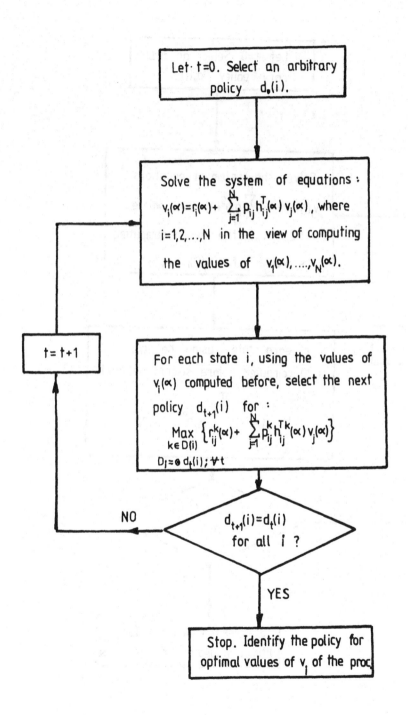

Figure 4.8(a). Flow chart for iterative policy with discounting coefficient α.

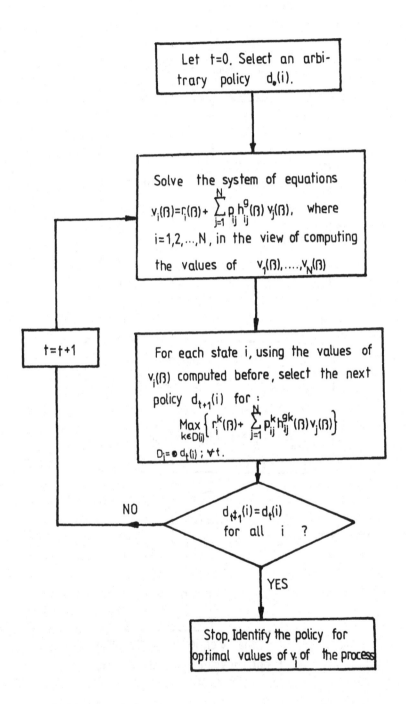

Figure 4.8(b). Flow chart for the iterative policy with discounting coefficient β.

DEMONSTRATION: Let us consider that a policy X has been evaluated and the next policy identified was Y in the policy iteration stage. The present value for the X and Y policies must satisfy equations (4.34) and (4.35).

$$v_i^y = q_i^y(\alpha) + \sum_{j=1}^{N} p_{ij}^y\, h_{ij}^{Ty}(\alpha)\, v_j^y; \quad i - 1, 2, \dots, N \tag{4.34}$$

$$v_i^x = q_i^x(\alpha) + \sum_{j=1}^{N} p_{ij}^x\, h_{ij}^{Tx}(\alpha)\, v_j^x; \quad i - 1, 2, \dots, N \tag{4.35}$$

Subtracting (4.35) from (4.34) we obtain:

$$v_i^y - v_i^x = q_i^y(\alpha) - q_i^x(\alpha) + \sum_{j=1}^{N} p_{ij}^y\, h_{ij}^{Ty}(\alpha)\, v_j^y - \sum_{j=1}^{N} p_{ij}^x\, h_{ij}^{Tx}(\alpha)\, v_j^x;$$

$$i = 1, 2, \dots, N. \tag{4.36}$$

Because we applied the policy improvement from X to Y, we obtain:

$$q_i^y(\alpha) + \sum_{j=1}^{N} p_{ij}^y\, h_{ij}^{Ty}(\alpha)\, v_j^y \geq q_i^x(\alpha) + \sum_{j=1}^{N} p_{ij}^x\, h_{ij}^{Tx}(\alpha)\, v_j^x;$$

$$i = 1, 2, \dots, N.$$

If:

$$\varepsilon_i = q_i^y(\alpha) - q_i^x(\alpha) + \sum_{j=1}^{N} p_{ij}^y\, h_{ij}^{Ty}(\alpha)\, v_j^x - \sum_{j=1}^{N} p_{ij}^x\, h_{ij}^{Tx}(\alpha)\, v_j^x;$$

$$i = 1, 2, \dots, N$$

and if $v_i^\delta = v_i^y - v_i^x$, then we find that:

$$v_i = \varepsilon_i + \sum_{j=1}^{N} p_{ij}^y\, h_{ij}^{Ty}(\alpha)\, v_j^\delta; \quad i = 1, 2, \dots, N.$$

If vectors v and ε have the components v_i^δ and ε_i, $i = 1, 2, \dots, N$, respectively, then:

$$v^\delta = C\varepsilon.$$

Because C and ε are positive values, therefore v is positive.

The value $v_i^\delta > 0$ is selected in such a way that the starting value in state i must be greater for Y policy than for the X policy, when it is possible to increase the test function in any state j which can be realized starting from state i under the conditions of policy Y. Under these circumstances we can, intuitively. prove that the algorithm does converge in a finite number of steps.

For the Markovian processes with transient behaviour, the total reward from the process before entering a recurrent state is the most important practical value for the algorithm. The general equation for v_i is , in this case:

$$v_i = (q_i - g)\,\tilde{\tau}_i + \sum_{j=1}^{N} p_{ij}\,v_j, \quad i = 1, 2, \dots, N .$$

We consider that all the states, except the N state, are transient states and state N is a trapping state for which $p_{NN} = 1$, $h_{NN}(1) = 1$. For $i = N$ the above equation can be written in the form:

$$g = qN .$$

If the system did start in state N the total reward up to time t would be $q_N t$. In this case v_N must be zero, and all other values can be computed using the relation:

$$v_i = (q_i - g)\,\tilde{\tau}_i + \sum_{j=1}^{N-1} p_{ij}\,v_j, \quad i = 1, 2, \dots, N - 1. \tag{4.36a}$$

By using the following notations:

$$\underline{e} = [e_i]; \quad i = 1, 2, \dots, N - 1$$

where $e_i = (q_i - g)\tilde{\tau}_i$, and:

$$\tilde{P} = [p_{ij}\,; \ i, j = 1, 2, \dots, (n - 1)],$$

relation (4.36) can be written in a vector form:

$$\underline{v} = \underline{e} + \tilde{P}\underline{v} \tag{4.37}$$

or

$$\underline{v} = [I - \tilde{P}]^{-1}\,\underline{e} .$$

For a transient process, the inverse matrix $[I - \tilde{P}]^{-1}$ does exist.

Iff:

$$\tilde{N} = \{\bar{v}_{ij}\} = [I - \tilde{P}]^{-1},$$

where $\tilde{v}_{ij}$ is the length of the mean time that the system will occupy state j in an infinite number of transitions, then:

$$\underline{v} = \tilde{N}\,\underline{e}\,;$$

or, successively:

$$v_i = \sum_{j=1}^{N-1} \bar{v}_{ij}\,(q_j - g)\,\bar{\tau}_j = \sum_{j=1}^{N-1} \bar{v}_{ij}\,q_j\,\bar{\tau}_j - g \sum_{j=1}^{N-1} \bar{v}_{ij}\,\bar{\tau}_j\,.$$

By analyzing the above equation we can see that the value for the process in state i is equal to the sum of the expected rewards for an unique occupancy of state j, multiplied by the mean duration that the system is in state j, minus the product of the value of the objective function g and the total expected time that the system will stay within the transient process.

It is considered that the policy iteration algorithm can be applied to this process.

4.3.2. MARKOV DECISION PROCESS

Next, we consider a Markov decition process characterized by the matrices $P = [p_{ij}]$ and $C = [c_{ij}]$, and β is the discounting coefficient $(0 \leq \beta \leq 1)$.

As in section 4.2.2 we shall consider the next $v_i(t)$, taking into consideration the influence of the discounting factor, such that:

$$v_i(t) = \sum_{j=1}^{N} p_{ij}\,[c_{ij} + \beta\,v_j\,(t-1)]; \qquad i = 1, 2, \ldots, N\,;\ t \geq 1.$$

Considering $q_i = \sum_{j=1}^{N} p_{ij}\,c_{ij}$ we can write:

$$v_i(t) = q_i + \beta \sum_{j=1}^{N} p_{ij}\,v_j\,(t-1); \qquad i = 1, 2, \ldots, N\,;\ t \geq 1.$$

In a vector form we can write:

$$\underline{v}\,(t+1) \;=\; q + \beta\,P\,\underline{v}\,(t).$$

The z-transform for $\underline{v}\,(t)$ is $\tilde{y}\,(z)$ and:

$$z^{-1}\,[\tilde{y}\,(z) - \underline{v}(0)] \;=\; (1-z)^{-1}\,q + \beta\,P\,\tilde{y}(z)$$

and in a successive manner:

$$\tilde{y}\,(z) - \underline{v}(0) \;=\; z\,(1-z)^{-1}\,q + \beta z\,P\tilde{y}\,(z)$$

$$(I - \beta z P)\,\tilde{y}\,(z) \;=\; \frac{z}{1-z}\,q + \underline{v}\,(0)$$

$$\tilde{y}\,(z) \;=\; \frac{z}{1-z}\;(I - \beta z P)^{-1}\,q + (I - \beta z P)^{-1}\,\underline{v}\,(0).$$

We can prove (see Howard (1960) [7]) that:

$$\tilde{y}\,(z) \;=\; \frac{z}{1-z}\,\Big[\frac{1}{1-\beta z}\,S + \mathfrak{I}(\beta z)\Big]\,q + \Big[\frac{1}{1-\beta z}\,S + \mathfrak{I}(\beta z)\Big]\,\underline{v}\,(0),$$

where S is the limit state probabilities matrix, and $\mathfrak{I}(\beta z)$ are values which tend to zero when $t \to \infty$.

If $\to \infty$, then:

$$\tilde{y}\,(z) \;=\; \Big\{\frac{z}{1-z}\;\cdot\;\frac{1}{1-\beta}\,S + \mathfrak{I}(\beta)\Big\}\,q$$

and:

$$\underline{v}\,(t) \;=\; \Big\{\frac{1}{1-\beta}\,S + \mathfrak{I}(\beta)\Big\}\,q\,.$$

Then we have:

$$\underline{v} \;=\; (I - \beta\underline{P}\,)^{-1}\,q,$$

or:

$$\underline{v}\,(t+1) \;=\; q + \beta P\,\underline{v}\,(t).$$

By an explicit enumeration of the vectors $\underline{v}(1)$, $\underline{v}(2)$, ... we can write:

$$\underline{v}(t) = \left[\sum_{j=0}^{n-1} (\beta P)^j\right] q + \beta^n \, P\underline{v}\,(0)$$

and because $0 \le \beta \le 1$, we can write:

$$\lim_{t \to \infty} \underline{v}(t) = \sum_{j=1}^{\infty} (\beta P)^j q = (I - \beta P)^{-1} q \,.$$

Within the value determination stage we must solve the following system of equations:

$$v_i = q_i + \beta \sum_{j=1}^{N} p_{ij} v_j \,; \quad i = 1, 2, \dots, N \,,$$

which corresponds to the general results given above.

The second stage refers to the policy improvement procedure. The optimal policy is that which realizes the largest present values corresponding to all states. For the case when an optimal policy up to time t has been identified, we must maximize the functional:

$$q_i^k + \beta \sum_{j=1}^{N} p_{ij}^k v_j(t) \,,$$

where k indicates an accessible decision alternative to a decision maker when the system is in state i. Because $\lim_{t \to \infty} v_j(t)$ exists and is equal to v_j, then it is necessary to maximize the functional:

$$q_i^k + \beta \sum_{j=1}^{N} p_{ij}^k v_j \,,$$

corresponding to all alternatives k in states i, $i = 1, 2, \dots, N$.

In this case it is necessary for each state i to identify the decision alternative which maximizes the test function:

$$\Gamma_i^k (\beta) = q_i^k + \beta \sum_{j=1}^{N} p_{ij}^k v_j \,,$$

using the values of the v_i 's determined in the previous stage.

Under these conditions, k becomes the new decision for the system when this is in state i.

The computational flow chart for the policy iteration algorithm for the case with discounting is given in Figure 4.9.

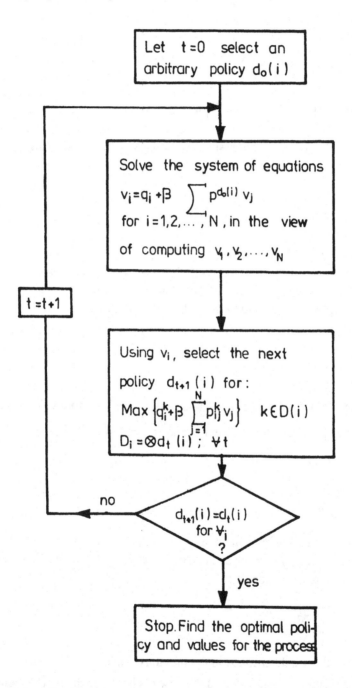

Figure 4.9. Computational flowchart for the policy iteration algorithm with discounting.

The convergence of this algorithm can be made in a similar way to the previous demonstration (see Howard (1960), Mine and Osaki (1970)) [7], [17].

4.4 Optimization algorithm using linear programming

4.4.1. SEMI-MARKOV DECISION PROCESS

In some situations applying linear programming (LP) for solving decision processes with a Markovian structure has special advantages from the computational point of view. A formulation for semi-Markov decision processes will be given.

The fundamental relations for semi-Markov decision processes without discounting ($\beta = 1$) are:

$$v_i + g\tilde{\tau}_i = q_i\tilde{\tau}_i + \sum_{j=1}^{N} p_{ij} v_j; \quad i = 1, 2, \dots, N,$$

or in an alternative representation:

$$g = q_i + (1/\tilde{\tau}_i) \left(\sum_{j=1}^{N} p_{ij} v_j - v_i \right); \quad i = 1, 2, \dots, N.$$

If g and v_i, for all i, correspond to the optimal policy, then for any sub-optimal policy k in state i we can write:

$$g \geq q_i^k + (1/\tilde{\tau}_i^k) \left(\sum_{j=1}^{N} p_{ij}^k v_j - v_i \right); \quad i = 1, 2, \dots, N.$$

If K_i represents the maximum number of decision alternatives in state i of the system, the total number of alternatives for the system is $\mathcal{K} = \otimes K_i = \sum_{i=1}^{N} K_i$.

The objective function for the semi-Markov decision process must satisfy the system of inequalities of the following form:

$$g \geq q_i^k + (1/\tilde{\tau}_i^k) \left(\sum_{j=1}^{N} p_{ij}^k v_j - v_i \right);$$

$$i = 1, 2, \dots, N; \quad k = 1, 2, \dots, K_i.$$

The desired objective function is the smallest value for g which satisfies the above given set of constraints.

Next we shall use the following notations and expressions:

a) $(1/\widetilde{\tau}_i^k) \sum\limits_{j=1}^{N-1} (\delta_{ij} - p_{ij}^k) v_j + g \geq q_i^k$,

$$i = 1, 2, \dots, N ; \quad k = 1, 2, \dots, K_i ;$$

b)   build the D matrix $(K \times N)$, with d_{ij} elements defined as:

$$d_{m(i, k), j} = (1/\widetilde{\tau}_i^k) (\delta_{ij} - p_{ij}^k) ; \quad j = 1, 2, \dots, N - 1,$$

$$d_{m(i, k), N} = 1,$$

where:

$$m(i, k) = k + K_0 + K_1 + \dots + K_{i-1},$$

$$k = 1, 2, \dots, K_i ; \quad i = 1, 2, \dots, N,$$

$$K_0 = 0.$$

In Figure 4.10 the general expression for matrix D is given.

Next, we shall define the column vectors by $(N \times 1)$ dimensions for $\underline{v}$ and t, and the column vector q by $(K \times 1)$ dimensions given by:

$$\begin{cases} v_i = v_i ; \ i = 1, 2, \dots, N - 1 \\ v_N = g, \end{cases}$$

$$t_i = 0 ; \ i = 1, 2, \dots, N - 1$$

$$t_N = 1,$$

$$q_{m(i, k)} = q_i^k .$$

The corresponding linear programming for the semi-Markov decision process with $\beta = 1$ is of the form:

$$\text{Min } \underline{t}^T \underline{v}$$

with the set of constraints:

$$D\underline{v} \geq q ; \quad \underline{v} \text{ without sign restrictions.}$$

$$
D = \begin{bmatrix}
(1-p_{11}^1)\dfrac{1}{\bar{\tau}_1^1} & -p_{12}^1\dfrac{1}{\bar{\tau}_1^1} & -p_{13}^1\dfrac{1}{\bar{\tau}_1^1} & \cdots & -p_{1,N-1}^1\dfrac{1}{\bar{\tau}_1^1} & 1 \\
\vdots & \vdots & \vdots & & \vdots & \vdots \\
(1-p_{11}^{K_1})\dfrac{1}{\bar{\tau}_1^{K_1}} & -p_{12}^{K_1}\dfrac{1}{\bar{\tau}_1^{K_1}} & -p_{13}^{K_1}\dfrac{1}{\bar{\tau}_1^{K_1}} & \cdots & -p_{1,N-1}^{K_1}\dfrac{1}{\bar{\tau}_1^{K_1}} & 1 \\
-p_{21}^1\dfrac{1}{\bar{\tau}_2^1} & (1-p_{22}^1)\dfrac{1}{\bar{\tau}_2^1} & -p_{23}^1\dfrac{1}{\bar{\tau}_2^1} & \cdots & -p_{2,N-1}^1\dfrac{1}{\bar{\tau}_2^1} & 1 \\
\vdots & \vdots & \vdots & & \vdots & \vdots \\
-p_{N1}^{K_N}\dfrac{1}{\bar{\tau}_N^{K_N}} & -p_{N2}^{K_N}\dfrac{1}{\bar{\tau}_N^{K_N}} & & \cdots & -p_{NN-1}^{K_N}\dfrac{1}{\bar{\tau}_N^{K_N}} & 1
\end{bmatrix}
$$

$$
\underline{v} = \begin{bmatrix} v_1 \\ v_2 \\ \cdot \\ \cdot \\ \cdot \\ v_{N-1} \\ g \end{bmatrix} \updownarrow N
\qquad
t = \begin{bmatrix} 0 \\ 0 \\ \cdot \\ \cdot \\ \cdot \\ 0 \\ 1 \end{bmatrix} \updownarrow N
$$

$$
q = \begin{bmatrix} q_1^1 & q_1^2 & \cdots & q_1^{K_1} & q_2^1 & q_2^2 & \cdots q_2^{K_2} & \cdots & q_N^1 & q_N^2 & \cdots & q_N^{K_N} \end{bmatrix}^T
$$

$$
|\!\!\xleftarrow{\quad K_1 \quad}\!\!|\ \ |\!\!\xleftarrow{\quad K_2 \quad}\!\!|\qquad |\!\!\xleftarrow{\quad K_N \quad}\!\!|
$$
$$
|\!\!\xleftarrow{\qquad\qquad\qquad K \qquad\qquad\qquad}\!\!|
$$

Figure 4.10. General expression for matrix D.

Next we shall define the column vectors by $(N \times 1)$ dimensions for v and t and the column vector q by $(K \times 1)$ dimensions given by:

$$\begin{cases} v_i = v_i; & i = 1, 2, \ldots, N-1 \\ v_N = g \end{cases}$$

$$\begin{cases} t_i = 0; & i = 1, 2, \ldots, N-1 \\ t_N = 1 \end{cases}$$

$$q_{m(i, k)} = q^k_i.$$

The corresponding linear programming for the semi-Markov decision process with $\beta = 1$ is of the form:

$$\text{Min } t^T v$$

with the set of constraints:

$$Dv > q$$

v without sign restrictions.

The above LP can be solved by using a standard technique as presented in Chapter 3.

The dual problem which corresponds to the LP model given above has the form:

$$\text{Max } q^T \phi$$

with the set of constraints:

$$D^T \phi = t$$

$$\phi > 0,$$

where the vector ϕ has $(K \times 1)$-dimensions.

Semi-Markov decision processes with $0 \leq \beta \leq 1$ can be formalized in an LP standard model. The basic equations are of the following form:

$$v_i = \rho_i(\alpha) + \sum_{j=1}^{N} p_{ij} h_{ij}^{T}(\alpha) v_j \; ;$$

$$i = 1, 2, \dots, N .$$

Next, we shall consider maximization of v_i's; taking into consideration the above assumptions we obtain:

$$v_i \geq \rho_i^{k}(\alpha) + \sum_{j=1}^{N} p_{ij}^{k} h_{ij}^{Tk}(\alpha) v_j \; ;$$

$$i = 1, 2, \dots, N$$

or:

$$\sum_{j=1}^{N} \left[\delta_{ij} - p_{ij}^{k} h_{ij}^{Tk}(\alpha) \right] v_j \geq \rho_i^{k}(\alpha) .$$

The following definitions will be taken into consideration:

$$d_{m(i,k),j} = \delta_{ij} - p_{ij}^{k} h_{ij}^{Tk}(\alpha) ,$$

$$i = 1, 2, \dots, N$$

$$\underline{v} = [v_i \; ; \; i = 1, 2, \dots, N]$$

$$\underline{S} = [s_i \; ; \; i = 1, 2, \dots, N] \; ; \; s_i = 1; \; \text{for all } i$$

$$\rho(\alpha) = [\delta_{m(i,k)}(\alpha) = \rho_i^{k}(\alpha)].$$

Finally, we obtain the LP formulated for semi-Markov decision processes with discounting:

Min $\underline{S}^{T} \underline{v}$

with the following set of constraints:

$D\underline{v} \geq \rho(\alpha)$

$\underline{v}$ without sign restrictions.

^{T} indicates the transpose value of the associated matrix.

The dual problem can be formulated by the following LP formulation:

$$\text{Max } \rho^T (\alpha) \underline{c}$$

with the following set of constraints:

$$D^T \underline{c} = \underline{S}$$

$$\underline{c} \geq 0,$$

where $\underline{c}$ is a column vector of $(1 \times K)$-dimensions.

4.4.2. MARKOV DECISION PROCESSES

Completely ergodic Markov decision processes can also be formulated by using a linear programming technique (see Mine and Osaki (1970) [17]). First, we shall consider the case with discounting ($\beta \neq 1$). The decision set included in the model has a stochastic structure (mixed strategies). For each moment t we shall define the probability $w_i^k(t)$ as the system being in state $i \in S$ (S is the state space of the system) and the selected decision is $k \in K_i$.

The problem is to identify an optimal strategy which maximizes the expected discounted reward. We can write that:

$$\sum_{k \in K_j} w^k(t) = w_j(0) = a_j \text{ for } t = 0$$

$$= \sum_{i \in S} \sum_{k \in K_i} p_{ij}^k w_i^k(t-1) \text{ for } t \geq 1, j \in S. \tag{4.39}$$

In the above relations, a_j is in state j at time 0 $\left(\sum_{j \in S} a_j = 1 ; 1 \geq a_j \geq 0 \right)$.

LEMMA 4.1. (Mine and Osaki (1970) [17]) *Any non-neagative solution $w_i^k(t)$ of (4.39) is a probability distribution, and the corresponding value of the discounted total reward is bounded.*

As a consequence of the above lemma, we can write the following objective function:

$$(P\,I) \text{ Max } \sum_{t=0}^{\infty} \beta^t \sum_{i \in S} \sum_{k \in K_i} c_i^k w_i^k(t)$$

with (4.39) as constraints and $w_i^k(t) \geq 0$, for $t \geq 0$, $i \in S$, $k \in K_i$ and c_i^k represents the reward generated by the system when it is found in state i and decision k is taken.

The variable x_j^k is defined as:

$$x_j^k = \sum_{t=0}^{\infty} \beta^t w_j^k(t) \; ; \; j \in S; \; k \in K_j \; ,$$

which is a z-transformed variable with $z = \beta$.

The problem can be formulated as an LP model with x_j^k as variable, se we can write:

$$(P \; II) \; \text{Max} \sum_{i \in S} \sum_{k \in K_i} c_i^k x_i^k$$

with the following constraints:

$$\sum_{k \in K_j} x_j^k - \beta \sum_{i \in S} \sum_{k \in K_i} p_{ij}^k x_i^k = a_j \; ; j \in S . \tag{4.40}$$

$$x_j^k \geq 0 ; \; j \in S; \; k \in K_j \; .$$

It can be proved that an optimal solution for this problem always exists.

The next theorems will be given without demonstration; they present additional properties for Markov decision processes with $\beta \neq 1$ (see Mine and Osaki (1970)).

THEOREM 4.4. *If the system of equations (4.40) is bounded only by the* x_i^k *variable, selected by any stationary strategy, then :*
    i) *system (4.40) has a unique solution ;*
    ii) *iff* $a_j \geq 0$ $(j \in S)$, $\exists \, x_j^k \geq 0$, $(j \in S)$;
    iii) *iff* $a_j > 0$ $(j \in S)$, $\exists \, x_j^k \geq 0$, $(j \in S)$.

THEOREM 4.5. *When* $a_j > 0$ *for* $j \in S$, *then there exists a corresponding mapping between stationary strategies and basic feasible solutions for (4.40) which are non-degenerate solutions.*

THEOREM 4.6. *When* $a_j, j \in S$ *is strictly positive, (P II) has an optimal basic solution and the dual problem as a unique optimal solution.*

COROLLARY 4.1. *For all values of* $a_j > 0, j \in S$, *there exists a basic solution with the property that for all i, there exists only one k such that:*

$$x_i^k > 0 \; and \; x_i^k = 0 \; for \; any \; other \; k.$$

The LP dual formulation is of the following form:

$$\text{Min} \sum_{i \in S} a_i v_i$$

with the following set of constraints:

$$v_i \geq c_i^k + \beta \sum_{j \in S} p_{ij}^k v_j; \ i \in S; \ k \in K_i$$

v_i without sign restrictions for all $i \in S$.

Markov decision processes with discountinng ($\beta \neq 1$) can also be formalized by using an LP formultion. In this case, we define $b_j^k, j \in S, k \in K_j$, as the mixed probability system will be in state j and a decision k is taken. The probability b_j^k does not depend on t because we deal only with stationary strategies. It is clear that:

$$\sum_{k \in K_j} b_j^k = 1; \ 0 \leq b_j^k \leq 1,$$

for $j \in S, k \in K_j$.

If π is the state probability vector with $\pi_j, j = 1, \dots, N$ as components, then the objective function for a Markov decision process with discounting is of the form:

$$\sum_{j \in S} \sum_{k \in K_j} \pi_j c_j^k b_j^k$$

with the following constraints:

$$\pi_j - \sum_{i \in S} \sum_{k \in K_j} \pi_j p_{ij}^k b_i^k = 0; \ j \in S$$

Iff we denote $x_j^k = \pi_j b_j^k \geq 0$, we shall write the LP model for the Markov decision process in the form:

$$\text{Max} \sum_{j \in S} \sum_{k \in K_j} c_j^k x_j^k$$

with the following constraints:

$$\sum_{k \in K_j} x_j^k - \sum_{i \in S} \sum_{k \in K_j} p_{ij}^k x_i^k = 0; \ j \in S,$$

$$\sum_{j \in S} \sum_{k \in K_j} x_j^k = 1,$$

$$x_j^k \geq 0 ; j \in S ; k \in K_j.$$

The LP dual formulation is of the form:

Min $\underline{v}_N$

with the following constraints:

$$v_N + v_i \geq c_i^k + \sum_{j=1}^{N-1} p_{ij}^k v_j ; \quad i \in S, k \in K_i.$$

v_i without sign restrictions for $i \in S$.

As it was proved for the policy iteration algorithm, there exists a solution for the LP primal which maximizes the objective function g such that LP dual can be written with the following form:

Max g

under the constraints:

$$g + v_i \geq c_i^k + \sum_{j=1}^{N-1} p_{ij}^k v_j ; \quad i \in S, k \in K_i.$$

g, v_i without sign restrictions for $i = 1, 2, \dots, N-1$

$v_N = 0$.

Markovian decision processes (with one or more chains) which do not obey the ergodicity property can also be solved by using an LP model such that the primal and dual have the following expressions:

Primal:

$$\text{Max} \sum_{j \in S} \sum_{k \in K_j} c_j^k x_j^k$$

under the constraints:

$$\sum_{i \in S} \sum_{k \in K_i} (\delta_{ij} - p_{ij}^k) x_i^k \neq 0 ; \quad j \in S$$

$$\sum_{k \in K_j} x_j^k + \sum_{i \in S} \sum_{k \in K_i} (\delta_{ij} - p_{ij}^k) x_i^k = 0 \,; \, j \in S,$$

where $y_j^k \geq 0$ for any transient state j and is given by the expression:

$$y_j^k = \sum_{i \in S} a_i \, d_{ij}(f) \, b_j^k \,; \, j \in S, \, k \in K_j,$$

where:

$$[d_{ij}(f)] = [I - P(f) + P^*(f)]^{-1} - P^*(f).$$

$$
P(f) =
\begin{bmatrix}
P_{11} & 0 & \cdot & \cdot & \cdot & & 0 \\
0 & P_{22} & 0 & \cdot & \cdot & & 0 \\
\cdot & \cdot & \cdot & \cdot & \cdot & & \cdot \\
\cdot & \cdot & \cdot & \cdot & \cdot & & \cdot \\
\cdot & \cdot & \cdot & 0 & P_{\alpha\alpha} & & 0 \\
P_{\alpha+1,1} & P_{\alpha+1,2} & \cdot & \cdot & P_{\alpha+1,\alpha} & & P_{\alpha+1,\alpha+1}
\end{bmatrix}
$$

where $P_{11}, P_{22}, \ldots, P_{\alpha\alpha}$ associated with each set $S\mu$, $\mu = 1, 2, \ldots, \alpha$ and

$$P^*(f) = \lim_{t \to \infty} \sum_{i=0}^{t-1} P^i(f)/t,$$

and $f \in K$ $(K = \bigotimes_{i \in S} K_i)$.

The LP dual for Markov decision processes has the following form:

$$\text{Max} \sum_{i \in S} a_i u_i$$

under the following constraints:

$$u_i \geq c_i^k + \sum_{j \in S} p_{ij}^k u_j; \quad i \in S; \quad k \in K_i,$$

$$u_i + v_i \geq c_i^k + \sum_{j \in S} p_{ij}^k v_j; \quad i \in S; \quad k \in K_i,$$

u_i, v_i without sign restrictions, $i \in S$,

where $u = [u_i]$; $\underline{v} = [v_i]$ are unique solution for $u = Pu$ and $u + \underline{v} = \underline{c} + P\underline{v}$.

4.5   Risk-sensitive decision processes

Risk-sensitivity phenomenon has been presented in Chapter 3. For an exponential utility function the "delta" property induces the following relations:

$$u(v) = -(\text{sgn } \gamma) e^{-\gamma v}$$

where $\gamma \in [-1, 1]$ is the risk aversion coefficient.

$$u^{-1}(x) = -\gamma^{-1} \ln (-(\text{sgn } \gamma)x),$$

$$u(v + \delta) = -(\text{sgn } \gamma) e^{-\gamma(v+\delta)} = e^{-\gamma\delta} u(v).$$

Next we shall introduce risk-sensitive Markovian decision processes for which $\gamma \neq 0$.

4.5.1. RISK-SENSITIVE FINITE HORIZON MARKOV DECISION PROCESSES

The Markov process is characterized by a finite number of states N, a probability transition matrix $P(t) = [p_{ij}(t)]$ at time t as well as the reward/losses matrix, $C(t) = [c_{ij}(t)]$, where elements $c_{ij}(t)$ can take negative and positive values, respectively. This kind of model has been investigated by Howard and Matheson (1972) and by Jacquette (1972) [10], [23].

Next we shall give the approach developed for the first time by Howard and Matheson.

Let $v_i(t + 1)$ denote the total reward generated by the process before termination and the certain equivalent is $\tilde{v}_i(t + 1)$. We make the hypothesis that the decision maker accepts an exponential utility function with the "delta" property. The Markov process is considered to be non-stationary and finite horizon.

The utility functional associated to the process (see Figure 4.11) can be written as:

$$u(\tilde{v}_i(t+1)) = \sum_{j=1}^{N} p_{ij}(t+1) \cdot u[c_{ij}(t+1) + \tilde{v}_j(t)], \quad t \geq 0.$$

Quantities $\tilde{v}_i(0)$, $j = ,1, 2, \dots , N$ can be directly allocated by the decision maker.

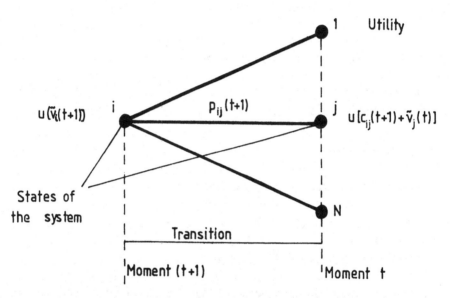

Figure 4.11. Utility function associated with the process.

Using the "delta" property given before, the relation written above can be written in the following form:

$$u(\tilde{v}_i(t+1)) = \sum_{j=1}^{N} p_{ij}(t+1) e^{-\gamma c_{ij}(t+1)} \cdot u(\tilde{v}_j(t)), \quad t \geq 0. \tag{4.41}$$

Next, define the state occupancy utility for the Markov decision process as being:

$$u_j(t) = u(\tilde{v}_j(t)) = -(\text{sgn}\,\gamma)\, e^{-\gamma \tilde{v}_j(t)}; \quad t \geq 0.$$

Equation (4.41) becomes:

$$u(\widetilde{v}_i(t+1)) = \sum_{j=1}^{N} p_{ij}(t+1) e^{-\gamma c_{ij}(t+1)} \cdot u_j(t)), \quad t \geq 0.$$

The notation $e_{ij}(n) = \exp(-\gamma c_{ij}(t))$, represents, in the content of this chapter, a negative utility for the reward $c_{ij}(t)$ associated with a transition from state i to state j. The expected value of the above quantity is given by:

$$q_{ij}(t) = p_{ij}(t) \exp(-\gamma c_{ij}(t))$$

$$= p_{ij}(t) e_{ij}(t),$$

or in a matrix form, $Q(t) = [q_{ij}(t)]$, where all the elements of $Q(t)$ are non-negative values.

The utility functional for the Markov decision process can be written in the following form:

$$u_i(t+1) = \sum_{j=1}^{N} q_{ij}(t+1)u_j(t), \quad t \geq 0. \tag{4.42}$$

The certain equivalent associated with the process is given by:

$$\widetilde{v}_i(t) = -(\gamma)^{-1} \ln[-(\text{sgn } \gamma) u_i(t)].$$

When a decision maker has the possiblity of applying a decision $k \in K_i$ when the system is in state i, and a transition to a state j takes place, (see Figure 4.12), the optimal policy for the finite horizon case can be identified by using the dynamic programming model.

$$u(\widetilde{v}_i(t+1)) = \max_{k \in K_i} \left\{ \sum_{j=1}^{N} p_{ij}(t+1) e^{-\gamma c_{ij}^{k}(t)} u_j(t)) \right\}; \quad t \geq 0.$$

where $d_i(t+1)$, is the maximal value for k, and $\underline{d}(t+1)$ is the corresponding decision vector.

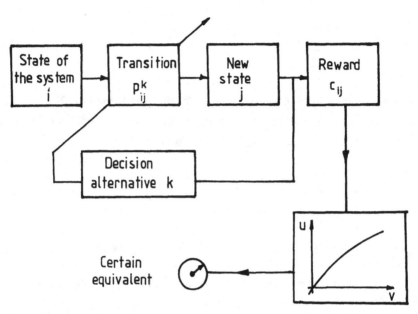

Figure 4.12. Application of decision k to a system in state i and a transition to state j takes place.

EXAMPLE. Let us consider a risk-sensitive Markov decision process ($\gamma \neq 0$) with $n = 3$; $K_1 = 3, K_2 = 2, K_3 = 2, P$ and C are as given in Table 4.1.

If $\gamma = 1.0$ (total risk aversion), then the optimal policy and the certain equivalent are given in Table 4.2 for $t = 1, 2, \ldots , 10$.

For the case when the decision maker is a totally risk preference person ($\gamma = -1$), the optimal policy and the certain equivalent are given in Table 4.3 for $t = 1, 2, \ldots , 10$.

State i	Alternative	p_{ij}^k			c_{ij}^k		
		$j=1$	$j=2$	$j=3$	$j=4$	$j=5$	$j=6$
	1	0.5	0.25	0.25	10	4	8
1	2				8	2	4
	3				4	6	4
	1	0.5	0	0.5	14	0	18
2	2				8	16	8
	1	0.25	0.25	0.5	10	2	8
3	2				6	4	2
	3				4	0	8

Table 4.1. Risk sensitive Markov decision process ($\gamma \neq 0$).

	Strategy d in state			Certain equivalent $\tilde{v}_i(t)$		
Stage t	1	2	3	$i = 1$	2	3
1	1	1	2	5.36329	14.675	3.47495
2	1	1	2	12.82038	19.94218	12.15518
3	1	1	1	21.29147	27.47853	20.73632
4	1	1	1	29.93613	36.04996	29.18795
5	1	1	1	38.4043	44.5913	37.66343
6	1	1	1	46.88206	53.05974	46.1496
7	1	1	1	55.3665	61.53781	54.63268
8	1	1	1	63.8495˙	70.0222	63.11497
9	1	1	1	72.33197	78.50518	71.59763
10	1	1	1	80.81462	86.98765	80.08033

Table 4.2. Total risk aversion ($\gamma = 1.0$).

	Strategy d in state			Certain equivalent $\tilde{v}_i(t)$		
Stage t	1	2	3	$i = 1$	2	3
1	1	1	2	9.37349	17.325	8.85351
2	3	2	2	21.2458	33.1947	21.03776
3	3	2	2	37.11204	49.05794	36.9038
4	3	2	2	52.9875	64.92441	52.77027
5	3	2	2	68.84497	80.79088	68.63673
6	3	2	2	84.71144	96.65735	84.5032
7	3	2	2	100.57791	112.52381	100.36967
8	3	2	2	116.44438	128.39028	116.23614
9	3	2	2	132.31085	144.25675	132.10261
10	3	2	2	148.17732	160.12322	147.96908

Table 4.3. Total risk preference ($\gamma = -1.0$).

4.5.2. RISK-SENSITIVE INFINITE HORIZON MARKOV DECISION PROCESSES

In the case of infinite horizon Markovian decision processes we make the assumption that the transition probabilities and rewards are the same for all the transition, such that:

$$p_{ij} = \lim_{t \to \infty} p_{ij}(t)$$

$$c_{ij} = \lim_{t \to \infty} c_{ij}(t).$$

The process is completely determined by matrices $P = [p_{ij}]$ and $C = [c_{ij}]$ as well as the disutility matrix $Q = [q_{ij}]$, where

$$q_{ij} = p_{ij} \exp(-\gamma c_{ij}).$$

Relation (4.42) can be written in a vector form such that:

$$u(t+1) = Qu(t); \quad t \geq 0$$

with a general solution of the form:

$$u(t) = Q^t u(0); \quad t \geq 0.$$

It has been proved by Howard and Matheson (1972) [10] that the Markov process is irreducible and acyclic and, therefore, matrix Q is irreducible and primitive, so that:

$$\lim_{t \to \infty} \lambda^{-t} Q^t u(0) = \lim_{t \to \infty} \lambda^{-t} u(t) = ku,$$

where λ is the largest eigenvalue for Q and u is the corresponding eigenvector to the λ eigenvalue. The constant k is chosen in such a way that $u_N = -\operatorname{sgn} \gamma$.

It is demonstrated that:

$$\lim \{\tilde{v}_i(t) - t(-\gamma^{-1} \ln \lambda)\} = \tilde{v}_i + c,$$

where $c = -\ln k/\gamma$, and $\tilde{v}_i$ is the certain equivalent in state i and is given by

$$\tilde{v}_i = -\gamma^{-1} \ln [-(\operatorname{sgn} \gamma) u_i].$$

The certain equivalent objective function $\tilde{g}$, for the risk-sensitive Markov process grows in a linear way, with ratio $(-\ln \lambda/\gamma)$, such that:

$$\tilde{g} = -\gamma^{-1} \ln \lambda,$$

and the asymptotic form for $\tilde{v}_i(t)$ can be given by the value $(t\tilde{g} + \tilde{v}_i + c)$.

The value of $\tilde{g}$ is bounded such that, regardless of the sign of γ:

$$\min_{i,j} c_{ij} \leq \tilde{g} \leq \max_{i,j} c_{ij}.$$

Starting from the utility functional for the process under investigation:

$$u_i(t+1) = \sum_{j=1}^{N} q_{ij} u_j(t),$$

we can divide the above equation by λ^t; when $t \to \infty$ we obtain the following relation:

$$u_i = \sum_{j=1}^{N} q_{ij} u_j.$$

Iff $u_i = (-(\text{sgn } \gamma) \exp(-\gamma\tilde{v}_i))$ and taking into consideration the computational relation for $\tilde{g}$, then $\lambda = \exp(-\gamma\tilde{g})$.

In turn, we can write that:

$$\exp(-\gamma(\tilde{g} + \tilde{v}_i)) = \sum_{j=1}^{N} q_{ij} \exp(-\gamma\tilde{v}_j)$$

or:

$$\exp(-\gamma(\tilde{g} + \tilde{v}_i)) = \sum_{j=1}^{N} p_{ij} \exp(-\gamma\tilde{v}_j). \tag{4.43}$$

COROLLARY 4.2. *If $\gamma \to 0$, when the decision maker becomes risk indifferent, then equation (4.43) can be written as*:

$$\tilde{g} + \tilde{v}_i = \sum_{j=1}^{N} p_{ij}(c_{ij} + \tilde{v}_j).$$

The general computational algorithm, for the risk-sensitive Markov decision process is that of the policy iteration. The test quantity within the policy improvement stage is of the form:

$$\Gamma_k^i = -\gamma^{-1} \ln \left[\sum_{j=1}^{N} p_{ij}^k \exp \left(-\gamma(c_{ij}^k + \tilde{v}_j) \right) \right].$$

$$i = 1, 2, \ldots, N ; \ k \in K_i.$$

The corresponding flow charts for model optimization are given in Figures 4.13 and 4.14.

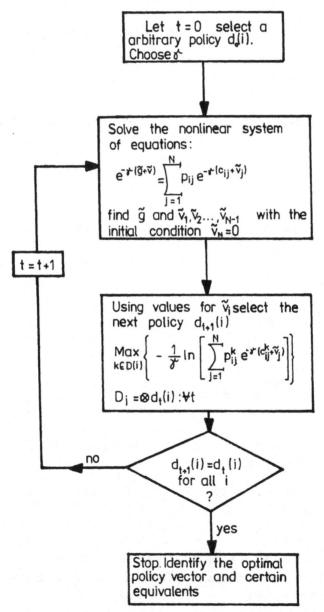

Figure 4.13. Policy iteration cycle for the risk-sensitive — certain equivalent form.

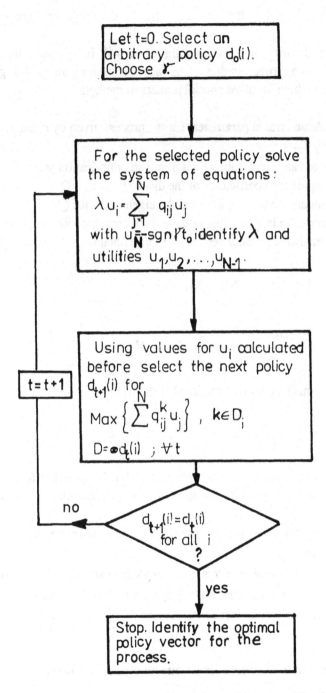

Figure 4.14. Policy iteration cycle for the risk-sensitive — utility form.

In Figure 4.13, the policy iteration cycle for the case risk-sensitive — certain equivalent form is given.

In Figure 4.14, the policy iteration cycle for the case risk-sensitive - utility is given.

Next, a revised form for the above algorithm will be given. The practical advantage for such an approach is due to the fact that it eliminates solving a set of non-linear equations which involves special numerical methods.

Step 0. Evaluate critical parameters for the process given by P and C matrices. Set $\gamma = -1$.

Step 1. Choose an initial policy for solving an optimization problem.

Step 2. Compute the coefficients of the disutility matrix, Q.

Step 3. Compute the eigenvalues and the eigenvetors for Q.

Step 4. Choose the largest eigenvalue and the corresponding eigenvector, u.

Step 5. Normalize the utility vector in order to obtain:

$$\hat{u} = [\hat{u}_i \; ; \; i = 1, 2, \dots , N], \quad \text{where}$$

$$\hat{u}_i = \frac{-(\text{sgn } \gamma)\, u_i}{u_N}; \; i = 1, 2, \dots , N .$$

Step 6. Minimize the utility functional of the form:

$$\tilde{r}_i^k = \sum_{j=1}^{N} q_{ij}^k \hat{u}_j \; ; \; i = 1, 2, \dots , N ; \; k \in K_i.$$

Step 7. If the condition for the policy iteration is statisfied (e.g. the same policy is obtained at two successive iterations), we obtain the optimal policy and go to Step 8.

Step 8. Print optimal and sub-optimal solutions corresponding to the risk aversion coefficient, γ.

Step 9. Increase the value of γ by an increment of size ε. Iff $\gamma = 1$, go to Step 10, iff $\gamma < 1$, $\gamma = \gamma + \varepsilon$ and then go to Step 0. Iff $\gamma = 0$ use a classical algorithm for the case of Markovian decision processes.

Step 10. Stop.

The above algorithm converges in a finite number of iterations.

EXAMPLE 1. A piece of automation equipment is characterized at any moment in time t $(t \geq 0)$ by one of the following states: initial, good, marginal and terminal (trapping state). We make the assumption that the technical equipment under investigation is totally observable to an external observer.

The deterioration (0)-maintenance (1) process for the equipment is approximated by a Markov chain with the transition probabilities:

$$P^0 = \begin{bmatrix} 0 & 0.80 & 0.15 & 0.05 \\ 0 & 0 & 0.97 & 0.03 \\ 0 & 0 & 0 & 1.0 \\ 0 & 0 & 0 & 1.0 \end{bmatrix}$$

$$P^1 = \begin{bmatrix} 0.4 & 0.3 & 0.2 & 0.1 \\ 0.85 & 015 & 0 & 0 \\ 0.75 & 0.20 & 0.05 & 0 \\ 0 & 0.5 & 0.4 & 0.1 \end{bmatrix}$$

The associated cost for the repair-replacement decisions are given in conventional units and are presented in Table 4.4.

A diagrammatic representation for the results using the risk -sensitive Markov decision model is indicated by Curve 1 in Figure 4.15 (also see Table 4.5). When we can choose a different repair policy, which finally affects the probability cost-structure for the process, the values for the certrain equivalent $\tilde{g}$ are represented by Curve 2 from Figure 4.15.

EXAMPLE 2. Consider a system with P and C parameters from the example given in section 4.5.1. When γ takes values between -1 and $+1$, we obtain the variation for $\tilde{g}$ and the optimal structure for the decision vector (see Figure 4.15).

Costs	State i			
	1	2	3	4
Cost occupancy for a state	1	2	3	7
Cost for repair (alternative 1)	2	5	10	25
Fixed cost for replacement of the system	5	10	20	30
Cost for repair (alternative 2)	3	6	9	20

Table 4.4. Cost for repair-replacement decisions.

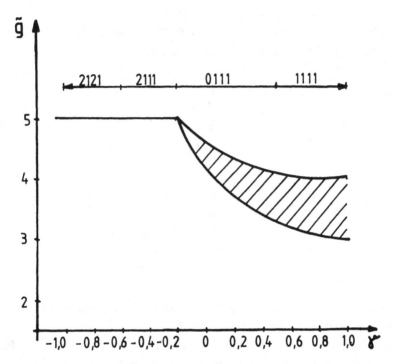

Figure 4.15. Variation for $\tilde{g}$ and the optimal structure for the decision vector.

γ	$\tilde{g}$	Optimal maintenance strategy				No. of iterations
-1.0	5.00058	2	1	2	1	3
-0.8	4.98539	2	1	2	1	3
-0.6	4.9999	2	1	2	1	3
-0.4	5.0000	2	1	1	1	2
-0.2	5.0000	2	1	1	1	2
0.2	2.50854	0	1	1	1	3
0.4	3.35222	0	1	1	1	3
0.6	3.17949	1	1	1	1	2
0.8	2.98928	1	1	1	1	2
1.0	2.84361	1	1	1	1	2

Table 4.5. Risk sensitive Markov decision model.

4.5.3. RISK-SENSITIVE FINITE HORIZON SEMI-MARKOV DECISION PROCESSES

The risk-sensitive semi-Markov decision process is characterized by a number N of states, the probability transition matrix $\underline{P} = [p_{ij}]$, the waiting time probability matrix $H(m) = h_{ij}(m)]$ as well as the costs given by $y_{ij}(\cdot)$ and $b_{ij}(\cdot)$. In Figure 4.16 a diagrammatic representation of the processes investigated in this section is given.

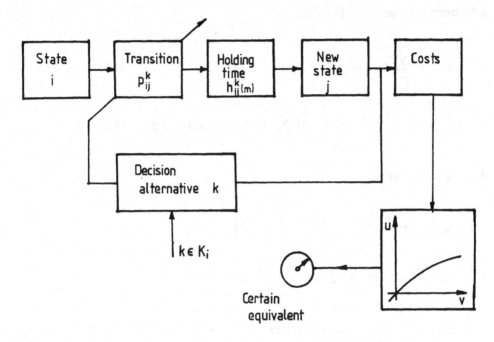

Figure 4.16. Diagrammatic representation of processes investigated in this section.

The process utility functional in accordance with the diagram, is of the following form:

$$u(\tilde{v}_i(t)) = \sum_{j=1}^{N} p_{ij} \sum_{m=1}^{t} h_{ij}(m) \cdot$$

$$\cdot \left(\sum_{l=0}^{m-1} y_{ij}(l) + b_{ij}(m) + \tilde{v}_j(t-m) \right) ; \ t \geq 1.$$

If the decision maker agrees on an exponential utility function, then the above functional is of the following form:

$$u(\tilde{v}_i(t)) = \sum_{j=1}^{N} p_{ij} \sum_{m=1}^{t} h_{ij}(m) \cdot$$

$$\cdot \exp\left(-\gamma\left(\sum_{l=0}^{m-1} y_{ij}(l) + b_{ij}(m)\right)\right) \cdot u(\tilde{v}_j(t-m)); \; t \geq 1,$$

or by using the alternative notation:

$$u_i(t) = \sum_{j=1}^{N} p_{ij} \sum_{m=1}^{t} h_{ij}(m) \cdot$$

$$\cdot \exp\left(-\gamma\left(\sum_{l=0}^{m-1} y_{ij}(l) + b_{ij}(m)\right)\right) \cdot u_j(t-m)); \; t \geq 1.$$

If we use the notation:

$$e_{ij} = \exp\left[-\gamma \cdot \left(\sum_{l=0}^{m-1} y_{ij}(l) + b_{ij}(m)\right)\right],$$

then the utility cost functional has the following form:

$$u_i(t) = \sum_{j=1}^{N} p_{ij} \sum_{m=1}^{t} h_{ij}(m) \, e_{ij}(m) \, u_j(t-m); \; t \geq 1. \tag{4.44}$$

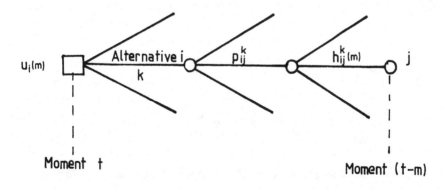

Figure 4.17. Diagrammatic representation of a process utility functional.

For the case when a decision maker is able to apply a decision alternative when the system is in state i, the optimal strategy for the finite horizon case is identified by using the dynamic programming technique, such that:

$$u_i(t) = \max_{k \in K_i} \left\{ \sum_{j=1}^{N} p_{ij}^k \sum_{m=1}^{t} h_{ij}^k(m) \cdot e_{ij}^k(m) \, u_j(t-m) \right\} ; \quad t \geq 1,$$

where $d_i(t)$ is a maximizing value for k and $\underline{d}(t)$ is the corresponding decision vector.

4.5.4. RISK-SENSITIVE INFINITE HORIZON SEMI-MARKOV DECISION PROCESSES

The utility functional given by (4.44) can be written in a vector form:

$$\underline{u}(t) = \sum_{m=0}^{t} \left[\left(P \,\square\, H(m) \right) \,\square\, E(m) \right] \_\underline{u}(t-m); \quad t \geq 1.$$

If we consider $Q(m)$ the disutility for the semi-Markov decision process with $\gamma \neq 0$, then:

$$Q(m) = \left(P \,\square\, H(m) \right) \,\square\, E(m),$$

and the corresponding utility functional for the process, in a matrix form, is:

$$\underline{u}(t) = \sum_{m=0}^{t} Q(m) \, \underline{u}(t-m); \quad t \geq 1.$$

When $t \rightarrow \infty$, then:

$$u_i(t) = \sum_{j=1}^{N} \sum_{m=0}^{t} q_{ij}(m) \, u_j(t-m); \quad t \geq 1,$$

$$\lim_{t \rightarrow \infty} \lambda^{-t} u_i(t) = \lim_{t \rightarrow \infty} \sum_{j=1}^{N} \sum_{m=0}^{n} \lambda^{-(t-m)} \lambda^{-m} \cdot q_{ij}(m) \, u_j(t-m);$$

$$\lim_{t \rightarrow \infty} \lambda^{-t} u_i(t) = \lim_{t \rightarrow \infty} \sum_{j=1}^{N} \lambda^{-(t-m)} q_{ij}^g \, \lambda^{-1} \cdot u_j(t-m) ;$$

$$u_i = \sum_{j=1}^{N} q_{ij}^g \, \lambda^{-1} \, u_j .$$

Howard's algorithm of the policy iteration considers, in the policy improvement phase, the use of a test quantity, which, for the case under investigation has the following form:

$$U_i^k = \sum_{j=1}^{N} q_{ij}^{g\,k} \lambda^{-1} u_j ; \; k \in K_i,$$

where $q_{ij}^{g(\cdot)} \lambda^{-1}$ is an adequate notation for the geometric transform by argument λ^{-1} for the disutility values.

The system of equations for the risk-sensitive semi-Markov decision process is of the following form:

$$\exp(-\tilde{w}_i) = \sum_{j=1}^{N} q_{ij}^{g} \left(\exp(\gamma\tilde{g}) \right) \exp(-\tilde{w}_j) ; \; i = 1, 2, \dots, N,$$

and the certain equivalent for the corresponding lottery (the test quantity for the policy improvement stage) is given by the expression:

$$V_i^k = -\gamma^{-1} \ln \sum_{j=1}^{N} q_{ij}^{g\,k} \left(\exp(\gamma\tilde{g}) \right) \exp(-\tilde{w}_j) ; \; i = 1, 2, \dots, N, \; k \in K_i,$$

where k is a possible decision to be applied when the system is in state i.

The computational flow charts for Howard's algorithm are shown in Figures 4.18 and 4.19.

In Figure 4.18 the policy iteration cycle for the case of the risk-sensitive-utility semi-Markov decision process is given.

In Figure 4.19 the policy iteration cycle for the case of the risk-sensitive-certain equivalent semi-Markov decision process is given.

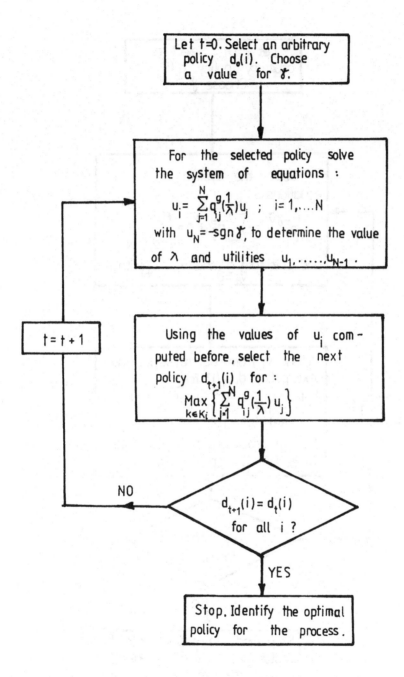

Figure 4.18. Policy iteration cycle for the case of the risk-sensitive-utility semi-Markov decision process.

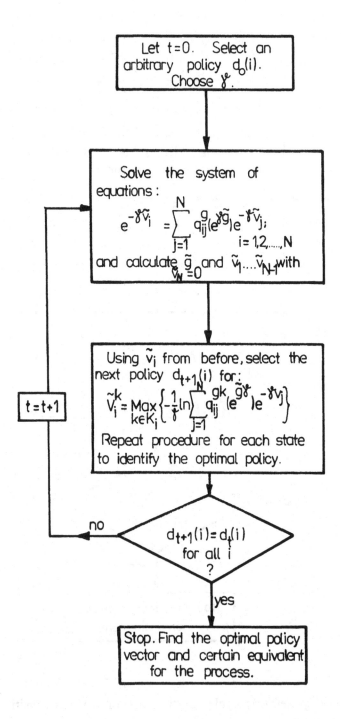

Figure 4.19. Policy iteration cycle for the case of the risk-sensitive-certain equivalent semi-Markov decision process.

4.6 On eliminating sub-optimal decision alternatives in Markov and semi-Markov decision processes

MacQueen (1967) [16] showed how sub-optimal decision alternatives in Markov process with discounting can be eliminated when the Markov chain has a finitie number of states. Later other theoretical and practical results were extended by Porteus (1971), Grinold (1973), Hastings and Mells (1973) etc. (see [22], [4], [6]).

In this way it is proved that we are able to identify sub-optimal alternatives in the case of Markov decision processes. Grinold presents upper and lower bounds for the corresponding objective function of the process when they are generated within the policy iteration or linear programming optimization models.

4.6.1. MARKOV DECISION PROCESS

The problem deals with maximizing the expected discounted reward when the Markov decision process has $\beta < 1$.

Let A be a current policy and v^A the corresponding value associated with this policy. For each state i the decision alternative k in this state is defined by γ_i^k with the following expression:

$$\gamma_i^k = q_i^k + \beta \sum_{j=1}^{N} p_{ij}^k v_j^A - v_i^A \ ; \ i = 1, 2, \dots, N \tag{4.44a}$$

and let $\gamma^* = \max_{k,i} \gamma_i^k$.

By using the policy iteration algorithm, we can find the following relation:

$$\gamma^* = \max_i \left\{ -v_i^A + \max_k \left[q_i^k + \beta \sum_{j=1}^{N} p_{ij}^k v_j^A \right] \right\}.$$

THEOREM 4.7. *Decision alternative k in state i is sub-optimal iff:*

$$\gamma_i^k < \frac{\gamma^* \beta}{(\beta - 1)} .$$

DEMONSTRATION. Let v_i^* be an optimal value function given by:

$$v_i^* = \max_k \left\{ q_i^k + \beta \sum_{j=1}^{N} p_{ij}^k v_j^* \right\}; \qquad i = 1, 2, \dots, N.$$

For any policy A there is the relation $v_i^A \leq v_i^*$ for $i = 1, 2, \dots, N$. It is accepted that for any v which satisfies the relation:

$$v_i > \max_k \left\{ q_i^k + \beta \sum_{j=1}^N p_{ij}^k v_j \right\}; \quad i = 1, 2, \dots, N,$$

an upper limit for v^* exists. (see MacQueen (1967) [16]).

It is also known that decision alternative k in state i is sub-optimal, if:

$$q_i^k + \beta \sum_{j=1}^N p_{ij}^k \, v_j^* < v_i^* \text{ for } i = 1, 2, \dots, N.$$

Because of the existence of v_i^* and equation (4.44) we can write:

$$q_i^k + \beta \sum_{j=1}^N p_{ij}^k \, v_j^* \leq q_i^k + \beta \sum_{j=1}^N p_{ij}^k \cdot [v_i^A + (1-\beta)^{-1} \gamma^*] =$$

$$= v_i^A + \gamma_i^k + (1-\beta)^{-1} \beta\gamma^*.$$

The lower bound for v^* as well as the condition $\gamma_i^k + (1-\beta)^{-1} \beta\gamma^* < 0$ lead to the inequality of the following form:

$$q_i^k + \beta \sum_{j=1}^N p_{ij}^k \, v_j^* \leq v_i^A + \gamma_i^k + (1-\beta)^{-1} \beta\gamma^* < v_i.$$

It is clear that iff:

$$\frac{\beta\gamma^*}{(\beta-1)} > \gamma_i^k; \quad i = 1, 2, \dots, N$$

$$k = 1, 2, \dots, K_i,$$

where alternative k in state i is sub-optimal.

Porteus (1980) proved the following theorem [102].

THEOREM 4.8. *For a Markov decision process with discounting, iff*

$$\gamma_i = \max_k \gamma_i^k, \quad \gamma_* = \min_i \gamma_i, \quad \gamma^* = \max_{k,i} \gamma_i^k \quad and \; then:$$

$$\gamma_i^k < \gamma_i + (\gamma_* - \gamma^*) \beta(1-\beta)^{-1},$$

indicates that decision alternative k is non-optimal for state i.

The above theorem does not have a very important practical application. Using the results given by this theorem, not more than 1% of non-optimal alternatives were eliminated when compared with other simpler rules which were applied.

4.6.2. SEMI-MARKOV DECISION PROCESSES WITH FINITE HORIZON

Consider a semi-Markov decision process with finite horizon where the rewards generated are discounted by means of the coefficient $\alpha > 0$. At time zero, in state i, there are accessible decisions $k \in K_i(0)$. The present value of the expected reward is c_i^k when the decision k is selected and the system is in state i. Transition probabilities are p_{ij}^k and the waiting time probabilities are $h_{ij}^k(m)$. The Cartesian product for all decision alternatives in the finite horizon semi-Markov process is $D(0) = \underset{i=1,N}{\otimes} K_i(0)$.

If d_{ij}^k represents the discounted transition probabilities which correspond to p_{ij}^k (see Hastings and Mello (1973)) then we can write [6]:

$$d_{ij}^k = p_{ij}^k \int_{m=0}^{\infty} h_{ij}^k(m) \exp(-\alpha m) \, dm < p_{ij}^k \leq 1.$$

For a decision policy $\delta \in D(0)$, let $B(\delta)$ be the probability transition matrix and $\underline{c}(\delta)$ be the column vector for the current rewards generated by the finite horizon semi-Markov decision process. We shall consider $\underline{v}(\delta)$ the column vector for the limiting present value using policy δ. A policy δ^* is optimal iff:

$$\underline{v}(\delta^*) > \underline{v}(\delta)$$

for each $\delta \in D(0)$.

The following inequality holds:

$$\underline{v}(\delta^*) \geq \underline{c}(\delta) + B(\delta)\,\underline{v}(\delta^*).$$

A decision alternative is sub-optimal if it is not among the optimal policies. As was already discussed, the methods and optimization algorithms for finite horizon semi-Markov decision processes are those of linear programming and policy iteration.

Let $f_i(t)$ be the maximal present value for the total mean reward, in a process with t stages, generated by state i.

Vector $f(t)$ is a column vector whose components are $f_i(t)$. We can write that:

$$f_i(t) = \max_k \left\{ c_i^k + b_i^{kT} f(t-1) \right\} ; \ k \in K_i(t-1); \ k \geq 1$$

and $f(0)$ has an arbitrary value.

$d_i^{kT} = [d_{ij}^k]$ is a raw vector of transition probabilities from state i to state j.
Let:

$$\mathcal{L}'(t) = \min_i \left\{ f_i(t) - f_i(t-1) \right\}$$

$$\mathcal{L}''(t) = \max_i \left\{ f_i(t) - f_i(t-1) \right\}$$

$$\eta_i^k = \sum_{j=1}^{N} d_{ij}^k ; \ \eta'(t) = \min_{i,k} [\eta_i^k]$$

$$\eta''(t) = \max_{i,k} [\eta_i^k] ; \ k \in K_i(t-1),$$

$$i = 1, 2, \dots, N ; \ t \geq 1.$$

MacQueen (1967) and Porteus (1971) estimated adequate limits for $v_i(\cdot)$ such that [16], [22]:

$$f_i(t) + \frac{\eta'(t) \ \mathcal{L}'(t)}{1 - \eta'(t)} \leq v_i(\delta_t) \leq v_i \leq f_i(t) + \frac{\eta''(t) \ \mathcal{L}''(t)}{1 - \eta''(t)} ,$$

where δ_t is the policy identified at iteration t by using the policy iteration algorithm.

Hastings and Mello (1973) considered a different test given by the above theorem [6].

THEOREM 4.9. *Decision alternative k in state i is sub-optimal iff*:

$$c_i^k + d_i^{kT} f(t-1) < f_i(t-1) + \frac{\eta'(t-1) \ \mathcal{L}'(t-1)}{1 - \eta'(t-1)} - \eta_i^k \eta''(t-1) \ \mathcal{L}''(t-1).$$

DEMONSTRATION. By adequate permutations of the above relation we can write:

$$c_i^k + d_i^{kT} \left\{ f(t-1) + \frac{\eta''(t-1)\, \mathcal{L}''(t-1)}{1-\eta''(t-1)} \mathbf{1} \right\} < f(t-1) + \frac{\eta'(t-1)\, \mathcal{L}'(t-1)}{1-\eta'(t-1)} \ ,$$

where $\mathbf{1} = [1 \ 1 \ \ldots \ 1]^T$.

Taking into consideration the results given by MacQueen and Porteus we can write:

$$c_i^k - d_i^{kY}\, v(\delta^*) < v(\delta^*)$$

which implies that alternative k is sub-optimal in state i of the semi-Markov process.

The above test is known as the Hastings-Mello (HM) test.

The quantities $\eta'(\,\cdot\,)$, $\eta''(\,\cdot\,)$, $\mathcal{L}'(\,\cdot\,)$, $\mathcal{L}''(\,\cdot\,)$ can be computed at step t and then used as computational elements in the test criteria at step $(t+1)$.

The MacQueen-Porteus test (see Hastings-Mello (1973)) for semi-Markov decision processes can be written in the following form [6]:

THEOREM 4.10. *Alternative k in state i is sub-optimal iff*:

$$c_i^k + d_i^{kT} f(t-1) < f_i(t-1) + \frac{\eta'(t-1)\, \mathcal{L}'(t-1)}{1-\eta'(t-1)} - \frac{\eta''(t-1)\, \mathcal{L}''(t-1)}{1-\eta''(t-1)}.$$

LEMMA 4.1. *(a)* $\mathcal{L}''(t) \le \eta''(t)\, \mathcal{L}''(t-1)$;

(b) $\mathcal{L}'(t) \ge \eta'(t)\, \mathcal{L}'(t-1)$.

COROLLARY 4.3. $\mathcal{L}''(t) - \mathcal{L}'(t) < \eta''(t) \left[\mathcal{L}''(t-1) - \mathcal{L}'(t-1) \right]$.

Let us consider a discrete semi-Markov process for which $d_{ij}^k = \beta\, p_{ij}^k$; for all i, j, k where $0 \le \beta \le 1$ is the discounting coefficient. The raw vector of transition probabilities from state i to state $j, j = 1, 2, \ldots, N$ under the decision alternative k, is p_i^{kT}. We shall also use the notation $\mathcal{D}_i^{HM}(t)$, and $\mathcal{D}_i^{MP}(t)$, respectively the set of alternatives for state i and transition t which were investigated by the Hastings-Mello (HM) test and by the MacQueen-Porteus (MP) test.

THEOREM 4.11. (a) *At iteration t, MP eliminates at least as many alternatives as HM and therefore $\mathfrak{D}_i^{HM}(t) \supseteq \mathfrak{D}_i^{MP}(t)$; and*
  (b) *HM at transition $(t+1)$ eliminates at least as many decision alternatives as does MP at transition t and therefore $\mathfrak{D}_i^{MP}(t) \supseteq \mathfrak{D}_i^{HM}(t+1)$,*

where $i = 1, 2, \ldots, N$; $t \geq 0$.

The corresponding demonstration is given in Hastings and Mello (1973) [6].

COROLLARY 4.4. *Each decision alternative in $\mathfrak{D}_i(0)$ which will not pass tests $HM_i(t)$ and $MP_i(t)$, respectively, will also not pass the tests $HM_i(t+1)$, $MP_i(t+1)$, $i = 1, 2, \ldots, N$; $t \geq 1$.*

In this respect the power of both tests is non-decreasing in t.

CHAPTER 5

PARTIALLY OBSERVABLE MARKOVIAN DECISION PROCESSES

5.1 Finite horizon partially observable Markov decision processes

The system dynamics for partially observable Markov processes have been presented in Chapter 1 of this book.

The cost of performing a transition from a state i to a state j in a Markov chain which produces, after a transition, an output θ is denoted by $c_{ij\theta}$. Next, we shall define γ_i as the expected immediate cost of making a transition and producing an observable output given $S(t) = i$, such that:

$$\gamma_i = \sum_{j=1}^{N} \sum_{\theta=1}^{R} c_{ij\theta}\, p_{ij}\, r_{j\theta}\,,$$

and in vector form $\gamma = [\gamma_i\,;\ i = 1, 2, \dots, N\,]$. If the process terminates in state i then γ_i^0 represents the expected cost incurred and in vector form $\gamma^0 = [\gamma_i^0\,]$.

For the analyzed process the expected cost of operating the process for n time periods is denoted by $\mathcal{C}^n(\pi)$, where the initial state of knowledge is π:

$$\mathcal{C}^n(\pi) = \pi\gamma + \sum_{\theta=1}^{R} \{\theta|\pi\}\, \mathcal{C}^{n-1}\, [T(\pi|\theta)],\ n \geq 1$$

and:

$$\mathcal{C}^0(\pi) = \pi\gamma^0\,.$$

PROPERTY 5.1. $\mathcal{C}^0(\pi)$ is linear in π.

Proof. The above property can be proved by induction starting from the fact that $\mathcal{C}^0(\pi)$ is linear in π.

$$\mathcal{C}^n(\pi) = \pi\gamma + \sum_{\theta} \{\theta|\pi\}\, \mathcal{C}^0\, [T(\pi|\theta)]$$

$$= \pi\gamma + \sum_{\theta} \{\theta|\pi\}\, T(\pi|\theta)\gamma^0$$

$$= \pi\gamma + \sum_{\theta} \{\theta|\pi\} \, \pi PR_\theta \, \gamma^0 \,/\{\pi|\theta\}$$

$$= \pi\gamma + \sum_{\theta} \pi PR_\theta \, \gamma^0$$

$$= \pi\gamma + \pi P\gamma^0 = \pi(\gamma + P\gamma^0)$$

$$\mathcal{C}^1(\pi) = \pi\alpha^1 \text{ where } \alpha^1 = (\gamma + P\gamma^0).$$

This proves that $\mathcal{C}^1(\pi)$ is linear in π. By induction, we can easily write that:

$$\mathcal{C}^1(\pi) = \pi(\gamma + P\gamma + ... + P\gamma^{n-1} + P\gamma^0),$$

which proves the above property.

The behaviour for large n is described by:

$$\mathcal{C}^n(\pi) \cong ng + \mathcal{C}(\pi), \tag{5.1}$$

where $g = \lim_{n\to\infty} \mathcal{C}^n(\pi)/n = \pi\bar{P}\gamma$, where $\bar{P}$ is the Cesaro limit of P^n which always exists [44], [66].

After some algebraic manipulation we can write:

$$ng + \mathcal{C}(\pi) = \pi\gamma + (n-1)\,g + \sum_{\theta} \{\theta|\pi\} \, \mathcal{C}\,[T(\pi|\theta)]$$

or:

$$g + \mathcal{C}(\pi) = \pi\gamma + \sum_{\theta=1}^{R} \{\theta|\pi\} \, \mathcal{C}\,[T(\pi|\theta)] .$$

If we consider that $\mathcal{C}(\pi) = \pi\alpha$ and if $\mathbf{1} = [1, 1, ... , 1]^T$, then:

$$g\,\mathbf{1} + \alpha = \pi[\gamma + \sum_{\theta} PR_\theta \, \alpha] .$$

Under the consideration that the above equality must hold for each vector $\pi \in \Pi$, then:

$$g\,\mathbf{1} + \alpha = \gamma + P\alpha.$$

The elements α_i of the vector α play the role of relative values in Howard's model. As in the policy iteration algorithm, by fixing one element of α and solving the above system of equations, we can have g and other $(N-1)$ elements of α.

In the case that an external observer of the partially observable Markov process does control it, he can select one of the K_a decision alternatives. The set of parameters which specify the process are $\{P^a, R^a, \gamma^a\}$. The dynamics of the decision process are represented in Figure 5.1 (see [18], [19], [20], [32], [44]).

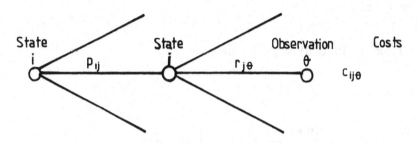

Figure 5.1. Dynamics of the decision process.

The decision maker is faced with the problem of minimizing the total expected cost, $C^n(\pi)$, of operating the process for n time periods with termination cost γ^0 and the initial state of knowledge $\pi(n)$.

Next, we emphasize the following:

a) π is a well-defined state for the controlled process;

b) the expected immediate reward in state π under alternative a is $\pi\gamma^a$;

c) the transition probability for entering the next state $T(\pi|\theta)$ is $\{\theta|\pi\}$.

The basic functional equation for a partially observable Markov decision process is:

$$C^0(\pi) = \pi\gamma^0$$

$$C^n(\pi) = \min_a \left\{ \pi\gamma^a + \sum_{\theta=1}^{R} \{\theta|\pi, a\} \, C^{n-1} \, [T(\pi|\theta, a)] \right\} \tag{5.2}$$

where:

$$\{\theta|\pi, a\} = P^a R_\theta^a \, 1$$

$$T(\pi|\theta, a) = P^a R_\theta^a / \{\theta|\pi, a\}.$$

An optimal solution can be found if $\mathcal{C}^{n-1}(\pi)$ is known. $\delta^n(\pi)$ is defined as an index that minimizes (5.2) as a function of π. As was proved by Sondik [20], the δ^n, $1 \leq n \leq L$, gives the complete solution of the decision process with L or fewer time periods remaining.

It is of interest to emphasize a few properties of the controlled expected cost $\mathcal{C}^n(\pi)$ [20].

LEMMA 5.1. $T(\pi|\theta, a)$ *preserves straight lines*; *if* $0 \leq \beta \leq 1$, $\bar{\beta} = 1 - \beta$ *then for* $\pi^1, \pi^2 \in \Pi$, *we have:*

$$T(\beta\pi^1 + \bar{\beta}\pi^2|\theta, a) = \mu e^1 + \bar{\mu} e^2$$

where:

$$\mu = \frac{\beta\{\theta|\pi^1, a\}}{\beta\{\theta|\pi^1, a\} + \bar{\beta}\{\theta|\pi^2, a\}} \ ; \ \bar{\mu} = 1 - \mu$$

and $e^i = T(\pi^i|\theta, a) \in \Pi$.

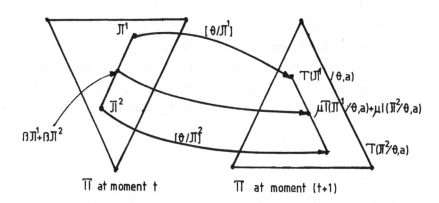

Figure 5.2. Transition mechanism.

Proof. In Figure 5.2 the transition mechanism for the analyzed process is represented. It is already known that:

$$T(\pi|\theta, a) = \pi P^a R_\theta^a / \{\theta|\pi, a\} = \pi P^a R_\theta^a \ \mathbf{1}.$$

If $H = P^a R_\theta^a$, then we have:

$$T(\beta\pi^1 + \bar{\beta}\pi^2|\theta, a) = \frac{(\beta\pi^1 + \bar{\beta}\pi^2) H}{(\beta\pi^1 + \bar{\beta}\pi^2) H\,1} =$$

$$= \frac{\beta\pi^1 H}{(\beta\pi^1 + \bar{\beta}\pi^2) H\,1} + \frac{\bar{\beta}\pi^2 H}{(\beta\pi^1 + \bar{\beta}\pi^2) H\,1}$$

With e^i defined as above, we can write that:

$$T(\beta\pi^1 + \bar{\beta}\pi^2|\theta, a) = \frac{1}{(\beta\pi^1 + \bar{\beta}\pi^2) H\,1}$$

$$[\beta e^1(\pi^1 H\,1) + \bar{\beta}\, e^2(\pi^2 H_1)] =$$

$$= \frac{1}{(\beta\pi^1 + \bar{\beta}\pi^2) H\,1}\ [e^1\beta\{\pi^1, a\} + e^2\bar{\beta}\,\{\theta|\pi^2, a\}] .$$

With the notation that:

$$(\beta\pi^1 + \bar{\beta}\pi^2) H\,1 = \beta\{\theta|\pi^1, a\} + \bar{\beta}\,\{\theta|\pi^2, a\}$$

and with μ as above, we can write:

$$T(\beta\pi^1 + \bar{\beta}\pi^2|\theta, a) = \mu e^1 + \bar{\mu} e^2.$$

THEOREM 5.1. *The basic functional equation* $C^n(\pi)$ *is piecewise linear and concave for all n.*

Proof. $C^0(\pi) = \pi\gamma^0$ is piecewise linear and concave in π.

It is known that the minimum of concave piecewise linear functions is also concave piecewise linear; the sum of such functions preserves these properties. It has to be shown that $\{\theta|\pi, a\}\ \mathcal{C}^{n-1}\ [T(\pi|\theta, a)]$ is concave piecewise linear.

We shall make the notation $f(\pi) = \{\theta|\pi, a\}\ \mathcal{C}^{n-1}\ [T(\pi|\theta, a)]$. Letting $\pi\beta = \beta\pi^1 + \bar{\beta}\pi^2$, we have:

$$f(\pi\beta) = \{\theta|\pi, a\}\ \mathcal{C}^{n-1}\ [T(\pi|\theta, a)] =$$

$$= \{\theta|\pi, a\}\ \mathcal{C}^{n-1}\ [\mu T(\pi^1|\theta, a) + \bar{\mu}\, T(\pi^2|\theta, a)]$$

$$\geq \{\theta|\pi\beta, a\}\big[\ \mu\mathcal{C}^{n-1}\ [T(\pi^2|\theta, a)] + \bar{\mu}\ \mathcal{C}^{n-1}\ [T(\pi^2|\theta, a)]\big]$$

$$= \big[\beta\{\theta|\pi^1, a\} + \bar{\beta}\,\{\theta|\pi^2, a\}\big]\big[\ \mu\mathcal{C}^{n-1}\ [T(\pi^1|\theta, a) + \bar{\mu}\ \mathcal{C}^{n-1}\ [T(\pi^2|\theta, a)]\big]$$

$$= \beta\{\theta|\pi^1, a\}\mathcal{C}^{n-1}\ [T(\pi^1|\theta, a)] + \bar{\beta}\,\{\theta|\pi^2, a\}\ \mathcal{C}^{n-1}\ [T(\pi^2|\theta, a)]$$

$$= \beta f(\pi^1) + \bar{\beta}\ f(\pi^2),$$

which indicates the concavity for $f(\ .\ )$.

$\mathcal{C}^{n-1}(\pi)$ is piecewise linear with $\alpha(\pi)$ ($\mathcal{C}^{n-1}(\pi) = \pi\alpha(\pi)$) and peicewise constant over Π, so that:

$$f(\pi) = \{\theta|\pi, a\}\mathcal{C}^{n-1}\ [T(\pi|\theta, a)]$$

$$= \{\theta|\pi, a\}\ \frac{P^a\ R_\theta^a\ \theta}{\{\theta|\pi, a\}}\quad \alpha\ [T(\pi|\theta, a)]$$

$$= P^a\ R_\theta^a\ \alpha\ [T(\pi|\theta, a)] = \pi\xi(\pi),$$

where ξ is piecewise constant over Π, and this proves the piecewise linearity.

Of real interest, is the computation of the expression $\mathcal{C}^n(\pi)$ so that:

$$\mathcal{C}^n(\pi) = \min_j\ [\pi\alpha_j^n],\quad j \geq 1. \tag{5.3}$$

Using the relation (5.2) we have:

$$\mathcal{C}^n(\pi) = \min_a \left[\pi\gamma^a + \sum_\theta \min_j \pi P^a R_\theta^a \alpha_j^{n-1}\right]. \tag{5.4}$$

The computational scheme of $\mathcal{C}^n(\pi)$ from $\mathcal{C}^{n-1}(\pi)$ gives the possibility of evaluating $\mathcal{C}^n(\pi)$ on a pointwise basis.

Next we shall define A^n as the set of values of $\alpha^n(\pi)$ so that $A^n = [\alpha_1^n, \alpha_2^n, \dots]$ and $\tilde{R}$ as the associated partition of Π, so that:

$$\tilde{R}^n = \bigcup_j \tilde{R}_j^n$$

$$\tilde{R}_j^n = [\pi : \mathcal{C}^n(\pi) = \pi\,\alpha_j^{n-1}].$$

The optimization result for the finite horizon partially observable Markov decision process indicates that the functional $\mathcal{C}^n(\pi)$ is determined by A^n.

For $\pi \in \tilde{R}_j^n$ we have:

$$\mathcal{C}^n(\pi) = \pi\alpha_j^n \; \min_a \left[\pi\gamma^a + \sum_\theta \min_k \pi P^a R_\theta^a \alpha_k^{n-1}\right]$$

$$= \pi\left[\gamma^{a_j} + \sum_\theta P^{a_j} R^{a_j} \alpha_{d_{j,\theta}^n}^{n-1}\right]$$

or:

$$\alpha_j^n = \gamma^{a_j} + \sum_\theta P^{a_j} R^{a_j} \alpha_{d_{j,\theta}^n}^{n-1}.$$

The optimal control for time n is denoted by $\delta^n(\pi)$ so that:

$$\delta^n(\pi) = a_j = \delta_j^n \quad \text{for } \pi \in \tilde{R}_j^n.$$

The optimization procedure for such a process is due to Sondik [20] and is known as "The One-Pass Algorithm". According to Sondik [20]:

"The one-pass algorithm is a systematic way of determining A from A^{n-1} (and consequently determining $\delta^n(\pi)$). The term "one-pass" refers to the fact that the optimal control for each state is found using a single computational pass over the state space Π for each time period the process is to operate. The algorithm is based on the fact that complete knowledge of A^{n-1} allows the computation of $\mathcal{C}^n(\pi)$ at some point π. Using the above equation for the calculation of $\mathcal{C}^n(\pi)$ gives the vector value of $\alpha^n(\pi) = \alpha^n_j$.

... when α^n_j is known, the boundaries of the region $\tilde{R}^n_j$ can be determined, and from these boundaries other values of $\alpha(\pi)$ can be determined.

.... The descriptive term one-pass refers to the fact that once a value of $\alpha^n(\pi)$, α^n_j, has been computed, it need not be computed".

The expected cost of using alternative a) for all states at time n is calculated as:

$$\mathcal{C}^n_a(\pi) = [\pi\gamma^a + \sum_\theta \{\theta|\pi, a\} \, \mathcal{C}^{n-1} \, [T(\pi|\theta, a)]$$

which is piecewise linear and concave.

$$\mathcal{C}^n_a(\pi) = \pi\alpha^{na}(\pi)$$

where:

$$\alpha^{na}(\pi) = \gamma^a + \sum_\theta P^a R^a_\theta \, \alpha^{n-1}[T(\pi|a, \theta)] .$$

The vector $\alpha^{na}(\pi)$ has α^{na}_j as components; the partition of Π induced by $\alpha^{na}(\pi)$ is given by $\tilde{R}^{na}$ with $\tilde{R}^{na}_j$ as components, so that:

$$\tilde{R}^{na}_j = [\pi : \mathcal{C}^n_a(\pi) = \pi\alpha^{na}_j] .$$

The component α^{na}_j can be written as:

$$\alpha^{na}_j = \gamma^a + \sum_\theta P^a R^a \, \alpha^{n-1}_{d^{na}_{j\theta}} , \tag{5.5}$$

where $d^{na}_{j\theta}$ is the subscript of the element A^{n-1}.

It is an equivalence between the vector d^{na}_j ($d^{na}_j = [d^{na}_{j1}, \ldots , d^{na}_{jk}]$) and the vector α^{na}_j; if d^{na}_j is known, then α^{na}_j can be calculated by relation (5.5).

The value of $\mathcal{C}^n_a$ can be computed as:

$$\tilde{C}_a^n(\pi) = \pi\gamma^a + \sum_\theta \min_{k_\theta} \pi P^a R_\theta^a \, \alpha_{k_\theta}^{n-1} = \pi\alpha_j^{na}. \tag{5.6}$$

where k_θ are variables equivalent to d_j^{na}, minimizing the relation (5.6).

LEMMA 5.2. [20]. *The set of points π in $\tilde{R}_j^{na}$ that lies on a boundary of $\tilde{R}_j^{na}$ obtained by transforming $\tilde{R}^{n-1}$ satisfies the functional relation :*

$$\pi P^a R_\theta^a \, (\alpha_k^{n-1} - \alpha_l^{n-1}) = 0.$$

Proof. The intersection between $\tilde{R}_k^{n-1}$ and $\tilde{R}_l^{n-1}$ $k \neq l$ is the set:

$$\tilde{R}_k^{n-1} \cap \tilde{R}_l^{n-1} = [\pi \in \Pi : \min_m \pi\alpha_m^{n-1} = \pi\alpha_k^{n-1} = \pi\alpha_l^{n-1}].$$

A set which contains the above set is:

$$[\pi \in \Pi : (\pi\alpha_k^{n-1} - \pi\alpha_l^{n-1}) = 0].$$

If $T(\pi|\theta, a)$ lies in $\tilde{R}_k^{n-1} \cap \tilde{R}_l^{n-1}$ at time $(n-1)$, then:

$$T(\pi|\theta, a) \, (\alpha_k^{n-1} - \alpha_l^{n-1}) = 0$$

$$\frac{\pi P^a R_\theta^a}{\{\theta|\pi, a\}} (\alpha_k^{n-1} - \alpha_l^{n-1}) = 0$$

which is equivalent to:

$$\pi P^a R_\theta^a (\alpha_k^{n-1} - \alpha_l^{n-1}) = 0 ; \ \pi \in \Pi .$$

THEOREM 5.2.

$$\tilde{R}_j^{na} = \left[\pi \in \Pi : \pi P^a R_\theta^a \big(\alpha_{d_{j\theta}^{na}}^{n-1} - \alpha_k^{n-1}\big) \leq 0, \text{ for all } \theta, \text{ for all } k \right].$$

Proof.

$$\mathcal{C}^{n-1} [T(\pi|\theta, a)] = T(\pi|\theta, a) \, \alpha_{d_{j\theta}^{na}}^{n-1} \leq T(\pi|\theta, a) \, \alpha_k^{n-1} ; \text{ for all } k .$$

If $\pi \in \tilde{R}^{na}_{j}$, then:

$$\pi \in \bigcap_{\theta} \left[\pi \in \Pi : \pi P^a R^a_{\theta} \left(\alpha^{n-1}_{d^{na}_{j\theta}} - \alpha^{n-1}_k \right) \leq 0, \text{ for all } k \right] =$$

$$= \left[\pi \in \Pi : \pi P^a R^a_{\theta} \left(\alpha^{n-1}_{d^{na}_{j\theta}} - \alpha^{n-1}_k \right) \leq 0, \text{ for all } \theta, \text{ for all } k \right] \equiv \hat{R} ,$$

where $\tilde{R}^{na}_{j} \subset \hat{R}$.

The next theorem will be given without proof.

THEOREM 5.3. [19], [20].

If $\pi P^a R^a_{\theta} \left[\alpha^{n-1}_{d^{na}_{j\theta}} - \alpha^{n-1}_k \right] \leq 0,$ *forms a boundary of* $\tilde{R}^{na}_{j}$ *where* α^{na} *is a value*

of $\alpha^{na}(\pi)$ *with d-representation* $d^{na}_{j} = [d^{na}_{j\theta} ...],$ *then over some region in* Π *the function* $C^n_a(\pi)$ *is given by* $C^n_a(\pi) = \pi \alpha^{na}_m,$ *where* α^{na}_m *has a d-representation of the form* $d^{na}_m = [d^{na}_{m1} = d^{na}_{j1}, ... , d^{na}_{m\theta} = l, ... d^{na}_{mN} = d^{na}_{jN}].$

The optimization model for a partially observable Markov process which operates over a finite horizon is known as the "one-pass algorithm".

The procedure for the above algorithm is given in Figure 5.3.

Next a numerical example will emphasize the practical way the algorithm is used to find a solution to a practical problem.

EXAMPLE 5.1. Suppose a technical system has its operational states described by three states and an external observer has access to two outputs. The parameters of the problem are given in Table 5.1.

It is known that $C^1(\pi) = \pi \gamma^0$, and we have:

$$C^1(\pi) = \min \begin{cases} \pi [\gamma^1 + \underline{P}^1 \gamma^0] \\ \pi [\gamma^2 + \underline{P}^2 \gamma^0] \end{cases}$$

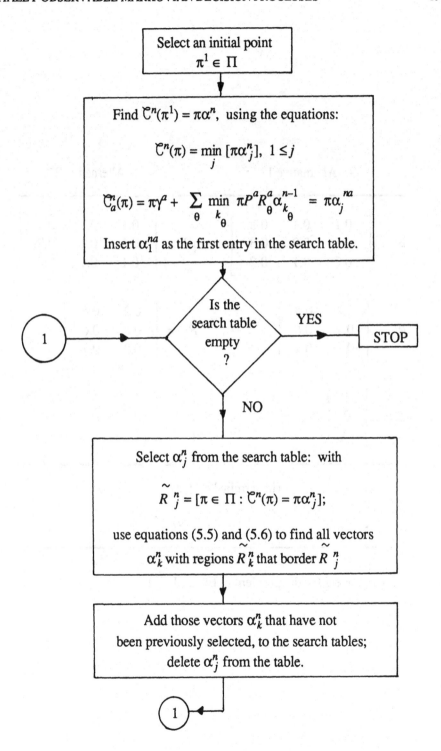

Figure 5.3. Procedure for the algorithm.

Alternative 1	Alternative 2

$$P^1 = \begin{bmatrix} 0.1 & 0.1 & 0.8 \\ 0.5 & 0.2 & 0.3 \\ 0.7 & 0.1 & 0.2 \end{bmatrix} \qquad P^2 = \begin{bmatrix} 0.1 & 0.8 & 0.1 \\ 0.7 & 0.1 & 0.2 \\ 0.1 & 0.9 & 0.0 \end{bmatrix}$$

$$R^1 = \begin{bmatrix} 0.2 & 0.8 \\ 0.4 & 0.6 \\ 0.3 & 0.7 \end{bmatrix} \qquad R^1 = \begin{bmatrix} 0.2 & 0.8 \\ 0.4 & 0.6 \\ 0.3 & 0.7 \end{bmatrix}$$

$$\gamma^1 = \begin{bmatrix} 1 \\ 0 \\ 0 \end{bmatrix} \qquad \gamma^2 = \begin{bmatrix} 0 \\ 1 \\ 0 \end{bmatrix}$$

The terminal cost is:

$$\gamma^0 = [0 \quad 3.5 \quad 0]$$

Table 5.1. Parameters for the problem in Example 5.1.

To start the "one-pass algorithm" an initial point $\pi^1 \in \Pi$ has to be chosen. When $\pi^1 = [1, 0, 0]$, by using equation $\mathcal{C}^n(\pi) = \min_a [\pi \gamma^a + \sum_\theta \pi P^a R_\theta^a \alpha_{k\theta}^{n-1}]$ to compute $\mathcal{C}_{a=1}^{n=1}(\pi)$ gives:

$$\mathcal{C}_1^1(\pi^1) = \pi^1 \alpha_1^{1,1} \quad \text{where} \quad \alpha_1^{1,1} = [1.35, \ 1.75, \ 0.32]^T,$$

$$\mathcal{C}_2^1(\pi^1) = \pi^1 \alpha_1^{1,2} \quad \text{where} \quad \alpha_1^{1,2} = [2.80, \ 0.35, \ 3.15]^T.$$

In this case $\mathcal{C}^1(\pi^1) = \pi^1 \alpha_1^{1,1}$; $\delta^1(\pi^1) = 1$; $\alpha_1^1 = \alpha_1^{1,1}$.

The boundary of $\hat{R}_1^1$ is calculated from the following set of constraints:

$$\pi_i \geq 0; \ \sum \pi_i = 1; \ \pi(\alpha_1^1 - \alpha_1^{1,2}) \leq 0.$$

In a graphical form, we can have the feature from Figure 5.4.

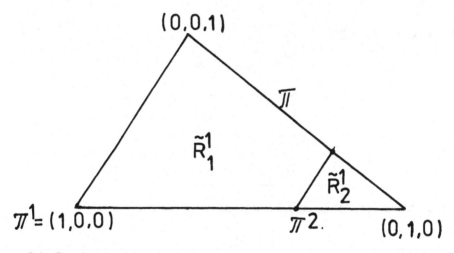

Figure 5.4. Graphical representation of Example 5.1.

To compute $\mathcal{C}^2(\pi)$ we have to choose $\pi^1 = (1, 0, 0)$ and calculate it using the equation:

$$\mathcal{C}^2(\pi^1) = \min_a \left[\pi^1 \gamma^a + \sum_\theta \{\theta \mid \pi^1, a\} \ \mathcal{C}^1 [T(\pi^1 \mid \theta, a)] \right]$$

$$= \pi^1 \alpha_1^2, \ \delta_1^2 = 2$$

where :

$$\alpha_1^{2,1} = [1.59; \ 1.25; \ 1.19]^T; \ \alpha_1^{2,2} = \alpha_1^2 = [1.56; 2.55; 1.6]^T.$$

The boundaries of $\tilde{R}_1^2$ are found by applying the following set of inequalities:

$$\pi_i \geq 0$$

$$\pi P^1 R_1^1 (\alpha_1^1 - \alpha_2^1) \leq 0$$

$$\sum_i \pi_i = 1$$

$$\pi P^1 R_2^1 (\alpha_1^1 - \alpha_2^1) \leq 0$$

$$\pi P^2 R_2^2 (\alpha_1^1 - \alpha_2^1) \leq 0$$

$$\pi P^2 R_1^2 (\alpha_2^1 - \alpha_1^1) \leq 0$$

$$\pi (\alpha_1^2 - \alpha_1^{2,1}) \leq 0$$

where for $n = 2, d_{j\theta}^{na}$ has the form:

a	j	θ	$d_{j\theta}^{2a}$
1	1	1	1
1	1	2	1
2	1	1	2
2	1	2	1

The graphical result is given in Figure 5.5.

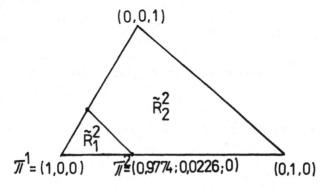

Figure 5.5. Graphical result of above.

The last constraint form the above set of inequalitites for $n = 2$ gives:

$$\alpha_1^2 = [1.56, \ 2.55, \ 1.60]^T$$

$$\alpha_2^2 = [1.59, \ 1.25, \ 1.19]^T$$

$$\delta_2^2 = 2 ; \ \delta_2^2 = 1.$$

The solution for $n = 3$ is found to be:

$$\alpha_1^3 = [1.28, \ 2.48, \ 1.28]^T$$

$$\alpha_2^3 = [2.23, \ 1.30, \ 1.48]^T$$

$$\delta_1^3 = 2 ; \ \delta_2^3 = 1;$$

and the graphical result is presented in Figure 5.6.

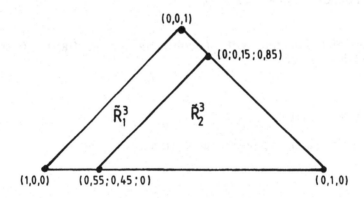

Figure 5.6. Graphical result for solution when $n = 3$.

Continuing in the way presented above, for $n = 10$, we have:

$$\alpha_1^{10} = [2.22, \ 3.78, \ 2.21]^T$$

$$\alpha_2^{10} = [3.17, \ 2.25, \ 2.31]^T$$

$$\alpha_3^{10} = [3.23, \ 3.86, \ 2.24]^T$$

$$\delta_1^{10} = 2 ; \ \delta_2^{10} = 1; \ \delta_3^{10} = 1,$$

and if

$$\min_{k} \pi_p^{10} = \pi\alpha_j^{10}, \text{ then } \delta^{10}(\pi) = \delta_j^{10}.$$

The graphical solution is given in Figure 5.7.

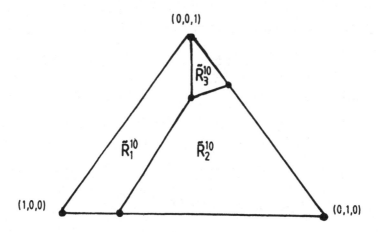

Figure 5.7. Graphical solution for $n = 10$.

5.2 The infinite horizon with discounting for partially observable Markov decision processes

5.2.1. MODEL FORMULATION

The discounting finite horizon problem can be written as:

$$\mathcal{C}^0(\pi) = \pi\gamma^0$$

$$\mathcal{C}^n(\pi) = \min_{a} \left[\pi\gamma^a + \beta \sum_{\theta} \{\theta|\pi, a\} \mathcal{C}^{n-1} [T(\pi \mid \theta, a)] \right]; \ n \geq 1$$

where it is already known that $\{\theta|\pi, a\} = \pi P^a R_\theta^a \mathbf{1}$ and $T(\pi|\theta, a) = P^a R_\theta^a / \{\theta|\pi, a\}$ and β is called the discount factor.

An optimal decision for the time n is denoted by δ^n and a sequence of such decision functions $(\delta^n, \delta^{n-1}, \delta^{n-2}, \dots, \delta^1)$ is called a policy. If the policy is $(\delta, \delta, \dots, \delta)$ then this is called a stationary policy, δ^∞.

The total expected cost for $0 < \beta < 1$, if $M = \max_{i,a} (\gamma_i^a)$, for $\gamma^0 = 0$ is:

$$|\mathcal{C}^n(\pi)| \leq M + \beta M + \ldots + \beta^n M$$

and:

$$\lim_{n \to \infty} |\mathcal{C}^n(\pi)| \leq \frac{M}{1 - \beta} < \infty.$$

According to Blackwell [1] an optimal expected cost $\mathcal{C}^*(\pi)$ exists so that:

$$\mathcal{C}^*(\pi) = \min_a \left[\pi \gamma^a + \beta \sum_\theta \{\theta | \pi, a\} \, \mathcal{C}^*[T(\pi|\theta, a)] \right].$$

The optimal control policy is stationary, so that:

$$\mathcal{C}^*(\pi) = \pi \gamma^{\delta^*} + \beta \sum_\theta \{\theta | \pi, \delta^*\} \, \mathcal{C}^*[T(\pi|\theta, \delta^*)],$$

where $\delta^*(\pi)$ is equal to the minimizing alternative for the previous equation.

The functional relationship between $\mathcal{C}^{n-1}$ and $\mathcal{C}^n$ is a contraction mapping which is an operator G such that $\|Gf_1 - Gf_2\| \leq \beta \|f_1 - f_2\|$ for some norm $\| \cdot \|$ and functions f_1 and f_2.

The contraction mapping is denoted by U such that:

$$U\mathcal{C}(\pi) = \min_a U_a(\pi, C)$$

where:

$$U_a(\pi, C) = \pi \gamma^a + \beta \sum_\theta \{\theta | \pi, a\} \, \mathcal{C}[T(\pi|\theta, a)]$$

with $\|C\| = \sup_{\pi \in \Pi} |\mathcal{C}(\pi)|$.

Considering $\mathcal{C}^n(\pi|\delta)$ as the n period expected cost of using control δ at each time period, then:

$$\mathcal{C}^n(\pi|\delta) = \pi \gamma^\rho$$

$$\mathcal{C}^n(\pi|\delta) = \pi \gamma^\delta + \beta \sum_\theta \{\theta | \pi, \delta\} \, \mathcal{C}^n[T(\pi|\theta, \delta)] \qquad \text{for } n \geq 0.$$

According to a previous remark, we have:

$$\mathcal{C}^n(\pi|\delta) = \pi\gamma^0$$

$$\mathcal{C}^{n-1}(\,\cdot\,|\,\delta) = U_\delta\,\mathcal{C}^n(\,\cdot\,|\,\delta),$$

where U_δ is the operator relating $\mathcal{C}^{n-1}$ to $\mathcal{C}^n$ and represents a contraction mapping with modulus β.

$\mathcal{C}(\pi|\delta^n)$ is an approximation to $\mathcal{C}^\infty(\pi)$ for large n.

Based on properties of contraction mappings [1] we have the following:

THEOREM 5.4 [20].

$$\|\mathcal{C}^n(\pi|\delta) - \mathcal{C}^*(\pi)\| \le \frac{2\beta^n}{1-\beta}\,K$$

where $K = \max_{a,i} |\gamma_i^a|$.

Proof.

$$\mathcal{C}(\pi|\delta^n) - \mathcal{C}^*(\pi) \le \|\mathcal{C}(\pi|\delta^n) - \mathcal{C}^n(\pi)\| + \|\mathcal{C}(\pi) - \mathcal{C}^*(\pi)\|,$$

but:

$$\|\mathcal{C}(\pi|\delta^n) - \mathcal{C}^n(\pi)\| \le (1-\beta)^{-1}\,\|U_\delta\mathcal{C}^n(\pi) - \mathcal{C}^n(\pi)\|$$

with $U_{\delta^n}\mathcal{C}^{n-1}(\,\cdot\,) = \mathcal{C}^n(\,\cdot\,)$, then:

$$\|\mathcal{C}(\pi|\delta^n) - \mathcal{C}^n(\pi)\| \le \beta\,\|\mathcal{C}(\pi|\delta^n) - \mathcal{C}^{n-1}(\pi)\| \le \beta(1-\beta)^{-1}\,\|\mathcal{C}^{n+1}(\pi) - \mathcal{C}^n(\pi)\|$$

$$\|\mathcal{C}(\pi|\delta^n) - \mathcal{C}^n(\pi)\| \le \frac{\beta^n}{1+\beta}\,\|\mathcal{C}^1(\pi) - \mathcal{C}^0(\pi)\|$$

---

[1] It is proved by Denardo and Blackwell (see comments in [20]) that:

(a) $U\mathcal{C}^* = \mathcal{C}^*$,   (b) $\|U^n\,\mathcal{C}^0 - \mathcal{C}^*\| \le \beta^n\|\mathcal{C}^0 - \mathcal{C}^*\|$, $n \ge 0$,

(c) $\|U^n\,\mathcal{C}^0 - \mathcal{C}^*\| \le (\beta^n/(1-\beta))\,\|U\mathcal{C}^0 - \mathcal{C}^0\|$, $n \ge 0$,

where U^n represents the iterated use of the operator U for n times.

$$\|\mathcal{C}^n(\pi) - \mathcal{C}^*(\pi)\| \leq \frac{\beta^n}{1 + \beta} \|\mathcal{C}^1(\pi) - \mathcal{C}^0(\pi)\|$$

which finally yields:

$$\|\mathcal{C}^n(\pi|\delta^n) - \mathcal{C}^*(\pi)\| \leq \frac{2\beta}{1 - \beta} \|\mathcal{C}^1(\pi) - \mathcal{C}^0(\pi)\|.$$

If $\mathcal{C}^0(\pi) = 0$, then

$$\|\mathcal{C}^1(\pi)\| = \min_{a} \|\pi\gamma^a\| \leq \min_{i,a} |\gamma_i^a|$$

which proves the above theorem.

A partially observable Markov decision process has to be considered as a Markovian decison problem whereby the states are defined by $\pi(t)$. In this case, the policy iteration model can be used to find an optimal decision policy. It consists of two steps known as value determination and policy improvement. In the value determination (VD) step it is computed as the cost of a stationary policy δ^∞. Next, in the policy improvement (PI) step, this cost is used to find a different stationary policy with a lower cost. As is already known, the cycle is repeated until a different policy cannot be found; this last policy is optimal.

It has already been shown that the total expected cost of a stationary policy δ^∞ is a well defined function given by $\mathcal{C}(\pi|\delta)$. Next, we shall prove that $\mathcal{C}(\pi|\delta)$ is used to identify a new policy $(\delta^1)^\infty$ such that:

$$\mathcal{C}(\pi|\delta^1) < \mathcal{C}(\pi|\delta), \quad \text{for all } \pi \in \Pi.$$

We shall consider the notation that if $f_1 \leq f_2$ this represents $f_1(\pi) - f_2(\pi) \leq 0$, for all $\pi \in \Pi$, and if $f_1 \leq f_2$, then $f_1(\pi) - f_2(\pi) \leq 0$, for all $\theta \in \Pi$, with the strict inequality holding for some π.

LEMMA 5.3. *For an infinite horizon partially observable Markov decision process with discounting, we have* :

Iff $f_1 \leq f_2$, then $U_\delta f_1 \leq U_\delta f_2$.

Next, an adapted Howard policy improvement for the class of processes that are dealt with in this chapter will be emphasized by the following theorem.

THEOREM 5.5. *If δ^∞, $C(\pi|\delta)$ represents the cost of a stationary policy and if $\delta^1(\pi)$ is the control function defined as the index minimizing $U_a(\pi, C(\cdot|\delta))$, where*

$$U_a\{\pi, C(\cdot|\delta)\} = \pi\gamma^a + \beta\sum_\theta \{\theta|\pi, a\} \cdot C[T(\pi > \theta, a)|\delta],$$

then we have:

$$C(\pi|\delta^1) < C(\pi|\delta), \quad \pi \in \Pi.$$

Proof.

$$\min_a U_a[\pi, C(\cdot|\delta)] \leq U_\delta\, C(\pi|\delta) = C(\pi|\delta)$$

$$U_{\delta 1} C(\pi|\delta)] = \min_a U_a[\pi, C(\cdot|\delta)]$$

and thus:

$$U_{\delta 1} C(\pi|\delta)] \leq C(\pi|\delta).$$

Using the above results, we have:

$$U_{\delta 1} U_{\delta 1} C(\cdot|\delta) \leq U_{\delta 1} C(\cdot|\delta) \leq C(\cdot|\delta)$$

$$\lim_{n\to\infty} U_{\delta 1}^n C(\cdot|\delta) = C(\cdot|\delta^1)$$

and thus:

$$C(\cdot|\delta^1) = C(\cdot|\delta).$$

From the following results from Sondik [20], we have the upper and lower bounds for the policy selection, so that:

$$\inf_\pi [C(\pi|\delta) - U C(\pi|\delta)] \leq \|C(\cdot|\delta) - C^*\| \leq (1-\beta)^{-1} \|C(\cdot|\delta) - U C(\cdot|\delta)\|.$$

From the above results, we can obtain adequate indications as to whether the cost $C(\pi|d)$ is sufficiently close to the optimal solution.

5.2.2. THE CONCEPT OF FINITELY TRANSIENT POLICIES.

Considering a stationary policy δ^{∞} let T_{δ} be the set function defined for any set $\Lambda \subset \Pi$ by:

$$T_{\delta}(\Lambda) = \text{closure } [T(\pi|\theta, \delta) : \pi \in \Lambda, \text{ for all } \theta].$$

If $S_{\delta} = \Pi$ represents the first set of succeeding elements S_{δ}^{n}, then we can have a recursive definition, so that:

$$S_{\delta}^{n} = T_{\delta}(S^{n-1}), \quad n \geq 1.$$

There is a different definition for the following set:

$$D_{\delta} = \text{closure } [\pi : \delta(\pi) \text{ is discontinuous in } \pi].$$

(D_{δ} is the smallest closed set of the discontinuities of δ.)

DEFINITION 5.1. A stationary policy δ^{∞} is finitely transient if and only if there exists an integer n $< \infty$, so that :

$$D_{\delta} \cap S_{\delta}^{n} = \varnothing,$$

where $\varnothing$ is the null set. We should call the *index* of the finitely transient policy (n_{δ}) the smallest integer.

LEMMA 5.4. *If δ^{∞} is finitely transient with index n_{δ} then $D_{\delta} \cap S_{\delta}^{n} = \varnothing$ for any* $n \geq n_{\delta}$.

LEMMA 5.5.

$$\mathcal{C}(\pi|\delta) = \pi \, \alpha(\pi|\delta)$$

where:

$$\alpha(\pi|\delta) = \gamma^{\delta(\pi)} + \sum_{\theta} P^{\delta(\pi)} \cdot R_{\theta}^{\delta(\pi)} \, \alpha[T(\pi|\theta, \delta) \mid \delta].$$

LEMMA 5.6. *If δ^{∞} is a finite transient policy for a partition $V = [V_1, V_2, \dots]$ associated with Π there exists a γ-mapping, so that:*

a) *V is an extension for $V^0 = [V_j^0]$, so that, if $\pi \in V_j^0$, then $\delta(\pi) = j$;*

b) *iff $\pi \in V_j$, then $T(\pi|\theta, \delta) \in V_{v(j,\theta)}$, where $v(j, \theta)$ defines a mapping for which $\pi \in V_j$, then $T(\pi|\theta, \delta) \in V_{v(j,\theta)}^0$.*

The set D_δ is defined as the smallest closed set of the discontinutities of δ. We can define the sequence of sets D^n as:

$$D^0 = D_\delta$$

$$D^{n+1} = [\pi : T(\pi|\theta, \delta) \in D^n \text{ for some } \theta], \ n \geq 0.$$

LEMMA 5.6. *D^n is the first empty set in the sequence $D^1, D^2, \ldots$ if and only iff δ^∞ is finitely transient with $n_\delta = n$.*

THEOREM 5.6. *If a policy δ^∞ is finitely transient, then $\mathcal{C}(\pi|\delta)$ is piecewise linear.*

DEMONSTRATION.   Since $\delta^\%$ is finitely transient, the partition $V = [V_j]$ and mapping v exist that satisfiy properties (a) and (b) given by the following relations:

   (a) if $\pi^1, \pi^2 \in V_j$ then $\delta(\pi^1) = \delta(\pi^2) \equiv v(j)$,

   (b) if $\pi \in V_j$ then $T(\pi|\theta, \delta) \in V_{v(j,\theta)}$.

In accordance with Lemma 5.5, $\mathcal{C}*\pi<\delta) = \pi\alpha(\pi|\delta)$, where $\alpha(\pi.\delta)$ is the unique bounded solution to:

$$\alpha(\pi|\delta) = \gamma^{\delta(\pi)} + \beta \sum_\theta P^{\delta(\pi)} R_\theta^{\delta(\pi)} \alpha[T(\pi|\theta, \delta) \mid \delta].$$

In the case that $\pi \in V_j$, $\alpha(\pi|\delta) = \alpha_{j.}$. Then for $\pi \in V_j$, $\delta(\pi) = v(j)$, so that the above equation can be transformed into the series of equations:

$$\alpha_j = \gamma^{v(j)} + \beta \sum_\theta P^{v(j)} R_\theta^{v(j)} \alpha_{v(j,\theta)}.$$

Because these equations are identical, their solutions are unique, and it follows that $\mathcal{C}(\pi|\delta) = \pi\alpha_j$ for $\pi \in V_j$ and the above theorem is proved.

5.2.3. THE FUNCTION $\check{C}(\pi|\delta)$ APPROXIMATED AS A MARKOV PROCESS WITH A FINITE NUMBER OF STATES.

Suppose $V = [V_1, V_2, \ldots , V_L]$ is a partition equivalent to the finitely transient policy δ^∞, which satisfies properties (a) and (b) from Lemma 5.6 (this can be written as $\pi^1, \pi^2 \in V_j$).

Suppose $\vec{\theta}$ represents some sequence of outputs $(\theta_1, \theta_2, \ldots , \theta_n) = \vec{\theta}$. We can write that:

$$\delta[T(\pi^1| \vec{\theta}, \delta)] = \delta[T(\pi^2| \vec{\theta}, \delta)].$$

The contribution to the cost functional from this sequence is of the following form:

$$\{ \vec{\theta} |\pi^1, \delta\} \, T(\pi^1| \vec{\theta}, \delta) \, \gamma^{\delta[T(\pi^1| \vec{\theta}, \delta)]} \equiv$$

$$\equiv \pi P^{v(\pi^1)} R^{v}_{\theta_1}(\pi^1) \, P^{v[T(\pi|\theta_1, \delta)]} \ldots R_{\theta_v}^{v[T(\pi^1| \vec{\theta}, \delta)]} \gamma^1$$

where it is assumed that $\delta[T(\pi^2| \vec{\theta}, \delta)] = 1$. The above sequence of matrices given by the previous relation is dependent only on the set V_j. Thus the difference in the contribution between π^1 and π^2, $\pi^1, \pi^2 \in V_j$ is linear and of the form:

$$(\pi^1 - \pi^2) \, P(j, \vec{\theta}) \gamma^1,$$

where $P(j, \vec{\theta})$ is the sequence of matrices.

Suppose that, for some k, the sequence $D^0, D^1, D^2, \ldots , D^k$ is found for an arbitrary policy δ^∞. If $D^k = \emptyset$, then δ^∞ is finitely transient with degree $n_\delta \leq k$. If $D^k \neq \emptyset$, we cannot define the nature of δ^∞. If $D^{k+1} \neq \emptyset$, then the partition V^k formed from $D^0, D^1, \ldots , D^k$ does not satisfy property (b) from Lemma 5.6.

Using the set V^k we can construct the mapping $\hat{v}$ that will approximate δ^∞. For this we select a point π^j in each set V^k_j and the $\hat{v}$ - mapping is defined by the fact that if $T(\pi^j|\theta, \delta) \in V^k_l$, then $\hat{v}(j, \theta) = l$.

The mapping is similar to that with v, but since V^k is not equivalent to the policy δ^∞ there is some set V^k_j, output θ and $\pi \in V^k_j$ such that $T(\pi|\theta, \delta) \notin V^k_{\hat{v}(j,\theta)}$.

The mapping $\hat{v}$ will now be used to construct a piecewise linear approximation to $\mathcal{C}(\pi|\delta)$ which has a bound on error proportional to β^k. This approximation to $\mathcal{C}(\pi|\delta)$, denoted by $\hat{\mathcal{C}}(\pi|\delta)$ is defined by:

$$\hat{\mathcal{C}}(\pi|\delta) = \pi\hat{\alpha}_j, \ \pi \in V^k_j,$$

where the linear segments satisfy the set of linear equations:

$$\hat{\alpha}_j = \gamma^{\delta(j)} + \beta \sum_\theta P^{\psi(j)} R^{\psi(j)}_\theta \hat{\alpha}_{\delta(j,\theta)}.$$

THEOREM 5.7.

$$\|\mathcal{C}(\pi|\delta) - \hat{\mathcal{C}}(\pi|\delta)\| \le \frac{\beta^k}{1-\beta^k} \frac{K}{1-\beta}$$

where:

$$K = \max_{a,i} \gamma^a_i - \min_{a,i} \gamma^a_i.$$

An ample demonstration of the above theorem is given in [20].

5.3 A useful policy iteration algorithm for discounted ($\beta < 1$) partially observable Markov decision processes

A methodology for the $\hat{v}$ construction for $n = 2$ and $n > 2$ is given.

5.3.1. THE CASE $\hat{v}$ FOR $N = 2$.

The transient policy δ^∞ will be approximated by a mapping $\Diamond$ constructed from the partition V^k. It is necessary to construct V^k so that the sequence of sets $D^0, D^1, \ldots$, D^k must be determined and then arranged to form the boundaries of the sets in V^k. Because of the special case of Markov processes that we are dealing with in this chapter (e.g. the state space Π is effectively the interval $[0, 1]$; the boundaries or edges of the sets of any partition Π are simply points), the set D^0 of discontinuities of δ is simply a set of points. The control function δ can be stored as the set D^0, which gives the set of intervals that comprise the partition V^0. The value of $\hat{v}$ can be stored as a set of vectors $[\alpha_1, \alpha_2, \dots]$ and an associated set $[a_j]$, so that if $\min_k \pi\alpha_k = \pi\alpha_j$, then $\delta(\pi) = a_j$.

The set $D^1 = [\pi = T^{-1}(\pi^0|\theta, \delta) \in \Pi : \pi^0 \in D^0]$ can be written as:

$$D^{-1} = [\pi = T^{-1}(\pi^0|\theta, \delta) \in \Pi : \pi^0 \in D^0],$$

where T^{-1} is the inverse mapping of T such that:

$$T^{-1}(\pi^0|\theta, \delta) = \pi Q_\theta^\delta / \pi Q_\theta^\delta 1,$$

where:

$$Q = [P^{\delta(\pi)} R_\theta^{\delta(\pi)}]^{-1}.$$

The relation which defines the set D^{-1} is equivalent to finding the set $B_{a,\theta}$ defined by:

$$B_{a,\theta} = [\pi = T^{-1}(\pi^0|\theta, a) \in \Pi : \pi^0 \in D],$$

and then eliminating those points in $B_{a,\theta}$ that do not lie in the regions of Π assigned control a by δ.

In a more general sense:

$$D^{n+1} = [\pi : T^{-1}(\pi|\theta, a) \in D^n \text{ for some } \theta].$$

It is clear that D^{n+1} is determined from D^n using the same steps as above for D^1. If the set D^k is completed, then $\hat{v}$ is constructed by combining the sets D^n into $\bigcup_{n=0}^{k} D^n$, and then ordering these points, forming a set of intervals given as V^k. Using the relation $T(\pi^1|\theta, \delta) \in V^k$, results in $\hat{v}(j, \theta) = l$, then we can calculate $\hat{v}$ and this can be constructed.

An algorithm for computing the optimal, or a nearly optimal, stationary policy is given next.

Policy Iteration Algorithm.

Step 0: Choose an arbitrary control function, $\delta(\pi) = a$ for any π.

Step 1: For k chosen to satisfy error requirements (i.e. $\|\mathcal{C}(\pi|\delta) - \hat{\mathcal{C}}(\pi|\delta)\| < \varepsilon_k$) find the partition V^k and the mapping $\hat{v}$.

Step 2: Calculate $\hat{\mathcal{C}}(\pi|\delta) = \pi\hat{\alpha}_j$, $\pi \in V^k_j$ from:

$$\hat{\alpha}_j = \gamma^{\hat{v}(j)} + \beta \sum_\theta \underline{P}^{v(j)} R_\theta^{v(j)} \hat{\alpha}_{\hat{v}(j,\theta)}.$$

Step 3:   Find the control function $\delta^1(\pi)$ where $\delta^1(\pi)$ is the alternative a that minimizes:

$$U_a[\pi, \bar{\hat{\mathcal{C}}}\ (\pi|\delta)] = \pi\gamma^a + \beta \sum_\theta \{\theta|\pi, a\} \hat{\mathcal{C}}[T(\pi|\delta)|\delta],$$

where $\hat{\mathcal{C}}(\cdot|\cdot)$ is the concave hull of $\hat{\mathcal{C}}(\pi|\delta)$.

Step 4: Compute the quantity

$$\|\bar{\hat{\mathcal{C}}}\ (\pi|\delta) - U\bar{\hat{\mathcal{C}}}\ (\pi|\delta)\|$$

where the operator $u = \min_a U_a$. If

$$\|\bar{\hat{\mathcal{C}}}\ (\pi|\delta) - U\bar{\hat{\mathcal{C}}}\ (\pi|\delta)\| < \varepsilon,$$

then Stop; the policy $(\delta)^\infty$ satisfies the relation

$$\|\mathcal{C}^*(\pi) - \hat{\mathcal{C}}(\pi|\delta)\| \le (1 - \beta)^{-1}$$

$$[\varepsilon + (1 + \beta)\ \varepsilon_k].$$

Otherwise return to Step 1 with δ replaced by δ^1.

5.3.2. THE CASE $\hat{V}$ FOR $N > 2$.

The major changes in the above algorithm due to $N > 2$ is for the mapping $\hat{V}$ For this case the sets $D^0, D^1, \ldots$ become surfaces rather than points and they must be arranged to form the regions of the partition V^k. For the case of $n = 3$, D_0^0 is a set of line segments. V^0 is defined as the partition induced by the control function δ. The space Π is broken into a set of sets V_j^0, each set associated with the vector α_j, where $\pi \in V_j^0$, if $\pi\alpha_j \le \pi\alpha_k$, for all k. It should be mentioned that each vector α_j is associated with a control alternative $a_j = \delta(\pi)$, $\pi \in V_j^0$.

Next we shall consider the partition V^1 which is constructed using the sets D^0 and D^1 as the boundaries of the sets in V^1. D^1 is itself constructed from D^0 as:

$$D^1 = [\pi = T^{-1}(\pi^0|\theta, \delta) : \pi^0 \in D^0].$$

The properties of a set V_j^1 of V^1 are:

(a) if $\pi \in V_j^1$ then:

$$\min_l \pi\alpha_l = \pi\alpha_j$$

and

(b) if $\delta(\pi) = a$ and if j_θ is defined so that:

$$\min_l \pi\alpha_{a,l,\theta}^1 = \pi\alpha_{j_\theta}^1$$

then V_j^1 can be said to be defined by the set of vectors $[\alpha_j, \alpha_{j_1}^1, \ldots \alpha_{j_M}^1]$.

Once V^1 is defined, V^2 can be determined by the vectors defining the sets in V^1. Once V^k is determined the mapping between regions can be defined and from this vectors $\hat{\alpha}_j$ can be computed using the following relation:

$$\hat{\alpha}_j = \gamma^{\phi(j)} + \beta\sum_\theta P^{\psi(j)} R_\theta^{\psi(j)} \hat{\alpha}_{\phi(j,\theta)}.$$

EXAMPLE 5.2. Let us consider the following partially observable Markov control process:

Control k	P^k	R^k	γ^k
1	$\begin{bmatrix} 0.8 & 0.2 \\ 0.5 & 0.5 \end{bmatrix}$	$\begin{bmatrix} 0.8 & 0.2 \\ 0.5 & 0.5 \end{bmatrix}$	$\begin{bmatrix} 4 \\ -4 \end{bmatrix}$
2	$\begin{bmatrix} 0.5 & 0.5 \\ 0.4 & 0.6 \end{bmatrix}$	$\begin{bmatrix} 0.9 & 0.1 \\ 0.4 & 0.6 \end{bmatrix}$	$\begin{bmatrix} 0 \\ -3 \end{bmatrix}$

ITERATION 1.

According to the policy iteration algorithm, an initial stationary policy must be chosen. First, we shall choose the policy that maximizes the expected immediate cost of operating the process. This policy $\delta^0(\pi)$ is simply the alternative k minimizing $\pi\gamma^k$ for each π:

$$\min_k \pi\gamma^k = \pi\gamma^{\delta^0(\pi)}.$$

With π defined as $\pi = [\pi_1, 1 - \pi_1]$, it is easily calculated that:

$$\delta^0(\pi) = \begin{cases} 1 & \text{if } \pi_1 \leq 0.2 \\ 2 & \text{if } \pi_1 > 0.2 \end{cases}$$

Next, we must compute the total expected cost of using this policy $\mathcal{C}(\pi|\delta^0)$. The set of discontinuities for this policy is $D_{\delta^0} = [0.2]$.

The transition diagram is given in Figure 5.8.

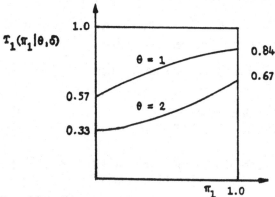

Figure 5.8. Transition diagram.

Following this graphical representation it is obvious that there is no point in Π which can reach D_{δ^0} using the control function δ^0. Thus $D^1 = \varnothing$ and (δ^0) is finitely transient.

This is equivalent to a partition $V = V_j^0 = [V_1^0, V_2^0]$. Selecting point π^j in each set V_j^0 of the partition and calculating $T(\pi^j|\theta, \delta^0)$ gives the mapping v^0 equivalent to the policy δ^0:

j	θ	$v^0(j,\theta)$	$v^0(j)$
1	2	1	1
1	2	2	1
2	1	2	2
2	2	1	2

Defining $\bar{\alpha} = \begin{bmatrix} \alpha_1 \\ \alpha_2 \end{bmatrix}$, where $\alpha_i = \begin{bmatrix} \alpha_{i1} \\ \alpha_{i2} \end{bmatrix}$, and α_{ij} is a scalar $\bar{\gamma} = \begin{bmatrix} \gamma^1 \\ \gamma^2 \end{bmatrix}$ and:

$$\bar{P} = \begin{bmatrix} 0 & P \\ P^2 R_2^2 & P^2 R_1^2 \end{bmatrix}$$

and solving:

$$\bar{\alpha} = \bar{\gamma} + \beta \bar{P} \, \bar{\alpha}$$

gives:

$$\alpha_1 = \begin{bmatrix} -9.86 \\ -18.78 \end{bmatrix}, \quad \alpha_2 = \begin{bmatrix} -14.76 \\ -18.14 \end{bmatrix}.$$

The expected cost of $(\delta^0)^\infty$ is represented in Figure 5.9. We notice that $\mathcal{C}(\pi|\delta^0)$ is not continuous at 0.2. In this case $\bar{\mathcal{C}}(\pi|\delta^0)$ must be used in place of $\mathcal{C}(\pi|\delta^0)$ in the policy improvement stage. In Figure 5.10 the following function is plotted:

$$U\mathcal{C}(\pi|\delta^0) = \min_k U_a[\pi, \mathcal{C}(\pi|\delta^0)].$$

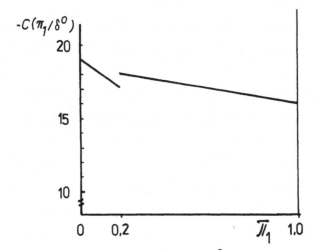

Figure 5.9. Representation of expected cost of $(\delta^0)^\infty$.

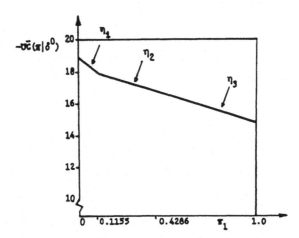

Figure 5.10. Plot of function $U\bar{C}(\pi|\delta^0)$.

The function consists of three linear segments labelled ξ_j, $j = 1, 2, 3$ where:

$$\xi_1 = \begin{bmatrix} -9.86 \\ -18.78 \end{bmatrix} ; \quad \xi_2 = \begin{bmatrix} -14.76 \\ -18.14 \end{bmatrix} ; \quad \xi_3 = \begin{bmatrix} -14.80 \\ -18.11 \end{bmatrix} .$$

The alternatives associated with these vectors are 1, 2, 2, respectively. It is clear that:

$$\|U\bar{C}(\pi|\delta^0) - \bar{C}(\pi|\delta^0)\| \leq 0.04.$$

In the case where this error was thought to be acceptable, the process could stop at this stage.

This policy $(\delta^1)^\infty$ associated with $U\bar{C}(\pi|\delta^0)$ is within $(1 - \beta)\ 0.04$.

We shall continue the steps within the algorithm.

ITERATION 2.

The policy defined by $U\bar{C}(\pi|\delta^0)$ is:

$$\delta^1(\pi) = \begin{cases} 1 & \text{if } \pi_1 \leq 0.1155 \\ 2 & \text{if } \pi_1 > 0.1155. \end{cases}$$

Thus $D_{\delta^1} = [0.1155]$. D^1 is calculated to be $D^1 = [0.396]$ and $D^2 = \emptyset$; in this case $(\delta^1)^\infty$ is finitely transient with $n_{\delta^1} = 2$. The associated V-partition consists of the three sets given in Figure 5.11.

The mapping v^1 is constructed as:

j	θ	$v^1(j,\theta)$	$v^1(j)$
1	2	3	1
1	2	3	1
2	1	3	2
2	2	1	2
3	1	3	2
3	2	2	2

Using the above results , the vector of linear segments is calculated to be:

$$\alpha_1 = \begin{bmatrix} -10.03 \\ -18.93 \end{bmatrix}; \ \alpha_2 = \begin{bmatrix} -14.89 \\ -18.27 \end{bmatrix}; \ \alpha_3 = \begin{bmatrix} -14.93 \\ -18.23 \end{bmatrix}.$$

After computing $\bar{C}(\pi|\delta^1)$, we must turn again to the policy improvement phase. Let δ^2 be the control function associated with $U\bar{C}(\pi|\delta^1)$ which is equal to $\bar{C}(\pi|\delta^1)$.

The cost of the optimal policy is given by the linear segments of $\bar{C}(\pi|\delta^1)$ since $\bar{C}(\pi|\delta^2) = U\bar{C}(\pi|\delta^1) = \bar{C}(\pi|\delta^1)$.

The control function δ^2 is found to be:

$$\delta^2(\pi) = \begin{cases} 1 & \text{if } \pi_1 \leq 0.1188 \\ 2 & \text{if } \pi_1 > 0.1188. \end{cases}$$

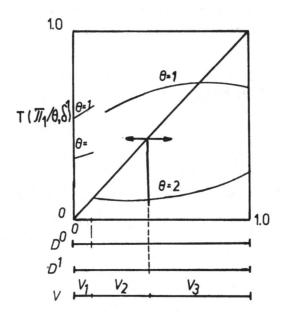

Figure 5.11. Associated partition for iteration 2.

The policy $(\delta^2)^\infty$ is finitely transient with mapping $\varpi^2 = \varpi^1$. In Figure 5.12 the block diagram of the optimal control is given. The alternatives are described solely by sequences of outputs (e.g. with 2 outputs $\theta = 2$ in a row, control $k = 1$ will be used).

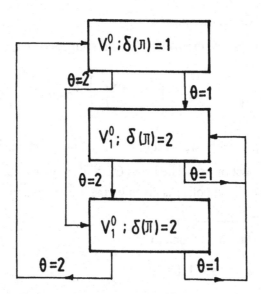

Figure 5.12. Block diagram of optimal control.

5.4 The infinite horizon without discounting for partially observable Markov processes

5.4.1. MODEL FORMULATIONS

For the case when $\beta = 1$, the costs of infinite horizon policies can become infinite, making it impossible to compare policies on a total expected cost basis.

When all the cost vectors γ^k, for each k, have values greater than 1, then:

$$\mathcal{C}^0(\pi) = \pi\gamma^0$$

$$\mathcal{C}^n(\pi) = \min_k \left[\pi\gamma^k + \sum_\theta \{\theta|\pi, k\} \, \mathcal{C}^{n-1}[T(\pi|\theta, k)] \right]$$

where $\mathcal{C}^n(\pi)$ is the minimum total expected cost with n time periods to operate, terminal cost vector γ^0, and current state of knowledge π, it is clear that $\mathcal{C}^n(\pi) \geq n$.

We can define $g^n(\pi)$ as:

$$g^n(\pi) = \frac{\mathcal{C}^n(\pi)}{n} .$$

It is not clear that $\lim_{n\to\infty} g^n(\pi)$, exists, in general, but we have the relations:

$$\min_{i,k} \pi\gamma_i^k \leq g^n(\pi) \leq \max_{i,k} \pi\gamma_i^k. \quad \text{for all } n \text{ and } \limsup_{n\to\infty} g^n(\pi) \text{ exists so that:}$$

$$g^*(\pi) = \limsup_{n\to\infty} g^n(\pi).$$

As Sondik has already emphasized:

"The optimal policy (which is not necessarily stationary) can be found at least conceptually by using the value iteration method ... , but this approach requires infinite computation time. ... It seems reasonable, comparing the partially observable case with the completely observable case, that such policies should exist! The fact that $g^\infty(\pi)$ does not depend on γ^0 is additional evidence that such policies may exist".

5.4.2. COST OF A STATIONARY POLICY.

A finitely transient policy is equivalent to a finite state Markov process.

A stationary policy δ^{∞} is finitely transient if the discontinuities of $\delta(\pi)$ (i.e. where $\delta(\pi)$ changes value) will never be encountered after some finite number of transitions, regardless of the initial state of knowledge:

(a) $D^0 = [\pi : \delta(\pi)$ is discontinuous at $\pi]$

(b) $V^0 = [V_1^0, V2, ...]$ as the partition of Π induced by δ, that is, if $\pi^1, \pi^2 \in V_j^0$, then:

$$\delta(\pi^1) = \delta(\pi^2).$$

The boundaries of the sets V_j^0 are contained in D^0.

A useful property of finitely transient policies is the simplified computation of their expected cost.

By the property of the partitions that $V^k = V^{k+1}$, the set containing $T(\pi|\theta, \delta)$ is the same for all $\pi \in V_j^k$; the index of this set is denoted by $v(j, \theta)$.

If $\mathcal{C}^n(\pi|\delta)$ is defined as:

$$\mathcal{C}^n(\pi|\delta) = \pi\alpha_j^n , \quad \pi \in V_j^k$$

then:

$$\mathcal{C}^0(\pi|\delta) = \pi\gamma^0$$

$$\mathcal{C}^n(\pi|\delta) = \pi\alpha_j^n , \quad \pi\gamma^{\delta(j)} + \sum_{\theta} \{\theta|\pi, \delta\} \, \mathcal{C}^{n-1}[T(\pi|\theta, \delta)|\delta]$$

where $\delta(j) \equiv \delta(\pi)$ for $\pi \in V_j^k$.

We can use an alternative notation of the form:

$$\alpha_j^0 = \gamma^0$$

$$\alpha_j^n = \gamma^{\delta(j)} + \sum_{\theta} P^{\delta(j)} R_{\theta}^{\delta(j)} \alpha_{\gamma(j,\theta)}^{n-1}.$$

This system of equations could be written in a matrix form:

$$\underline{\alpha}^0 = \underline{\gamma}^0$$

$$\underline{\alpha} = \underline{\gamma} + P\underline{\alpha}^{n-1}$$

where:

$$\underline{\alpha}^n = [\alpha_1^n, \alpha_2^n, \dots]^T$$

$$\underline{\gamma}^0 = [\gamma^0, \gamma^0, \dots]^T$$

$$\underline{\gamma}^0 = [\gamma^{\delta(1)}, \gamma^{\delta(2)}, \dots]^T$$

and the matrix P with $(N \times N)$ sub-matrices, where the sub-matrix in the position (j, k), $P_{j,k}$ is defined as:

$$P_{j,k} = \begin{cases} \sum_\theta P^{\delta(j)} R_{\gamma(j,\theta)}^{\delta(j)} & \text{for all } \theta \text{ such that } (j, \Theta) = k \\ 0 & \text{otherwise.} \end{cases}$$

The number of equations in the above representation is equal to the number of sets in V^k.

Next, we shall approximate a stationary policy δ^∞ in the following way.

First construct the sequence of partitions $V^0, V^1, \dots$ up to V^k for some value of k. If $V^k = V^{k+1}$ then δ^∞ and, according to Sondik, δ^∞ is a finitely transient Markov process.

Next, we shall consider that $V^k \neq V^{k+1}$. This does not preclude the possibility that the policy is finitely transient. It may have this property but the partition associated with the property is a degree greater than k. Under these circumstances, we construct an approximation to δ^∞.

Construct the mapping $\hat{V}(j, \theta)$ as follows:

(a) arbitrarily select a point $\pi^j \in V_j^k$; for all j,

(b) set $\hat{V}(j, \theta) = l$ if $T(\pi^j | \theta, \delta) \in V_l^k$.

Define:

$$\hat{C}^n(\pi|\delta) = \pi \hat{\alpha}_j^n, \quad \pi \in V_j^k$$

where:

$$\hat{\alpha}_j^0 = \underline{0}$$

$$\hat{\alpha}_j^n = \gamma^{\delta(j)} + \sum_\theta P^{\delta(j)} R_\theta^{\delta(j)} \hat{\alpha}_{\hat{V}(j,\theta)}^{n-1}.$$

Defining:

$$\hat{\alpha}^n = [\hat{\alpha}_1^n \, ... \,]^T$$

$$\underline{\gamma} = [\gamma^{\delta(1)} ... \,]^T$$

and P appropriately, the above equation can be written as:

$$\underline{\hat{\alpha}} = \underline{0}$$

$$\underline{\hat{\alpha}}^n = \underline{\gamma} + P \underline{\alpha}^{n-1} \, .$$

The system is of the form that permits the computation of the asymptotic properties of $\underline{\hat{\alpha}}^n$.

5.4.3. POLICY IMPROVEMENT PHASE

Stationary monodesmic policies δ^{∞} are characterized by a gain g_δ and a relative cost function $\mathcal{C}(\pi|\delta)$ so that:

$$g_\delta + \mathcal{C}(\pi|\delta) = \pi\gamma^{\delta(\pi)} + \sum_\theta \{\theta|\pi, \delta\} \, \mathcal{C}[T(\pi|\theta, \delta)|\delta].$$

Sondik proved that if a monodesmic policy δ_1 is chosen so that:

$$g_\delta + \mathcal{C}(\pi|\delta) \geq \pi\gamma^{\delta_1(\pi)} + \sum_\theta \{\theta|\pi, \delta_1\} \cdot \mathcal{C}[T(\pi|\theta, \delta_1)|\delta]$$

then $g_{\delta_1} \geq g_\delta$. This policy improvement involves finding a control function δ_1 satisfying the above equation.

The function δ_1 can be found by setting $\delta_1(\pi) = a^*(\pi)$ where $a^*(\pi)$ the index a that minimizes:

$$\left[\pi\gamma^a + \sum_\theta \{\theta|\pi, a\} \, \mathcal{C}[T(\pi|\theta, a)|\delta]\right].$$

With δ_1 chosen in this way, it is clear that the above inequality is satisfied since:

$$\min_a \left[\pi\gamma^a + \sum_\theta \{\theta|\pi, a\} \, \mathcal{C}[T(\pi|\theta, a)|\delta] \right] \le$$

$$\le \pi\gamma^\delta + \sum_\theta \{\theta|\pi, \delta\} \, \mathcal{C}[T(\pi|\theta, \delta)|\delta]] = g + \mathcal{C}(\pi|\delta).$$

The proof that $g_{\delta^1} \le g_\delta$ is similar to Howard's proof of the policy iteration algorithm for the discrete state problem.

5.4.4. POLICY ITERATION ALGORITHM.

In the completely observable problem there are only a finite number of alternatives and in this case the policy iteration algorithm must converge in a finite number of iterations. In the case of a partially observable problem, because of its continuous state space which admits an uncountable number of stationary controls, the above statements are no longer valid. In this case it is necessary to have a measure of closeness to the optimal policy control at each point in the iteration.

If we define a measure as:

$$\Delta(\pi) = g + \mathcal{C}(\pi|\delta) - \left[\pi\gamma^{\delta_1} + \sum_\theta \{\theta|\pi, \delta_1\} \, \mathcal{C}[T(\pi|\theta, \delta_1)|\delta] \right],$$

then for the case of optimal gain, we can write:

$$\min_\pi \Delta(\pi) \le g_\delta - g^* \le \max_\pi \Delta(\pi).$$

The policy iteration algorithm for $\beta = 1$ can be given as follows:

Step 0: Pick an arbitrary policy, say $(\pi) = a$, for all π.
Step 1: For k chosen to satisfy error requirements, find the partition V^*.
Step 2: Construct the mapping v from V^*.
Step 3: Calculate $\underline{\alpha}_\delta$ and g_δ from the system of equations:

$$g_\delta \mathbf{1} + \underline{\alpha}_\delta = P_\delta \underline{\alpha}_\delta + \underline{\gamma}.$$

Step 4: Find the policy $\delta_1(\pi)$ where $\delta_1(\pi)$ minimizes:

$$\pi\gamma^a + \sum_\theta \{\theta|\pi, a\} \, \mathcal{C}[T(\pi|\theta, \delta)|\delta]$$

over a, and where $\tilde{C}(\pi|\delta) = \pi\alpha_{\psi(\pi)}$.

Step 5: Evaluate $g_\delta - g^*$ from $\Delta(\pi)$ where:

$$\Delta(\pi) = [g_\delta + \pi\alpha_{\psi(\pi)}] - \left[\pi\gamma^{\delta_1(\pi)} + \sum_\theta \{\theta|\pi, \delta_1\} \; \tilde{C}[T(\pi|\theta, \delta_1)|\delta]\right]$$

and $\min \Delta(\pi) \leq g_\delta - g^* \leq \max \Delta(\pi)$.

Step 6: If $|g_\delta - g^*| < \varepsilon$ then Stop; the optimal policy (within ε) is δ, otherwise return to Step 1 with δ replaced by δ_1.

EXAMPLE 5.3. (a case for $N = 3$). The parameters of the problem are presented in Table 5.2. The terminal cost vector of the finite horizon problem is also given.

	P^a			R^a		γ^a
$a = 1$	0.1	0.1	0.8	0.7	0.3	1
	0.2	0.5	0.3	0.1	0.9	0
	0.7	0.1	0.2	0.4	0.6	0
$a = 2$	0.1	0.8	0.1	0.2	0.8	0
	0.7	0.1	0.2	041	0.6	1
	0.1	0.9	0.0	0.3	0.7	0

Table 5.2. Parameters of the problem in Example 5.3.

The initial control is considered to be the optimal control with 10 operating periods. The optimal policy δ^∞ for 10 operating periods based on γ^0 is $\delta(\pi) = a_j$ where j is the index minimizing $\pi\alpha_j$ for:

$$\alpha_1 = [2.22 \quad 3.78 \quad 2.21]^T \quad a_1 = 2$$

$$\alpha_2 = [3.17 \quad 2.25 \quad 2.31]^T \quad a_2 = 1$$

$$\alpha_3 = [3.23 \quad 2.86 \quad 2.24]^T \quad a_3 = 1.$$

To compute the gain $g\delta$ and relative costs $\bar{C}(\pi|\delta)$ for δ^{∞} it is necessary to calculate a partition V^k for some k and the associated mapping v. The partition induced by δ, V^0, is given below (see Figure 5.13).

The partition V^1 involves finding all the vectors of the form:

$$\alpha^1_{a,j,\theta} = P^a R^a_\theta \, \alpha_j .$$

Each set in V^1, V^1_k is then defined by $(M + 1)$ vectors of the form $[\alpha_k, \alpha_{k_1}, \dots , \alpha_{k_M}]$ where for $\pi \in V^1_k$, $\min \pi\alpha_j = \pi\alpha_k$ and if $\delta(\pi) = a$ then we have:

$$\min_l \pi\alpha^1_{a,l,\theta} = \pi\alpha^1_{k_\theta}.$$

The partition V^1 consists of the following sub-sets:

$$V^1_j = [\alpha_{j1}, \alpha^1_{j1}, \alpha^1_{j2}]$$

where:

j	α_{j1}	α^1_{j1}	α^1_{j2}
1	α_1	$P^2 R^2_1 \alpha_2$	$P^2 R^2_2 \alpha_2$
2	α_2	$P^1 R^1_1 \alpha_1$	$P^1 R^1_2 \alpha_2$
3	α_3	$P^1 R^1_1 \alpha_1$	$P^1 R^1_2 \alpha_1$
4	α_4	$P^1 R^1_1 \alpha_1$	$P^1 R^1_2 \alpha_1$

Next we must calculate the partition V^2 defined as:

$$V^2_j = [\alpha_{j1}, \alpha^1_{j1}, \alpha^1_{j2}, \alpha^2_{j1}, \alpha^2_{j2}],$$

where we have the following relations:

$$\min_k \pi\alpha_k = \pi\alpha_{j1} ; \ \pi \in V^2_j$$

$$\min_k \pi P^a R^a_\theta \alpha_k = \pi\alpha^1_{j\theta}$$

$$\min_{k,\theta_1} \pi P^a R^a_{\theta} \alpha^1_{k_{\theta_1}} = \pi\alpha^2_{j\theta}$$

for $\delta(\pi) = a$, and for every θ:

$$\alpha^2_{j_\theta} = P^1 R^1_\theta \alpha^1_{2_2}, \quad 2 \le j \le 4$$

$$\alpha^2_{1_\theta} = P^2 R^2_\theta \alpha^1_{2_2}.$$

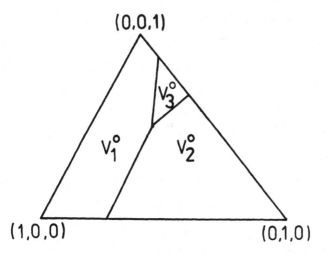

$$\delta(\pi) = 2; \quad \pi \in V^0_1$$
$$\delta(\pi) = 1; \quad \pi \in V^0_2$$
$$\delta(\pi) = 1; \quad \pi \in V^0_3$$

Figure 5.13. Partition induced by δ, V^0.

The policy δ^∞ is finitely transient ($V^2 = V^1$) and mapping v associated with V^1 is an adequate description of the dynamics of the state of knowledge for δ^∞ (if $\pi \in V_j$ then $T(\pi|\theta, \delta) \in V_{v(j,\theta)}$).

The mapping v is of the following form:

j	θ	$v(j,\theta)$	$v(j)$
1	1	2	
			2
1	2	2	
2	1	1	
			1
2	2	2	
3	1	2	
			1
3	2	2	
4	1	2	
			1
4	2	1	

and $v(j)$ is the control defined by δ for all points $\pi \in V_j^1$.

The matrix P is of the following form:

$$P = \begin{bmatrix} 0 & P^2 & 0 & 0 \\ P^1 R_1^1 & P^1 R_2^1 & 0 & 0 \\ P^1 & 0 & 0 & 0 \\ P^1 & 0 & 0 & 0 \end{bmatrix}$$

By solving the matrix linear equation:

$$g_\delta \, \underline{1} + \underline{\alpha} = \gamma + P\underline{\alpha},$$

where:

$$\underline{\alpha} = [\alpha_1 \ \alpha_2 \ \alpha_3 \ \alpha_4]^T$$

$$\underline{\gamma} = [\gamma_1 \ \gamma_2 \ \gamma_3 \ \gamma_4]^T.$$

By setting $\alpha_{23} = 0$, we can have:

$$g_\delta = 0.133 \, ; \ \alpha_1 = [-0.080, \ 1.463, \ 0.104]^T$$

$$\alpha_2 = [\, 0.861, \ -0.0633, \ 0.0]^T$$

$$\alpha_3 = \alpha_4 = [-0.080, \ 0.547, \ -0.077]^T.$$

The next step consists of improving the policy δ using $\mathcal{C}(\pi|\delta)$.
By setting $\delta(\pi)$ equal to the index minimizing the equation:

$$\pi\gamma^a + \sum_\theta \{\theta|\pi, a\} \, \mathcal{C}[T(\pi|\theta, a)|\delta] \ = \ \pi\gamma^a + \sum_\theta \min_j \pi \, P^a R_\theta^a \, \alpha_j,$$

for each π, we can identify a new policy δ^∞.
Defining:

$$\mathcal{C}(\pi) = \min_a [\pi\gamma^a + \sum_\theta \min_\theta \pi \, P^a R_\theta^a \, \alpha_j]$$

we can have:

$$\hat{\mathcal{C}}(\pi) = \bar{\mathcal{C}}(\pi|\delta) + g_\delta.$$

The following results are given:

(a) the PI (policy improvement) performed policy improvement on δ_1^∞ since $\bar{\mathcal{C}}(\pi|\delta) = \mathcal{C}(\pi|\delta_1)$, and

(b) the PI resulted in policy δ_1^∞.

Policy δ_1^∞ is the optimal one. δ_1^∞ is a finitely transient policy and can be implemented using the block diagram shown in Figure 5.14 where V_2, V_3 implies the use of control 1 and V_1 implies the use of control 2.

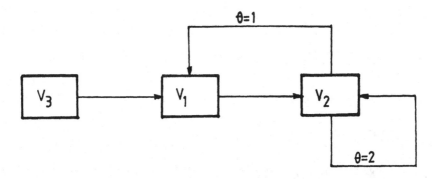

Figure 5.14. Block diagram for implementation of a finitely transient policy.

5.5   Partially observable semi-Markov decision processes

The above class of processes will be developed through a dynamic decision model for clinical diagnosis. This unified approach combines partially observable Markovian (semi-Markov) decision processes with cause-effect models as a probabilistic representation of the diagnostic process. Pattern recognition techniques are used in a first stage of system state identification.

The methodology is given for combining the patient state of health, the clinician's state of knowledge of the cause-effect representation from the observation space (measurements), feature selection using pattern recognition techniques and, finally, the treatment decisions with which to restore the patient to a more desirable state of health. A cost functional for the process has then to be optmized according to some pre-assigned obective function, when the process has an infinite time horizon.

5.5.1. MODEL FORMULATION.

Quantitative approaches to clinical diagnosis and decision making offer the promise of an improvement in the understanding of relationships between process variables and an improved quality of patient care. Examining the real decision process of clinical diagnosis and its associated infrastructure, it can be easily seen that the clinician, with his *a priori* medical knowledge, has to make observations on the patient who behaves as a dynamic system under conditions of uncertainty. The clinician has to "recall" his professional knowledge, identify patterns and classify features as a result of making these observations and measurements (see Chapter 1).

The whole process is not a "one-stage" operation but rather a hierarchical system of pattern recognition, identification and decision making. In an advanced conceptual expert system for medical diagnosis (see Figure 5.15) we can identify the dynamic state space of the patient (i.e. states of health for a particular class of disease), a cause-effect model, and clinical decisions bringing about a change in the patient's state of health in accordance with the new information vector (state of health - measurements - *a priori* knowledge).

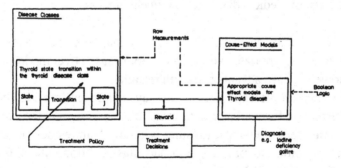

Figure 5.15. An advanced conceptual expert system for medical diagnosis.

The model presented here uses hierarchical modelling techniques and faces two distinct stages in the medical diagnosis-treatment process. (See Figure 5.16.) Firstly, the pattern recognition techniques can be used to identify the state space. Then a semi-Markov model can be used when the states of the process are only partially observable and we have a set of cause-effect models available which describe the probabilistic relationship between the observable effects and the causes to which these observations can be ascribed.

The cause-effect model provides a mapping from a set of input measurements to a diagnosis/primary event.

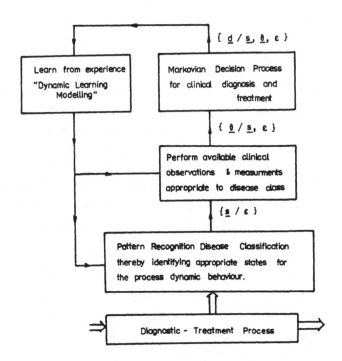

Figure 5.16. Model of medical diagnosis-treatment process.

A pattern recognition system provides a mapping from a set of input measurements to the appropriate pattern class. Both cause-effect models and pattern recognition, however, have advantages worth retaining.

Pattern recognition techniques, as already mentioned, are ideally suited to extracting information from a large data base. Such a system is therefore used to operate on the raw measurements, together with any features derived from those measurements, in order to produce a broad classification into classes. This classification is then passed to the cause-effect model, together with the raw measurements and derived features, for a finer classification. Effectively, the pattern recognition system guides the cause-effect model to an appropriate area of uncertainty.

5.5.2 STATE DYNAMICS.

Let us consider that, after careful analysis, it has been decided that there are N states within the appropriate disease class, and that a finite set of M observations is always available to the clinician. For example, in the case of thyroid disease, the patient states would be hyper-thyroid, eu-thyroid and hypo-thyroid, with the observations corresponding to the levels of the thyroid hormones, thyroxine and tri-iodo-thyonine and thyroid stimulating hormone.

The dynamics of patient health evolution are considered to be a semi-Markov process with the probabilistic rate transition:

$$p_{ij}^k = \Pr\{S(t+1) = j | S(t) = i; \ u(t) = k, \ \mathcal{E}(t)\}$$

$$i = 1, 2, \ldots, N; \ j = 1, 2, \ldots, N$$

$$k = 1, 2, \ldots, K_i,$$

where K_i is the maximum number of decision alternatives available in state i, k is a pertinent treatment alternative, S is the patient sate, $\mathcal{E}(t)$ is the current state of knowledge of the system and $u(t)$ is the decision at time t.

$$p_{ij}^k > 0; \ \sum_{j=1}^{k} p_{ij}^k = 1; \ i = 1, 2, \ldots, N; \ k = 1, 2, \ldots, K_i.$$

Letting the random integer valued variable τ_i be the holding time in state i, the cumulative probability distribution function for the semi-Markov process, (the holding-time mass function), can be defined as:

$$h_{ij}(m) = \Pr\{\tau_i \geq m\}$$

considering the process to be completely observable, $f_{ij}(t)$ is defined as the first passage time for the semi-Markov discrete process. This is the probability that the first passage from state i to state j will require at least t time units and, as such, is the measure of how long it takes to reach a given state from another state. The analytical relationship between p_{ij}, h_{ij}, and f_{ij} is given by:

For a time-dependent Markov process, the initial state vector can be written as:

$$f_{ij}(t) = \sum_{\substack{r=1 \\ r \neq j}}^{N} \sum_{m=0}^{t} p_{ir} h_{ir}(m) f_{rj}(t-m) + p_{ij} h_{ij}(t)$$

$$f_{ij}(0) = \begin{cases} 1 & \text{if } i = j \\ 0 & \text{otherwise.} \end{cases}$$

$$\pi(t) = [\pi_i(t), \ i = 1, 2, \ldots, N]$$

$$= [\Pr\{S(t) = i | \mathcal{E}(t)\}; \ i = 1, 2, \ldots, N].$$

The initial state vector depends on the past history of decisions. A set of all possible initial state vectors can be defined, given by:

$$\Pi = \text{set } \pi_i : \pi_i \geq 0 ; \sum_{i=1}^{N} \pi_i = 1).$$

5.5.3. THE OBSERVATION SPACE.

By performing a set of independent observations, $\theta = 1, 2, \dots , M$, the clinician obtains information about the patient's state. These observations may be classified as cheap or expensive. The random process of clinical observation is described by the probability:

$$r^k_{j\theta} = [\text{Pr}\{Z(t+1) == \theta | S(t+1) = j, u(t) = k, \mathcal{E}(t)\}$$

$$i = 1, 2, \dots , N ; \quad \theta = 1, 2, \dots , M ; \quad k = 1, 2, \dots , K_i$$

and:

$$0 \leq r^k_{j\theta} \leq 1; \quad \sum_{\theta=1}^{M} r^k_{j\theta} = 1 ; \quad j = 1, 2, \dots , N; \quad k = 1, 2, \dots , K_i ,$$

where $Z(t + 1)$ is an event representing an observable output at time $(t+1)$. Thus $r^k_{j\theta}$ defines the probability of achieving observation θ, given that state j results from a particular treatment k.

Following an observation and clinical judgement a cause-effect model can be built in which all the logical combinations (such as AND, OR, TRUE, FALSE) of the causes are correlated with the observation outputs. The pattern recognition model used in the first step of the hierarchical modelling process assists in building up a more complete and accurate cause-effect model. If $Y(t+1)$ is an event representing a cause and ξ_n is a primary event in a cause-effect model, the model is then characterized by:

$$l_{\theta\xi_n} = \text{Pr}\{Y(t+1) = \xi_n | Z(t+1) = \theta, S(t+1) = j, \mathcal{E}(t)\}$$

$$\theta = 1, 2, \dots , M ; \quad \xi_n = 1, 2, \dots , N_{A_n}$$

where N_{A_n} is the number of causes which could give rise to a given observation and:

$$0 \leq l_{\theta\xi_n} \leq 1; \; l_{\theta\xi_n} + \mathcal{C}(l_{\theta\xi_n}) = 1,$$

where $\mathcal{C}$ is the complementary function.

In the two equations above $l_{\theta\xi_n}$ specifies the probability of a particular cause, given an observation θ and the present state being achieved as a result of a specific treatment history. If the case of thyroid disease, causes would include thyro-toxicosis, toxic nodule, iodine deficiency goitre and Hashimoto's disease.

5.5.4. OVERALL SYSTEM DYNAMICS.

From the above we notice that the partially observable , semi-Markov diagnosis decision process is given by the quadruplex:

$$w = \{S, \mathcal{O}, \mathcal{E}, \mathcal{D}\},$$

where S is the state space, $\mathcal{O}$ the observation space, $\mathcal{E}$ represents the primary events (causes) of the process and $\mathcal{D}$ is the decision space.

As input parameters for the model we can have:

$$w^* = \{P, H(m), R_\theta, L_{\xi_n}, k\}$$

where

$$P = [p_{ij}], H(m) = [h_{ij}(m)], R_\theta = \mathrm{diag}_{\theta=\mathrm{const}} R = \mathrm{diag}_{\theta=\mathrm{const}}[r_{j\theta}], L_{\xi_n} = [l_{\theta\xi_n}]$$

and k is a treatment decision.

Extending previous results, given in Chapter 1 for the semi-Markov processes, using the first passage time, we can write the new system dynamics as:

$$T_j(\pi|\theta, \xi_n, t, k) = \frac{\sum\limits_{i=1}^{N} \pi_i f_{ij}^k(t)\, r_{j\theta}^k\, l_{\theta\xi_n}^k}{\sum\limits_{i=1}^{N}\sum\limits_{j=1}^{N}\sum\limits_{\theta=1}^{M} \pi_i f_{ij}^k(t)\, r_{j\theta}^k\, l_{\theta\xi_n}^k}$$

or in vector form:

$$T(\pi|\theta, \xi_n, t, k) = [T_j(\pi|\theta, \xi_n, t, k) \,;\; j = 1, 2, \dots, N]$$

$$= \frac{\pi F^k(t)\left(R_\theta^k \nabla L_{\xi_n}^k\right)}{\{\theta, \xi_n|\pi, t, k\}} = \frac{\pi F^k(t)\, X_d^k}{\{d\,|\pi, t, k\}} \;; \tag{5.7}$$

where $X^k = (R_\theta^k \nabla L_{\xi_n}^k)$, ∇ being in effect a logical multiplication operator. X^k is therefore the probability of a particular cause having given rise to a specific state which was achieved as a result of decision alternative k.

$$\{\theta, \xi_n|\pi, t, k\} = \{d|\pi, t, k\} = \pi F^k(t)\, X_d^k\, \mathbf{1}$$

is the probability that the observed output is θ, given that the process started under the initial probability state vector π was holding for t time units in state i, before making the transition to state j due to treatment decision k; d is an index for the cause-effect model. For example, if there were two causes and two observations, d would have four elements corresponding to each combination of cause and observation.

As X^k is a non-square matrix, a square matrix x_d^k is defined as $\mathrm{diag}_{d\text{-const}} X^k$.

Equation (5.7) is valid since dimensionally $\{\theta, \xi_n\,|\pi, t, k\} = \{d\,|\pi, t, k\}$ involves the matrix multiplication $(1 \times N)\,(N \times N)\,(N \times N)\,(N \times 1)$ leading to a scalar quantity and $\mathbf{1} = [1, \dots, 1]^T$.

Consideration as a Markov process leads to the following results:

a) if $R_\theta = I$, the process degenerates to a completely observable semi-Markov decision process;

b) if $H(1) = I$, then we have a partially observable Markov process with auxiliary cause-efect models.

A pictorial representation of the patient state dynamics for the case $N = 3$ is given in Figure 5.17.

If the process could be represented using conditional transition probabilitites (first the system selects a new state j, before making any real transition), then the first time passage is given by the relation:

$$f_{ij}(t) = \sum_{r=1}^{N} \sum_{m=1}^{t} c_{ir}(m)\, f_{rj}(t - m) + c_{ij}(t)$$

where $c_{ij}(m) = p_{ij}\, h_{ij}(m)$ for $i, j = 1, 2, \dots, N$ and the general equation of the system dynamics is given by equation (5.7).

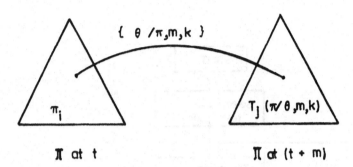

Figure 5.17. Pictorial representation of patient state dynamics when $N = 3$.

Figure 5.18 provides a logical diagram for the process represented above and Figure 5.19 shows an extension of this in a computer-aided diagnosis-treatment scheme, a step toward expert systems engineering.

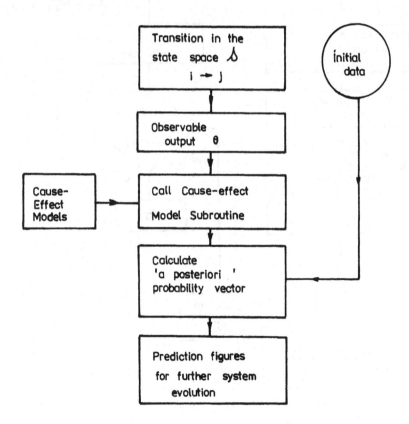

Figure 5.18. Logical diagram for the process described above.

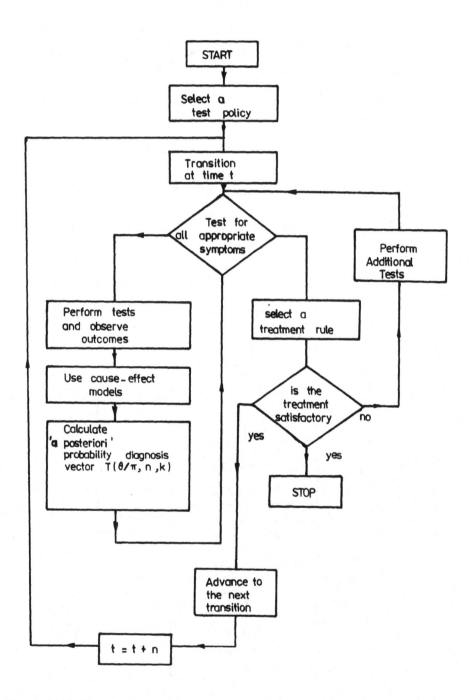

Figure 5.19. Extension of process in a computer-aided diagnosis-treatment scheme.

5.5.5. DECISION ALTERNATIVES IN CLINICAL DISORDERS.

To improve the patient's state of health or even to maintain him in an acceptable state, it is necessary to carry out a set of treatment decisions. From the viewpoint of the model, we can ascribe rewards to the implementing of alternative treatments to patients in each state, given the current observations.

If, in the above decision process, y_{ij}^k is the variable reward (per unit of holding time) and b_{ij}^k is the fixed reward, then the expected value of the process for the case of a completely observable semi-Markov process is:

$$q(\pi, k) = \sum_i \sum_j \pi_i p_{ij}^k \sum_{m=0}^{t} h_{ij}^k(m) \, (my_{ij}^k + b_{ij}^k).$$

However, the reward functional for the clinical treatment-diagnosis process which has to be maximized using a discount factor β, is given by:

$$g(\pi, t) = \max_{k \in \mathcal{D}} \left[q(\pi, k, t) + \beta \sum_{\theta=1}^{m} \left(\sum_{j=1}^{N} \sum_{i=1}^{N} \cdot \right.\right.$$

$$\left.\left. \cdot \, \pi_i p_{ij}^k \, r_{j\theta}^k \, l_{\theta\xi_n}^k \sum_{m=1}^{t} h_{ij}^k(m) \, g(T(\cdot \mid \cdot), (t-m)) \right].$$

Adapting an explicit cost $\tilde{c}_{j\theta}^k$ for the observation process under a treatment alternative, k the reward functional which has to be maximized can be expressed as:

$$v_j(t) = \max_k \left(\gamma^k_{ij\theta\xi_n}(t) + \beta \sum_{i,j,\theta,\xi_n} \pi_i c^k_{ij}(m) r^k_{j\theta} l^k_{\theta\xi_n} \cdot v_{T_j(\pi|\theta,m,k)}(t-m) \right)$$

$$j = 1, 2, \ldots, N$$

where:

$$\gamma^k_{ij\theta\xi_n}(t) = \sum_{i=1}^N \pi_i \sum_{j=1}^N \sum_{m=0}^t c^k_{ij}(m) \sum_{\theta=1}^M r^k_{j\theta} \sum_{\zeta_n}^{N_{A_n}} 1_{\theta\xi_n} (my^k_{ij} + b^k_{ij} + \tilde{c}^k_{j\theta}). \tag{5.8}$$

The reward functional given by (5.8) is piecewise linear and convex; the proof being similar to that given for the case of partially observable Markov processes.

Provided that the treatment process for the partially observable system with Markovian dynamics can be presented as a decision tree, an optimization algorithm (Branch and Bound algorithm) can be built up to find the optimal treatment decision, k, under conditions postulated by the model [18], [32], [54], [70].

5.6   Risk-sensitive partially observable Markov decision processes

5.6.1. MODEL FORMULATION AND PRACTICAL EXAMPLES.

In Chapter 4 partially observable Markov processes were introduced. When the system makes a transition an independent observer is able to see a change in the "core process", but he cannot identify for sure the state of that process. After making an inspection, the decision maker has a set of finite independent observations 1, 2, ... , M, probabilistically correlated with the internal performance states of the system. However, both completely observable and partially observable processes   are considered to have a finite number of states, labelled 1, 2, ... , N.

The decision maker has a choice of decision that depends on the information about the system dynamics.

An important limitation of such Markovian decision processes is that there has been little research into incorporating the risk-sensitivity of the decision maker. In the literature, a first attempt to formulate risk-sensitive Markov decision processes when the system is completely observable is the work of Howard and Matheson [10].

In this chapter, a model of a risk-sensitive Markovian decision process with probabilistic observation of states is formulated and developed. It shows how risk sensitivity may be treated in such processes if the decision maker is characterized as a constant risk aversion person, when the utility function is exponential in form and follows the "delta property".

Several systems applications for the above process are emphasized. A model of a risk-sensitive Markovian decision process with probabilistic observation of states is formulated for the case of an infinite horizon. In the present chapter, a Branch and Bound algorithm is also used which performs maximization of the certain equivalent reward generated by the partially observable MDP with constant risk sensitivity.

5.6.1.1. *Maintenance Policies for a Nuclear Reactor Pressure Vessel.*

To show the practical value of the model given above, we shall first consider an example from nuclear engineering. A BWR (Boiling Water Reactor) system is under observation. Let us suppose that the pressure vessel could be in a finite number of states of deterioration. Each state in the Markov chain is defined as an operating state or a failure state, where the type of failure is cracking of the pressure vessel. The major index description for the state of the vessel is given by the depth of the crack, so that:

state 1— no detectable crack,
state 2 — crack size 0.23622 - 0.3556 mm.,
state 3 — crack size 0.3556 - 0.5588 mm.,
state 4 — crack size 0.5588 - 1.0717 mm.,
state 5 — crack size 1.0717 - 2.54 mm.,
state 6 — crack size 2.54 - 12.7 mm.,
state 7 — crack size > 12.7 mm., failure.

The deterioration process is given by the rate transition matrix:

$$P = \begin{bmatrix} q & p & 0 & . & . & . & 0 \\ 0 & q & p & 0 & . & . & . \\ . & . & q & p & . & . & . \\ . & . & . & . & . & . & . \\ . & . & . & . & . & . & . \\ . & . & . & . & . & . & . \\ . & . & . & . & 0 & q & p_1 \end{bmatrix}$$

where $q = \exp(-\lambda \Delta t)$ and $p = 1 - q$. In the above equation λ is the failure rate constant associated with the deterioration process of the pressure vessel. The system states are arranged in the order of increasing deterioration.

A set if independent observations such as tensometric measurements, noise level, X-ray, on-line TV pictures is available and correlates, probabilistically, the deterioration state of the pressure vessel and its structure with information given by the set of independent observations.

The decision maker (i.e. maintenance engineer) could take decisions such as "do nothing" or "repair" to bring the system into a more desirable operational state. The sensitiviry to risk is described by an exponential utility function. We are interested in how the spectrum of optimal decisions and the process gain changes when the attitude towards risk of the the decision maker changes from risk preference to risk aversion.

5.6.1.2. Medical Diagnosis and Treatment as applied to Physiological Systems. An example from physiological systems as applied to the respiratory system is presented. It is assumed that the decision maker (i.e. the clinician) wishes to choose an action from a finite set of treatment alternatives. The states of the patient (in this case his respiratory system) are considered to follow Markovian dynamics, describable in terms of a performance index based upon deterioration of the system with time or due to a fault such as infection. The system is complex and not completely observable, so any change of state can only be assessed on the basis of clinical observations which may be cheap or expensive to perform.

In addition, care must be taken to ensure that any measurements performed on the patient do not lead to a deterioration in his state of health. It is on the basis of these observations that decisions are made.

In carrying out the diagnosis, the clinician aims to ensure that the outcome of his decision is satisfactory both to the pattient and to himself.

During this process, three stages can be considered:

a) taking measurements, (e.g. pulse rate, temperature),

b) reaching a conclusion as to the state of the process; whether it is a case of heart disease, lung disease, poor environment, etc.,

c) taking a decision regarding any course of action (e.g. do nothing, apply medicine, change environment).

For the respiratory system the measeurement space may consist of heart rate, respiratory rate, temperature, tidal volume, noise level in lungs and oxygen and carbon dioxide concentrations in venous and arterial blood. Any set of observations could result form a number of different malfunctions. For example, a high ventilation rate could be due to low perfusion ventilation or a slow rate of blood flow through the system.

The clinician has also to decide at what point to stop making further observations and start making decisions about medication of some alternative treatment. Over the treatment process outcomes, a utility value could be assessed; the decision maker should be interested in maximizing the utility of the process over a long time horizon.

5.6.2. THE STATIONARY MARKOV DECISION PROCESS WITH PROBABILISTIC OBSERVATIONS OF STATES.

Let us consider a stationary Markov process with a finite number N of states.

The probability rate transition matrix $P^k = [p^k_{ij}]$ of the "core process" is given by:

$$p^k_{ij} = \Pr\{S(t+1) = j \mid S(t) = i, u(t) = k, \mathcal{E}(t)\},$$

$$1 \geq p^k_{ij} \geq 0 ; \quad i, j = 1, 2, \ldots, N ; \quad k = 1, 2, \ldots, K_i$$

$$\sum_{j=1}^{N} p^k_{ij} = 1 ; \quad i = 1, 2, \ldots, N .$$

The matrix P^k has $(N \times N)$ dimensions. K_i represents the maximum number of decision alternatives for the process in state i.

The above probabilistic measure gives information on the system's behaviour and shows that the present state of the system depends only on the immediate past transition and our knowledge about the system evolution. For a time dependent Markov process the state probability vector π is given by:

$$\pi(t) = [\pi_i(t) ; \; i = 1, 2, \ldots, N]$$

$$= [\Pr\{S(t) = i \mid \mathcal{E}\}, \text{ for all } i].$$

We can define a set of all possible initial state vectors by:

$$\Pi = \text{set} \, [\pi_i : \pi_i \geq 0 ; \quad \sum_{i=1}^{N} \pi_i = 1].$$

The state of a completely observable process is known at any time for an independent observer. When the process cannot be continuously observed, we have to use past knowledge, $\mathcal{E}$, about the process for encoding probabilities. We shall refer to $\pi(t)$ as the state of knowledge of the partially observable Markov process when $\mathcal{E}$ contains all past information ($\pi(t)$ is only a partial encoding of that information). The state space Π is a continuum and the conditional probability density function of the next state depends on any pertinent observed output and is given as a discrete probability distribution, referred to here as the system's dynamics.

After a transition in the state space of the core process, an independent observer is able to observe an output θ which is probabilistically correlated with the internal performance state of the "core process", such as:

$$r^k_{j\theta} = \Pr\{z(t+1) = \theta \mid S(t) = j, u(t) = k, \mathcal{E}(t)\},$$

$$1 \geq r^k_{j\theta} \geq 0 ;$$

$$\sum_{\theta=1}^{M} r^k_{j\theta} = 1 ; \quad j = 1, 2, \dots , N ,$$

or in matrix form $R^k = [r^k_{j\theta}]$ (we shall use the diagonal matrix $R^k_\theta = \operatorname*{diag}_{\theta=ct} R^k$).

The system has a dynamic behaviour. The reward associated with making a transition from state i to state j in the chain, and producing an output θ after the transition, is c_{ij}. A utility value associated with the reward is then generated by the process (see Figure 5.20).

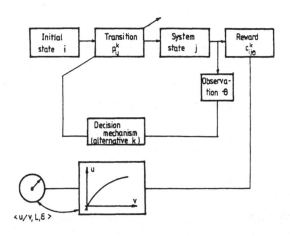

Figure 5.20 . Step in decision process.

LEMMA 5.8. *For a risk-sensitive decision process, if x is the value of an outcome and u(x) is the utility associated with it, and if the utility function u(·) is an exponential one, then u(x) → x, with some appropriated scale change if the risk-aversion coefficient tends to zero (γ → 0).*

Prrof. Let $u(\cdot)$ given by the relation:

$$u(x) = k[1 - e^{-\gamma x}]$$

where k is a constant for an appropriate scale change. (See, also, Figure 5.21).

Let us set k such that $u(1) = 1$. Then we can have $k = [1 - e^{-\gamma}]$ and we want to show that $u(x) \to x$, as $\gamma \to 0$.

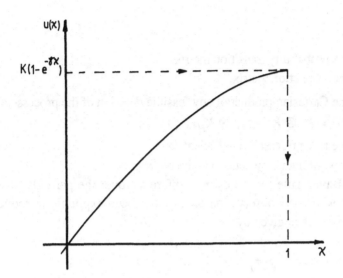

Figure 5.21. Illustration of Lemma 5.8.

By expanding the above relation in a Taylor series, we can write:

$$u(x) = [1 - e^{-\gamma x}]^{-1} [1 - e^{-\gamma x}]$$

$$= \frac{1 - (1 - \gamma x + \frac{(\gamma x)^2}{2!} + \dots)}{1 - (1 - \gamma + \frac{\gamma^2}{2!} + \dots)},$$

and if $\gamma \cong 0$, then we can write:

$$u(x) = \frac{\gamma x}{\gamma} = x,$$

which proves the above lemma.

From the above lemma, we can see that a risk-sensitive partially observable Markov process for $\gamma = 0$ is the case given in the first part of this chapter. If $R_\theta = I$, the identity matrix, then we have a completely observable Markov process.

Now we are ready to define a risk-sensitive Markovian decision process with probabilistic observation of states as a quintuple:

$$\Gamma = \{P, R_\theta, \Omega, \gamma, \mathcal{C}\}$$

where:

P — the probability transition matrix,
R_θ — the inspection matrix,
Ω — the Cartesian product of any feasible decision of the process
$(\Omega = K_1 \otimes K_2 \otimes ... \otimes K_N)$,
γ — the risk-preference coefficient,
$\mathcal{C}$ — reward matrix structure for the process.

By using Bayes' rule for the dyamic inference over the partially observable Markov process where the observation process has been explicitly introduced, the new state of knowledge is given by:

$$\pi'_j = T_j(\pi|\, \theta, k) = \frac{\sum_i \pi_i p_{ij}^k r_{j\theta}^k}{\sum_{i,j} \pi_i p_{ij}^k r_{j\theta}^k}$$

if at the starting time the initial state vector was given by π and k represents a pertinent decision alternative in state i before the process visits a new state j.

With this as a background, the remainder of this section will formulate and examine an implicit enumeration (Branch and Bound) solution to calculate the optimal control policy for a partially observable, risk-sensitive, Markov process.

We define $u[\tilde{v}_t(\pi)]$ as the utility functional for the lifetime of the process if its current information is π and there are t control intervals remaining before the process terminates. Then, we can write:

$$u[\tilde{v}_t(\pi)] = \sum_{i=1}^{N} \pi_i \sum_{j=1}^{N} p_{ij} \sum_{\theta=1}^{M} r_{j\theta} \cdot u\{c_{ij\theta} + \tilde{v}_{t-1}[T(\pi|\,\theta)]\}, \quad t \geq 0.$$

Because it is a dynamic process, the utility of accepting the certain equivalent must be the same as the utility of continuing and the quantities $\tilde{v}_0[T(\pi|\,\theta)]$ can be assigned directly by the decision maker.

Using the "delta property" given in Chapter 3, we can write, for the above equation:

$$u[\tilde{v}_t(\pi)] = \sum_{i=1}^{N} \pi_i \sum_{j=1}^{N} p_{ij} \sum_{\theta=1}^{M} r_{j\theta} e^{-\gamma c_{ij\theta}} \cdot u[\tilde{v}_{t-1}[T(\pi|\,\theta)]], \quad t \geq 0.$$

Using an alternative notation, we can write:

$$u_t(\pi) = \sum_{i=1}^{N} \pi_i \sum_{j=1}^{N} p_{ij} \sum_{\theta=1}^{M} r_{j\theta} e^{-\gamma c_{ij\theta}} u_{t-1} [T(\pi| \theta)], \quad t \geq 0. \qquad (5.9)$$

Following Howard and Matheson [10], the "disutility" for the partially observable process is $e^{-\gamma c_{ij\theta}}$ and the alternative notation used here is $e_{ij\theta}$.

We shall let q_θ be the symbol for the product of the disutility measure, the probability state vector transition rate and the inspection, such as:

$$q_\theta = \sum_{i,j} \pi_i p_{ij} r_{j\theta} e_{ij\theta} .$$

Equation (5.9), using an alternative notation, can be written as:

$$u_t(\pi) = \sum_{\theta=1}^{M} q_\theta u_{t-1} [T(\pi | \theta)], \quad t \geq 0.$$

The certain equivalent of the partially observable process can be written as:

$$\tilde{v}_t(\pi) = -\gamma^{-1} \ln[- (\text{sgn } \gamma) u_t(\pi)].$$

When a decision alternative k is available at time t then we have:

$$u_t(\pi) = \underset{k}{\text{Max}} \left[\sum_{\theta=1}^{M} q_\theta^k u_{t-1} [T(\pi | \theta)] \right], \quad t \geq 0.$$

In this model the upper bound on the utility reward functional is obtained by assuming that the system's states are completely observable ($R_\theta = I$) resulting in a modified set of decisions. The lower bound on the utility cost functional is given by the expected utility reward of any reasonable policy. We shall consider that if the process has an infinite time horizon, then $u_i = \lim_{t \to \infty} u_i(t)$ where $u_i(t)$ are successive solutions for the completely observable process (see Chapter 4):

$$u_i(t) = \text{Max} \left\{ \sum_{j=1}^{N} p_{ij}^k y_{ij}^k u_j(t-1) \right\} \quad i = 1, 2, \dots, N ; \quad t \geq 0.$$

$$u_j(0) = 1 \quad j = 1, 2, \dots, N$$

$$y^k_{ij} = \exp(-\gamma c^k_{ij})$$

and

$$e^k_{ij} = p^k_{ij} y^k_{ij}.$$

LEMMA 5.9. *An upper bound for a risk-sensitive MDP with probabilistic observation of states exists and is given by:*

$$g(\pi) \le \sum_{i=1}^{N} \pi_i u_i.$$

Proof. By induction, for the utility functional $g(\pi, \cdot)$, we can write:

$$g(\pi, 1) = \underset{k}{\text{Max}} \left\{ \sum_i \sum_j \pi_i p^k_{ij} e^k_{ij} \right\}$$

$$\le \sum_{i=1}^{N} \pi_i \underset{k}{\text{Max}} \left\{ \sum_{j=1}^{N} p^k_{ij} e^k_{ij} \right\}$$

$$\le \sum_{i=1}^{N} \pi_i u_i(1).$$

Suppose that we have $g(\pi, t) \le \sum_{i=1}^{N} \pi_i u_i(t)$. Then we can write:

$$g(\pi, t+1) = \underset{k}{\text{Max}} \left\{ \sum_{i=1}^{N} \pi_i \sum_{j=1}^{N} p^k_{ij} r^k_{ij} \cdot u(\pi_i p^k_{ij} r^k_{j\theta} g(\pi', t)) \right\}$$

$$\le \sum_{i=1}^{N} \pi_i \underset{k}{\text{Max}} \left\{ \sum_{j=1}^{N} p^k_{ij} e^k_{ij} u(\pi_i p^k_{ij} r^k_{j\theta} v_j(t)) \right\}$$

$$\le \sum_{i=1}^{N} \pi_i \underset{k}{\text{Max}} \left\{ \sum_{j=1}^{N} p^k_{ij} e^k_{ij} u(\widetilde{v}_j(t)) \right\}$$

$$\le \sum_{i=1}^{N} \pi_i u_i(t+1)$$

$$g(\pi, t+1) \le \sum_{i=1}^{N} \pi_i u_i(t+1).$$

Then, we can write:

$$g(\pi) = \lim_{t\to\infty} g(\pi, t) \le \lim_{t\to\infty} \sum_{i=1}^{N} \pi_i \, u_i(t)$$

$$\le \sum_{i=1}^{N} \pi_i \lim_{t\to\infty} u_i(t) \le \sum_{i=1}^{N} \pi_i \, u_i \; .$$

LEMMA 5.10. *A lower bound for a risk-sensitive MDP with probabilistic observation of states for* $\gamma = $ *constant, exists and is given by*:

$$g(\pi) \ge \mathop{\mathrm{Max}}_{k} \pi[I - \lambda^k P^k]^{-1} \hat{u}^{\,k},$$

where:

$$u^k(t) = [u_i^k(t)]$$

and:

$$u_i^k(t) = \sum_{i=1}^{N} p_{ij}^k \, u(c_{ij}^k(t))$$

$$= \sum_{j=1}^{N} p_{ij}^k \, \exp\left(-\gamma c_{ij}^k(t)\right)$$

$$= \sum_{j=1}^{N} q_{ij}^k(t)$$

$$u^k = \lim_{t\to\infty} u^k(t),$$

and λ^k is the largest eigenvalue for the disutility matrix Q^k, over a completely observable RSMDP (risk-sensitive Markov decision process).

REMARK 5.1. u^k is the eigenvector corresponding to λ. In a normalized form, we can write:

$$\hat{u}_i^k = \frac{-(\mathrm{sgn}\,\gamma)\, u_i}{u_N} \; ; \quad i = 1, 2, \dots, N$$

and

$$\hat{u}^k = [\hat{u}_i^k] \qquad \text{(see Gheorghe [32].}$$

Proof. A policy that can be used is to make the same decision for every period in the future, when $t \to \infty$. Let us call the decision k. Prior to using the results given by Howard and Matheson [10] for large t, we need to multiply the vector by λ (the largest eigenvector of Q), when the process moves from one step to the other. At the next transition we need to multiply the transition probability matrix P^k by λ^k, for any feasible decision k.

Then:

$$g(\pi) \geq \pi \hat{u}^k + \pi \lambda^k P^k \hat{u}^k + \pi \lambda^k \lambda^k P^k P^k \hat{u}^k + \dots$$

$$\geq \pi [I - \lambda^k P^k]^{-1} \hat{u}^k$$

$$\geq \underset{k}{\text{Max }} \pi [I - \lambda^k P^k]^{-1} \hat{u}^k .$$

Next, piecewise linear and convexity properties for the above partially observable Markovian models will be given.

THEOREM 5.7. $u_t(\pi)$ *is piecewise linear and convex in* π *for all t.*

Proof. Let us consider:

$$y_i = \sum_{j=1}^{N} p_{ij} \sum_{\theta=1}^{M} r_{j\theta} e_{ij\theta}$$

which in matrix form is given by:

$$u_0(\pi) = \pi y_0,$$

where y_0 is the expected disutility vector of making a transition at time zero. From the above equation, we can see that $u_0(\pi)$ is not only linear in π but is piecewise linear and convex.

LEMMA 5.11. $T(\pi|\theta, k)$ *preserves straight lines. That is, if* $0 \leq \beta \leq 1$, $\bar{\beta} = 1 - \beta$, *and* $\pi^1, \pi^2 \in \Pi$, *then* $\beta \pi^1 + \bar{\beta} \pi^2$ *is a straight line in* π *with end points* π^1, π^2.
In the transformation of this line fixed θ *and k are considered, and then:*

$$T(\beta \pi^1 + \bar{\beta} \pi^2 | \theta, k) = \mu T(\pi^1 | \theta, k) + \bar{\mu} T(\pi^2 | \theta, k)$$

where β *and* μ *range between* 0 *and* 1.

$$\mu = \frac{\beta[\theta \mid \pi^1]}{\beta[\theta \mid \pi^1] + \beta[\theta \mid \pi^2]} \; ; \; \bar{\mu} = 1 - \mu .$$

Prrof. Let $g(\pi) = [\theta \mid \pi, k] \, U_{t-1}[T(\pi \mid \theta, k)]$ be considered convex, piecewise linear and $[\theta \mid \pi, k]$ is the disutility value of the next observed output θ if the current information vector is π and the next alternative selected is k.

Then letting:

$\pi\beta = \pi^1\beta + \pi^2 \bar{\beta}$, for β as defined above, we can write:

$g(\pi\beta) = [\theta \mid \pi\beta, k] \, U_{t-1}[T(\pi\beta \mid \theta, k)]$

$\qquad = [\theta \mid \pi\beta, k] \, U_{t-1}[\mu T(\pi^1 \mid \theta, k) + \bar{\mu} \, T(\pi^2 \mid \theta, k)]$

$\qquad \leq [\theta \mid \pi\beta, k] \cdot \left[\mu \, U_{t-1}[T(\pi^1 \mid \theta, k) + \bar{\mu} \, U_{t-1}[T(\pi^2 \mid \theta, k)]\right]$

$\qquad = \left[\beta \, [\theta \mid \pi^1\beta, k] + \bar{\beta} \, [\theta \mid \pi^2\beta, k]\right] \left[\mu U_{t-1}[T(\pi^1 \mid \theta, k)] + \right.$

$\qquad\qquad\qquad\qquad\qquad\qquad\qquad\qquad \left. + \bar{\mu} \, U_{t-1}[T(\pi^2 \mid \theta, k)]\right]$

$\qquad = \beta \, [\theta \mid \pi^1\beta, k] \, U_{t-1}[T(\pi^1 \mid \theta, k)] + \bar{\beta} \, [\theta \mid \pi^2\beta, k] \, U_{t-1}[T(\pi^2 \mid \theta, k)]$

$\qquad = \beta g(\pi^1) + \bar{\beta} g(\pi^2),$

which preserves concavity.

Let $U_{t-1}(\pi) = \pi\alpha(\pi)$, that is let $U_{t-1}(\pi)$ be piecewise linear, with $\alpha(\pi)$ being piecewise constant over Π.

$$\bar{g}(\pi) = \{\theta \mid \pi, k\} \frac{P^k R^k_\theta}{\{\theta \mid \pi, k\}} E^k_\theta \alpha[T(\pi \mid \theta, k)]$$

$$= \pi P^k R^k_\theta E^k_\theta \propto [T(\pi \mid \theta, k)] = \pi\xi(\pi),$$

where $\xi(\pi)$ is piecewise constant over Π (a continuum for the state space of the process).

5.6.3. A BRANCH AND BOUND ALGORITHM.

Suppose that the decision maker has available a finite set of decision alternatives for the process when the system could be in any of its states. He has to select the best policy for the process, that is to say, to find the optimal decision vector which is selected after an optimization procedure has been applied.

The Branch and Bound algorithm to find the optimal policy for the partially observable Markov decision process when the decision maker is a risk-sensitive person has also been investigated by Satia et al. [18] for a similar Markov model when the decision maker is risk-neutral ($\gamma = 0$).

It would seem that the analysis using a decision tree approach (see Chapter 3) may have certain advantages of branching and bounding in the decision tree.

In this section, a definition of the above procedure is given from Lawler and Wood [13]:

"Among the most general approaches to the solution of constrained optimization problems is that of "branching-and-bounding", or in the words of Bertier and Roy, "separation et evaluation progressive". Like dynamic programming, branching and bounding is an intelligently structured search of the space of all feasible solutions. Most commonly, the space of all feasible solution is repeatedly partitioned into smaller and smaller sub-sets and a lower bound (in the case of minimization) is calculated for the cost of the solutions within each sub-set. After partitioning, those sub-sets with a bound that exceeds the cost of the known feasible solution are excluded from all further partitionings. The partitioning continues until a feasible solution is found such that its cost is no greater than the bound of any sub-set".

Lemmas 5.10 and 5.11 provide the computational procedure for the bounds of the partially observable RSMDP.

The utility of the upper bound on the return is obtained by using a modified decision set assuming that the states are observed with certainty ($R_\theta = I$). The utility of the lower bound on the return is given by the utility of the total expected return of any reasonable decision policy.

Suppose that the whole process can be described as a decision tree as given in Figure 5.22. If the node has no continuation, we can write:

$$u\left(\mathrm{UPB}(I\,I_k\,O_\theta)\right) \;=\; \sum_j T_j(\pi \mid k, \theta)\, u(\tilde{v}_j),$$

where $\tilde{v}_j$ is the certain equivalent of the lottery in the case of the completely observable state, j.

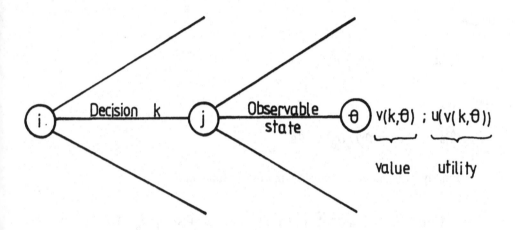

Figure 5.22. Decision tree of partially observable RSMDP.

Similarly, we can write:

$$u\big(\mathrm{LOB}(I\,I_k\,O_\theta)\big) \;=\; \max_m \,[T_j(\pi \mid m,\,\theta) \cdot (I - \lambda^m\,P^m)^{-1}\,u^m].$$

If the node has a branch originating from it, the above computational routine must be performed on all the terminal nodes.

In this case, we can write:

$$u\big(\mathrm{UPB}(I)\big) \;=\; \max_k \Big\{ \sum_\theta \,\{\theta \mid \pi,\,k\} \cdot u\big[w_{I\,I_k O_\theta} + \mathrm{UPB}(I\,I_k\,O_\theta)\big]\Big\}$$

$$\{\theta \mid \pi, k\} = \sum_{i,j} \pi_i p_{ij}^k r_{j\theta}^k$$

$$= \underset{k}{\text{Max}} \left\{ \sum_{\theta} \{\theta \mid \pi, k\} \cdot \exp(-\gamma w_{I I_k O_\theta}) \cdot u\big(\text{UPB}(I I_k O_\theta)\big) \right\}$$

and:

$$u\big(\text{LOB}(I)\big) = \underset{k}{\text{Max}} \left\{ \sum_{\theta} \{\theta \mid \pi, k\} \cdot u\big[w_{I I_k O_\theta} + \text{LOB}(I I_k O_\theta)\big] \right\}$$

$$= \underset{k}{\text{Max}} \left\{ \sum_{\theta} \{\theta \mid \pi, k\} \cdot \exp(-\gamma w_{I I_k O_\theta}) \cdot u\big(\text{LOB}(I I_k O_\theta)\big) \right\},$$

where:

$$\{\theta \mid \pi, k\} = \sum_{i,j} \pi_i p_{ij}^k r_{j\theta}^k$$

and:

$$e_{ij\theta}^k = \exp(-\gamma w_{I I_k O_\theta}) = e_\theta^k.$$

Finally, we can write the following utility functionals:

$$u\big(\text{UPB}(I)\big) = \underset{k}{\text{Max}} \left\{ \sum_{\theta} \{\theta \mid \pi, k\} \cdot e_\theta^k \cdot u\big(\text{UPB}(I I_k O_\theta)\big) \right\}$$

$$u\big(\text{LOB}(I)\big) = \underset{k}{\text{Max}} \left\{ \sum_{\theta} \{\theta \mid \pi, k\} \cdot e_\theta^k \cdot u\big(\text{LOB}(I I_k O_\theta)\big) \right\}.$$

The decision branching criteria for further tree extension has considerable effect on the efficiency of the Branch and Bound algorithm convergence and computation time. We can use a Fibonacci search method to find potential nodes in the decision tree to be extended.

Following Satia et al. [18], as a value branching criteria for a risk-neuter decision maker ($\gamma = 0$), we can have any node I:

$$\text{VALB} = (\beta^m) [\text{UPB}(I) - \text{LOB}(I)] \cdot \text{Pr}\{I\},$$

or entering the above results using Lemma 5.10, we can have, for the case $\gamma \neq 0$:

$$VALB = (\lambda)^n \cdot \left[u(UPB(l)) - u(LOB(l))\right] \cdot Pr\{l\}$$

for $\gamma = 0$, $\alpha \to \beta$, where β is a discount factor for convergence reasons, and n represents the stage of the computation procedure (see Figure 5.23).

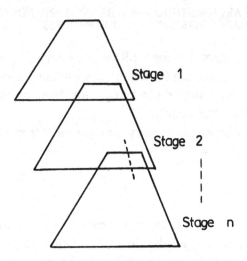

Figure 5.23. Stages of computation procedure.

As a bounding evaluation function for the optimal decision in any state we shall also consider:

"The utility of the lower bound for the decision $k \in K_i$, $i = 1, 2, \dots, N$, is greater than the utility of the upper bound for any other decision l, for that branch ($l \neq k$) (lexicographical ordering)".

Description of Algorithm.

The above steps of the algorithm are as follows:

    Step 1:   Find all labelled nodes in the tree.

    Step 2:   Perform the above routine (Lemmas 5.9, 5.10, 5.11) on the labelled nodes.

    Step 3:   Starting from the highest level, scan all nodes.

    Step 4:   If at level l, a dominating decision k is foud, then the procedure terminates. Otherwise, repeat the procedure with a new tree.

Better upper and lower bounds on the utility functional will increase the computational efficiency. If $u(L^A_i) \geq u(L^B_i)$, then $L^A_i \geq L^B_i$ or $A \succ B$ (policy A

preferred to policy B), then policy A dominates B and policy B need not be considered for any further analysis. First of all we need to find policies which are not dominated.

The algorithm is sequential in nature. As the process makes the transition, part of the computation done previously, can be used for the next stage.

5.6.4 A FIBONACCI SEARCH METHOD FOR A BRANCH AND BOUND ALGORITHM FOR A PARTIALLY OBSERVABLE MARKOV DECISION PROCESS

In a Branch and Bound algorithm for the RSMDP, we must have a decision criteria for further branching in gathering for the optimal solution. Two criteria have been given. A Fibonacci search method can be used for the selection of new branching trees (Step 4 in the above algorithm),

Let us consider that we can identify a number of H terminal nodes and define the Fibonacci number as:

$$F_N = 1 + H. \tag{5.10}$$

Suppose that in the range of interest, h cases must be compared to determine the optimum node to be expanded, where H satisfies the above equation. The ordering for search of those nodes can be done in the increasing value of the nodes cardinality. Once the system has been ordered, it is only necessary to label the various possible cases from 1 to H, consecutively. Then the range of interest can be viewed as being $(H + 1)$ in length. Any given case will be identified by an integer h, where:

$$1 \leq h \leq H + 1.$$

For X experiments (where X is defined by equation $F_X = (1 + H)$), each experiment will correspond to an integer value of H, and hence to a particular case in the decision tree. The position of the first experiment (the first searching procedure) is given by:

$$l_1 = \frac{F_{N-2}}{F_N}(H + 1),$$

or using (5.10), we can write:

$$l_1 = F_{N-2}.$$

The experiments to be performed correspond to $h = l_1$ and $h = 1 = H - l_1$. For the general case, the node to be searched is given by the integral distance in the interval $(1 + H)$:

$$l_j = F_N - (j + 1) \frac{1 + H}{F_N}$$

or, using results from above, we can write:

$$l = F_N - (j + 1),$$

and l_j is always an integer.

We have to use artificial locations for the case when

$$1 + h \neq F_N,$$

which will be added at one end of the sequence, to bring the total to a Fibonacci number such that:

$$1 + H + F = F_N,$$

where now $F_N > 1 + H > F_{N-1}$, and F is the number of pseudo experiments which have to be added. The search should then be applied as in the previous description.

For the RSMDP, as a criteria for comparison, we can use:

$$y(\cdot) = \text{UPB}(\cdot) - \text{LOB}(\cdot).$$

where UPB$(\cdot)$ and LOB$(\cdot)$ have been calculated in earlier stages of the Branch and Bound algorithm (Step 2).

5.6.5 A NUMERICAL EXAMPLE

Let us consider a system under maintenance control alternatives. For the sake of simplicity, we shall consider that $N = 2$ and $M = 3$. The decision maker is a risk-sensitive person and follows an exponential utility function.

The data for the problem are given in Table 5.3.

In the table v^1 and v^2 represent the loss from the system when a "do nothing" or a "do repair" decision is taken, respectively.

	Alternative 1 "do nothing"		Alternative 2 "do repair"

$$P^1 = \begin{bmatrix} 0.4 & 0.6 \\ 0.7 & 0.3 \end{bmatrix} \qquad P^2 = \begin{bmatrix} 0.5 & 0.5 \\ 0.6 & 0.7 \end{bmatrix}$$

$$R^1 = \begin{bmatrix} 0.3 & 0.3 & 0.4 \\ 0.5 & 0.2 & 0.3 \end{bmatrix} \qquad R^2 = \begin{bmatrix} 0.4 & 0.2 & 0.4 \\ 0.3 & 0.5 & 0.2 \end{bmatrix}$$

$$v^1 = \begin{bmatrix} 3 \\ 2 \end{bmatrix} \qquad v^2 = \begin{bmatrix} 2 \\ 1 \end{bmatrix}$$

$$\pi = \begin{bmatrix} 0.6 & 0.4 \end{bmatrix}$$

Table 5.3. Data for the numerical example.

First Iteration

For a total risk-aversion person ($\gamma = 1$), we have the following data:

$$c^1_{ij} = \begin{bmatrix} 1 & 2 \\ 3 & 5 \end{bmatrix} \qquad c^2_{ij} = \begin{bmatrix} 3 & 4 \\ 2 & 4 \end{bmatrix}$$

$$(v^1) = \begin{bmatrix} 0.04978 \\ 0.13533 \end{bmatrix} \qquad (v^2) = \begin{bmatrix} 0.13533 \\ 0.36787 \end{bmatrix}$$

$$k = 1 \qquad\qquad k = 2$$

$$w_{I I_k O_\theta}: \begin{bmatrix} 3 & 2 & 2 \end{bmatrix} \qquad \begin{bmatrix} 4 & 2 & 1 \end{bmatrix}$$

Let us suppose that the policy which we are going to follow has given us the utility vector:

$$L^A = \begin{bmatrix} 4 \\ 2 \end{bmatrix} \Rightarrow u(L^A) = \begin{bmatrix} 0.01831 \\ 0.13533 \end{bmatrix}$$

The upper and lower bounds of the nodes from the decision tree are given in Table 5.4.

Node	Upper bound	Lower bound
1,1,1	0.0882775	0.070969
1,1,2	0.0711675	0.047565
1,1,3	0.073734	0.0510756
1,2,1	0.2027666	0.0522458
1,2,2	0.2702032	0.0861816
1,2,3	0.1841634	0.0428842
1,1	$2.8219 \cdot 10^{-3}$	$1.9968 \cdot 10^{-3}$
1,2	0.01604	$4.1916 \cdot 10^{-3}$
1	0.01604	$1.9968 \cdot 10^{-3}$

Table 5.4. Upper and lower bounds of nodes from the decision tree.

We need to proceed to a second iteration such that a new tree needs to be built up (see Figure 5.24). We choose, as a new branching node (1, 2, 3).

At the second iteration, the probability state vector for alternatives one and two are given in Table 5.5 and 5.6.

Alternative 1				Alternative 2		
$T(\pi \mid \theta, 1)$	State			$T(\pi \mid \theta, 2)$	State	
	1	2			1	2
$\theta = 1$	0.49	0.51		$\theta = 1$	0.80	0.20
$\theta = 2$	0.86	0.14		$\theta = 2$	0.27	0.73
$\theta = 3$	0.82	0.18		$\theta = 3$	0.90	0.10

Table 5.5. Probability state vector for Alternative 1 (second iteration).　　Table 5.6. Probability state vector for Alternative 2 (second iteration).

The bounds for the nodes are given in Table 5.7.

Node	Upper bound	Lower bound
1,1	$2.99 \cdot 10^{-3}$	$2.125 \cdot 10^{-3}$
1,2	0.0182770	$4.7696 \cdot 10^{-3}$

Table 5.7. Bounds for the nodes at the second iteration.

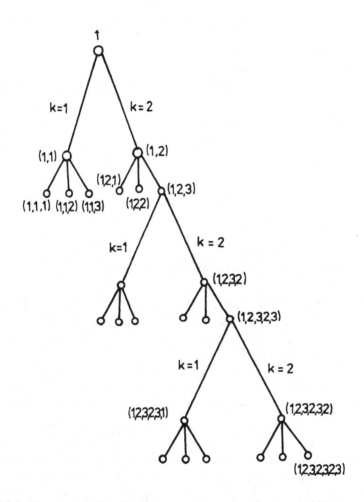

Figure 5.24. New tree following the second iteration.

At the second iteration the process does not reach the operational criterion. Then, we need to proceed to the third iteration, The probability state vector at the third iteration is given in Tables 5.8 and 5.9.

	Alternative 1	
$T(\pi \mid \theta, 1)$	State	
	1	2
$\theta = 1$	0.43	0.57
$\theta = 2$	0.92	0.08
$\theta = 3$	0.89	0.11

Table 5.8. Probability state vector for Alternative 1 (third iteration).

	Alternative 2	
$T(\pi \mid \theta, 2)$	State	
	1	2
$\theta = 1$	0.87	0.13
$\theta = 2$	0.16	0.84
$\theta = 3$	0.96	0.04

Table 5.9. Probability state vector for Alternative 2 (third iteration).

The bounds for the nodes for the third iteration are given in Table 5.10.

Node	Upper bound	Lower bound
1,1	$3.1404 \cdot 10^{-3}$	$2.21533 \cdot 10^{-3}$
1,2	0.1974659	$5.1597 \cdot 10^{-3}$

Table 5.10. Bounds for the nodes at the third iteration.

At the fourth iteration the probability state vectors are given in Tables 5.11 and 5.12. The bounds for the nodes in the new tree at the fourth iteration are given in Table 5.13.

	Alternative 1	
$T(\pi \mid \theta, 1)$	State	
	1	2
$\theta = 1$	0.38	0.62
$\theta = 2$	0.96	0.04
$\theta = 3$	0.93	0.07

Table 5.11. Probability state vector for Alternative 1 (fourth iteration).

	Alternative 2	
$T(\pi \mid \theta, 2)$	State	
	1	2
$\theta = 1$	0.92	0.08
$\theta = 2$	0.87	0.13
$\theta = 3$	0.98	0.02

Table 5.12. Probability state vector for Alternative 2 (fourth iteration).

Node	Upper bound	Lower bound
1,1,1	0.10282696	0.0908624
1,1,2	0.0532089	0.01811892
1,1,3	0.0557754	0.02655014
1,2,1	0.1539388	0.0276716
1,2,2	0.16556602	0.02788587
1,2,3	0.13998616	0.0206504
1,1	$2.68105 \cdot 10^{-3}$	$1.4499 \cdot 10^{-3}$
1,2	0.105869	$2.2737 \cdot 10^{-3}$
1	0.105869	$1.4499 \cdot 10^{-3}$

Table 5.13. Bounds of the nodes of the new tree at the fourth iteration.

At the fourth iteration LOB(1,2) > UPB(1,1) and then decision 2 ("do repair"), is preferred to decision 1 when the system is in state 1

In the case of a totally risk-averse decision maker, repair action should always be followed when the system is in state 1.

CHAPTER 6

POLICY CONSTRAINTS IN MARKOV DECISION PROCESSES

For the cases analyzed in previous chapters (mainly Chapter 4) no restrictions are made on the way in which alternatives are selected in different states of the Markov process. Research on Markov decision processes with policy constraints has been done by Nafeh [103]. He stated that:

"It is assumed that the selection of an alternative in a given state has no effect on alternative selection in any other state. In other words, there is no interaction, or "coupling" between alternatives in different states. This is the feature that allows each state to be considered separately ... , it is an idealized situation that might or might no hold in real life. With the profusion of rules and regulations governing economic activities in this day and age, it might very well turn out that the alternative selection is not as "coupling-free" as the idealized situation envisages.

Introducing inter-state dependence in alternative selection, in effect, imposes constraints on the policies. Some policies become "infeasible" when satisfying the constraints. ... for policy constrained problems, we start with the optimal policy in the absence of constraints and go backwards in the ordering, checking the feasibility, until we hit the feasible policy yielding the highest gain. The problem with this brute force method is that once we get beyond the second-to-optimum policy we have to evaluate an increasingly large number of policies for each step in the ordering process. Moreover, we first have to solve the unconstrained policy problem before we can solve the constrained policy one".

Within the same framework the above author mentioned:

"Also of interest in a constrained policy problem is the "sensitivity" of optimal policies to the policy constraints. This "sensitivity" can best be expressed in terms of value, i.e. how much a rational decision maker would be willing to pay to remove a constraint".

6.1 Methods of investigating policy constraints in Markov decision processes

The main model used to investigate and solve the above problems is the LP formulation. A quantity d_i^k is introduced (e.g. the conditional probability that, given the process is in state i, alternative k is selected). In an LP approach, for each state i only one $d_i^k = 1$ and the rest are zero. The policy constraints and the Markov decision process are formulated in terms of d_i^k.

A "two-alternative-coupling" case explains the interaction between an alternative k in state i and an alternative l in some other state j (e.g. d_i^k and d_j^l will be denoted by a and b, respectively).

There can only be five different types of constraints:

$$a + b \leq 1 \tag{6.1a}$$

$$a + b \geq 1 \tag{6.1b}$$

$$a - b \geq 0 \tag{6.1c}$$

$$a + b = 1 \tag{6.1d}$$

$$a - b = 0 \tag{6.1e}$$

Assume that we have a policy $\mathcal{P}$ whose first R components ($R > 1$) are $a, b, c, \dots, r$, respectively.

If we want to make $\mathcal{P}$, and all policies that differ from $\mathcal{P}$ in exactly one of the first R components, infeasible, we can do it by the constraint:

$$d_1^a + d_2^b + \dots + d_R^r \leq R - 2.$$

For the risk-indifferent case, the objective function to be maximized is the gain. As is considered in the literature, the constraints are the equations defining the limiting state probabilities and those requiring the d_i^k to sum to unity in each state. We can also consider additional relationships between d_i^k of different states.

According to Nafeh [103]:

"The realization that we are dealing with a constrained optimization problem leads to the Lagrange multiplier rule which enables us to reduce the problem to two iteratively coupled problems defined on the associated Lagrangian".

For the case of a risk-sensitive Markov decision process a Lagrange multiplier formulation is also applied; the objective function is the maximal eigenvalue of the matrices with the feasible policies.

The constraints considerd here are the eigenvalue problem defining the maximal eigenvalue, plus the policy constraints. It is to be mentioned that, according to the relevant literature, "as in the risk-indifferent case, Howard's and Matheson's ... algorithm is shown to be the transformation of the original problem via Lagrange multipliers, to two problems", [103].

Within the framework of this chapter, a sensitivity analysis will be considered.

6.2 Markov decision processes with policy constraints

For such a process we can write the following non-linear problem:

$$\max \sum_{i=1}^{N} \pi_i \sum_{k=1}^{K_i} q_i^k d_i^k \qquad (6.2)$$

subject to:

$$\pi_j - \sum_{i=1}^{N} \pi_i \sum_{k=1}^{K_i} p_{ij}^k d_i^k = 0 \; ; \; j = 1, 2, \dots , N \qquad (6.3)$$

$$\sum_{i=1}^{N} \pi_i \sum_{k=1}^{K_i} d_i^k = 1 \qquad (6.4)$$

$$\sum_{k=1}^{K_i} d_i^k = 1 \; ; \quad i = 1, 2, \dots , N \qquad (6.5)$$

$$d_i^k \geq 0 \; ; \qquad i = 1, 2, \dots , N \qquad (6.6)$$

where $q_i^k = \sum_{j=1}^{N} p_{ij}^k r_{ij}^k \; ; \; i = 1, 2, \dots , N$.

In the solution of (6.2) to (6.6) the d_i^k are either zero or unity, with $d_i^k = 1$ for only one k in each state i.

Next, constraints on policies will be introduced (e.g. interaction, or coupling, between alternatives in different states). Selecting alternative j, when the system is in state i, prevents the selection of alternative l in state k. This leads to the fact that any policy having $P(i) = j$ and $P(k) = l$ is non-feasible.

The quantity d_i^k is zero or one for only one k in each i. The constraint is expressed as:

$$d_i^j + d_k^l \leq 1. \qquad (6.7)$$

Quantity $d_i^k = 1$ or 1 plus relation (6.7) imply that no more than one of d_i^j and d_k^l can be unity.

THEOREM 6.1. [103]. *Any policy constraint consisting of an interaction between alternative j in state i and alternative l in state k can be expressed as one of the relations (6.1a) to (6.1e), where $a = d_i^j$ and $b = d_k^l$ and both are either zero or unity.*

Proof. An extensive demonstration of this theorem is given in [103].

DEFINITION 6.1.   Let X be an event. Its associated algebraic variable is a real number x restricted to the values 1 or 0 such that X is true if $X = 1$. Values true and false will be denoted by T and F.

PROPOSITION 6.1. *In the case that X is an event whose associated algebraic variable is x, then the algebraic variable associated with $\overline{X}$ (the complement of X) is $(1 - x)$.*

Proof. Let $Y = X, y = 1 - x$.

Then: $Y = T \Leftrightarrow = X = F$

$\qquad X = T \Leftrightarrow x = 1$

Assume: $Y = T$

Then: $X = F \Rightarrow x \neq 1 \Rightarrow x = 0 \Rightarrow y = 1 - x = 1$

$\qquad Y = T \Rightarrow y = 1$

Assume: $y = 1$

Then: $x = 1 - y = 0 \neq 1 \Rightarrow X = F \Rightarrow Y = T$

i.e. $y = 1 \Rightarrow Y = T$.

We have proved that $y = T \Leftrightarrow y = 1$, which is the definition of the associated algebraic variable.

PROPOSITION 6.2. $B = ABC \dots Z$ *is true* $\Leftrightarrow abc \dots z = 1$.

PROPOSITION 6.3. $B = B_1 + B_2 + ... + B_N$ is true $\Leftrightarrow b_1 + b_2 + ... + b_n \geq 1$ where B_i is a Boolean product of events, and b_i is the corresponding product of the associated algebraic variables.

COROLLARY 6.1. Let :
    (a) $B = B_1 + B_2 + ... + B_N$, where B_i is the Boolean product of events ;
    (b) $I = \{1, 2, ... , N\}$.
If $B_i B_j = F$; for all $i, j \in I$.
Then :

$$B = T \Leftrightarrow \sum_{i=1}^{N} b_i = 1 .$$

Proof. Assume that $\exists i, j \ni B_i = T$, and $B_j T$.
Then $B_i B_j = T$.
    This is a contradiction. $B_i B_j = F$; for all i, j . There cannot exist more than one $i \ni B_i = T$, and in this case:

$$B = T \Leftrightarrow \exists_i \ni B_i = T \text{ and } B_j = F \text{ for } j \neq i$$

$$\Leftrightarrow \exists_i \ni b_i = 1 \text{ and } b_j = 0 \text{ for } j \neq i$$

$$\Leftrightarrow \sum_{i=1}^{N} b_i = 1.$$

It has been proved by Nafeh [103] that the MDP with policy constraints can be formulated with extra constraints on a sub-set of d_i^k.

We shall give, without proof, the following proposition.

PROPOSITION 6.4. For the problem :

$$\max \sum_{i=1}^{N} \pi_i \sum_{k=1}^{K_i} q_i^k d_i^k \tag{6.8}$$

subject to :

$$\pi_j - \sum_{i=1}^{N} \pi_i \sum_{k=1}^{K_i} p_{ij}^k d_i^k = 0 \ ; \ j = 1, 2, ... , N \tag{6.9}$$

$$\sum_{i=1}^{N} \pi_i \sum_{k=1}^{K_i} d_i^k = 1 \tag{6.10}$$

$$\sum_{k=1}^{K_i} d_i^k = 1 \; ; \quad i = 1, 2, \dots, N \tag{6.11}$$

$$d_i^k \geq 0 \; ; \quad i = 1, 2, \dots, N \tag{6.12}$$

$$C_l(d_i^k) \leq 0 \; ; l = 1, 2, \dots, m \quad (i, k) \in S_1$$

$$C_p(d_i^k) = 0 \; ; p = 1, 2, \dots, q \quad (i, k) \in S_2, \tag{6.13}$$

where S_1 and S_2 are sub-sets of $S = \{(i, k)\}$.

In the case where d_i^k is assumed to take values of 0 or 1 for $(i, k) \in S_1 \cup S_2$, then d_i^k will only take on values of 0 or 1 for $(i, k) \in S_1$.

6.2.1. A LAGRANGE MULTIPLIER FORMULATION

This method as applioed to the constrained maximization problem will reduce the constrained problem to a number of unconstrained ones.

The Lagrange multiplier rule is defined as:

(a) Let $X^* \in E^n$ maximize $g(x)$ subject to $C_i(x) = 0 \; ; i = 1, 2, \dots, m$.

(b) Then there exist real numbers $\lambda_i^*, i = 1, 2, \dots, m$ (Lagrange multipliers) such that the point $(x_1^*, x_2^*, \dots, x_n^*, \lambda_1^*, \lambda_2^*, \dots, \lambda_m^*) = (x^*, \lambda^*) \in E^{n+m}$ is a critical point of the function

$$L(x, \lambda) = f(x) + \sum_{i=1}^{m} \lambda_i C_i(x) \tag{6.14}$$

i.e.

$$\nabla L(x^*, \lambda^*) = 0$$

and, moreover, $L(x^*, \lambda^*) = f(x^*)$.

The Lagrangian offers the possibility of an adequate iterative algorithm. Such a procedure will be used to solve the Markov decision process.

Next the form of the general problem which will include the policy constraints will be written as:

$$\max \sum_{i=1}^{N} \pi_i \sum_{k=1}^{K_i} q_i^k d_i^k$$

subject to:

$$\sum_{i=1}^{N} \pi_i \sum_{k=1}^{K_i} p_{ij}^k d_i^k - \pi_j = 0 ; \quad j = 1, 2, \ldots, N$$

$$1 - \sum_{i=1}^{N} \pi_i \sum_{k=1}^{K_i} d_i^k = 0$$

$$\sum_{i=1}^{K_i} d_i^k - 1 = 0.$$

$$C_l(d_i^k) \leq 0 ; l = 1, 2, \ldots, m \qquad (i, k) \in S_1$$

$$C_p(d_i^k) = 0 ; p = 1, 2, \ldots, q \qquad (i, k) \in S_2,$$

where S_1 and S_2 are sub-sets of the set $S = \{(i, k)\}$ of (i, k) pairs defining each alternative in each state.

The Lagrangian associated with the above optimization problem is given by:

$$L = \sum_{i=1}^{N} \pi_i \sum_{k=1}^{K_i} q_i^k d_i^k + \sum_{j=1}^{N} v_j \left[\sum_{i=1}^{N} \pi_i \sum_{k=1}^{K_i} p_{ij}^k d_i^k - \pi_j \right] +$$

$$+ g \left[1 - \sum_{i=1}^{N} \pi_i \sum_{k=1}^{K_i} d_i^k \right] + \sum_{i=1}^{N} \beta_i \left[\sum_{k=1}^{K_i} d_i^k - 1 \right] \qquad (6.15)$$

where v_j, g, β_i are the Lagrange multipliers, $(2n + 1)$ in number.

This relation is a function of the limiting state probabilities π_i, the conditional probabilities d_i^k of selecting alternative k given state i. The dimension of the Euclidian space of L is $(3N + \sum_{i=1}^{K} K_i + 1)$, and the objective is to find a critical point for L.

The iterative algorithm implies:

Step 1: Starting with d_i^k 's that satisfy (6.11) and (6.12) we can set the spatial derivatives of L with respect to the π 's equal to zero.

Step 2: Use the result which tells us that $d_i^k = 0, 1$.

The partial derivatives of L with respect to the π 's are:

$$\frac{\partial L}{\partial \pi_i} \sum_{k=1}^{K_i} q_i^k d_i^k - v_i \sum_{j=1}^{N} v_j \sum_{k=1}^{K_i} p_{ij}^k d_i^k - g \sum_{k=1}^{K_i} d_i^k \; ; \; i = 1, 2, \dots, N,$$

$$(6.16)$$

Given a policy P, we can get k for each i and relation (6.16) can be written as:

$$\frac{\partial L}{\partial \pi_i} = q_i(P) - v_i + \sum_{j=1}^{N} p_{ij}(P) v_j - g$$

or by setting the derivatives equal to zero we can write:

$$q_i + \sum_{j=1}^{N} p_{ij} v_j = g + v_i \; ; \; i = 1, 2, ./.. , N.$$

$$(6.17)$$

Relation (6.17) is, in fact, the value determination algorithm within Howard's algorithm.

In matrix form, we can write the following:

$$\begin{bmatrix} 1-p_{11} & -p_{12} & \cdot & \cdot & -p_{1N} & 1 \\ -p_{21} & 1-p_{22} & \cdot & \cdot & -p_{2N} & 1 \\ \cdot & \cdot & \cdot & \cdot & \cdot & \cdot \\ -p_{N-1,1} & -p_{N-1,2} & \cdot & \cdot & -p_{N-1,N} & 1 \\ -p_{N,1} & -p_{N,2} & \cdot & \cdot & 1-p_{NN} & 1 \end{bmatrix} \begin{bmatrix} v_1 \\ v_2 \\ \cdot \\ v_N \\ g \end{bmatrix} = \begin{bmatrix} q_1 \\ q_2 \\ \cdot \\ \cdot \\ q_N \end{bmatrix}$$

For $v_N = 0$ (e.g. we can eliminate the N-th column to obtaine a $N \times N$ system of equations); we can write:

$$\begin{bmatrix} 1-p_{11} & -p_{12} & \cdot & \cdot & -p_{1N-1} & 1 \\ -p_{21} & 1-p_{22} & \cdot & \cdot & -p_{2N-1} & 1 \\ \cdot & \cdot & \cdot & \cdot & \cdot & \cdot \\ \cdot & \cdot & \cdot & \cdot & \cdot & 1 \\ -p_{N,1} & -p_{N,2} & \cdot & \cdot & p_{N,N-1} & 1 \end{bmatrix} \begin{bmatrix} v_1 \\ v_2 \\ \cdot \\ v_{N-1} \\ g \end{bmatrix} = \begin{bmatrix} q_1 \\ q_2 \\ \cdot \\ \cdot \\ q_N \end{bmatrix}$$

Or we have:

$$
\begin{bmatrix} v_1 \\ v_2 \\ \cdot \\ v_{N-1} \\ g \end{bmatrix} = [\tilde{P}]^{-1} \begin{bmatrix} q_1 \\ q_2 \\ \cdot \\ \cdot \\ q_N \end{bmatrix} ,
$$

where (see Equation (6.17)):

$$
[\tilde{P}] = \begin{bmatrix} 1 - p_{11} & -p_{12} & \cdot & \cdot & 1 \\ & \cdot & & \cdot & \cdot & \cdot \\ \cdot & & \cdot & \cdot & \cdot \\ \cdot & & \cdot & \cdot & \cdot \\ -p_{N,1} & -p_{N,2} & \cdot & \cdot & 1 \end{bmatrix} .
$$

By differentiating the Lagrangian with respect to v's and g, we can have:

$$
\frac{\partial L}{\partial v_j} = \sum_{i=1}^{N} \pi_i \sum_{k=1}^{K_i} p_{ij}^k \, d_i^k - \pi_i \, ,
$$

$$
\frac{\partial L}{\partial g} = 1 - \sum_{i=1}^{N} \pi_i \sum_{k=1}^{K_i} d_i^k \, .
$$

For a given policy, by setting the partial derivative equal to zero, we obtain the definition of limiting probabilities:

$$
\pi_j - \sum_{i=1}^{N} p_{ij} \pi_i = 0 \; ; \quad j = 1, 2, \dots , N
$$

$$
\sum_{i=1}^{N} \pi_i = 1 \, .
$$

In matrix form, we have the following relations:

$$
\begin{bmatrix}
1-p_{11} & -p_{12} & \cdot & \cdot & -p_{1N-1} \\
-p_{21} & 1-p_{22} & \cdot & \cdot & -p_{2N-1} \\
\cdot & \cdot & \cdot & \cdot & \cdot \\
-p_{N,1} & -p_{N,2} & \cdot & \cdot & p_{N,N-1} \\
1 & 1 & \cdot & \cdot & 1
\end{bmatrix}
\begin{bmatrix}
\pi_1 \\
\pi_2 \\
\cdot \\
\pi_{N-1} \\
\pi_N
\end{bmatrix}
=
\begin{bmatrix}
0 \\
0 \\
\cdot \\
0 \\
1
\end{bmatrix}
$$

By proceeding to some algebraic manipulations, we have:

$$
\begin{bmatrix}
\pi_1 \\
\pi_2 \\
\cdot \\
\\
\pi_N
\end{bmatrix}
= ([\tilde{P}]^{-1})^T
\begin{bmatrix}
0 \\
0 \\
\cdot \\
\cdot \\
1
\end{bmatrix} .
$$

The π_i, $i = 1, 2, \dots, N$ are the last column $([\tilde{P}]^{-1})^T$; these values are calc-ulated from $\tilde{P}^{-1}$ in the first phase of the Howard's algorithm. This has a special importance when we consider policy constraints.

Concerning the importance of Lagrange multipliers, the values of v_i, $i = 1, 2, \dots, N$ are involved in L in the form:

$$
\sum_{j=1}^{N} v_j \frac{\partial L}{\partial v_j} ,
$$

where $(\partial L/\partial v_j)$ gives the amount by which the j-th constraint is violated; it is necessary that t is zero for all j. Where this is not the case we shall not have a critical point for L. The values of v_j give the cost of violating the j-th constraint by one unit, which, in turn, represents the equilibrium of probabilistic flows in the steady state. If in state j, the equilibrium of probabilistic flows does not hold, v_j gives the cost per unit of "disequilibrium" for that state. There is no guarantee that by computing the new π_i and v_i, $i = 1, 2, \dots, N$ we can get a net increase in the Lagrange functional, L.

For g, we have:

$$
g = \sum_{i=1}^{N} \pi_i q_i ,
$$

which is the value of the original function we are trying to maximize. It is clear that, for a given policy, the value of g is one of the Lagrange multipliers, which makes the partial derivatives equal to zero.

In this case, the value determination phase of the Howard's algorithm results in equating some of the partial derivatives of L to zero, for a given policy. It is known, from this chapter, that d_i^k, $i = 1, 2, \ldots, N$ are 0 or 1 for the case of unconstrained policies. Now, we can treat the maximization process as a discrete problem by trying to increase the value of L.

The equation (6.15) can be written in such a form as to bring out the dependence on d_i^k:

$$L = \sum_{i=1}^{N} \pi_i \sum_{k=1}^{K_i} \left(q_i^k + \sum_{j=1}^{N} p_{ij}^k v_j \right) d_i^k - \sum_{j=1}^{N} v_j \pi_j +$$

$$+ g\left(1 - \sum_{i=1}^{N} \pi_i \sum_{k=1}^{K_i} d_i^k\right) + \sum_{i=1}^{N} \beta_i \left[\sum_{k=1}^{K_i} d_i^k - 1\right]. \tag{6.18}$$

The main objective here is to select the d_i^k, $i = 1, 2, \ldots, N$ so as to obtain the largest possible increase in L.

Due to the fact that:

(a) $d_i^k = 0$ or 1, for unconstrained policies, according to our prior knowledge,

(b) the values of π_i are held constant during improvement of the policy,

we must be concerned only with the first term in relation (6.18).

In this case, we must maximize the quantity:

$$\sum_{i=1}^{N} \pi_i \sum_{k=1}^{K_i} \left(q_i^k + \sum_{j=1}^{N} p_{ij}^k v_j \right) d_i^k , \tag{6.19}$$

which is the inner product of two N-component vectors.

The first component is the vector $(\pi)\mathcal{P}$, as determined by the current policy $\mathcal{P}$. Its components are non-negative. The second vector is a variable. For each policy $\mathcal{P}'$, where $\mathcal{P}'(i) = k'$, the $d_i^{k'}$ satisfy (6.18), and $t_i^{k'} = [q_i^{k'} + \sum_{j=1}^{N} p_{ij}^{k'} v_j]$ is the i-th component of the vector selected by $\mathcal{P}'$. We shall denote the vector by $T(\mathcal{P}')$. Hence, the maximization of (6.19) is reduced to the selection of vector $T(\mathcal{P}')$. (policy $\mathcal{P}'$ which yields the largest value of inter-product with the constant vector $\Pi(\mathcal{P})$). This is a combinatorial problem.

For the case of no policy constraints, all possible vectors $T(\mathcal{P}')$ are feasible. There are $\prod\limits_{i=1}^{N} K_i$ such vectors and among them is that vector for which:

$$t_i = \max_{k=1,2,\ldots,K_i} \left[q_i^k + \sum_{j=1}^{N} p_{ij}^k v_j \right], \quad i = 1, 2, \ldots, N . \tag{6.20}$$

The maximization of (6.20) yields, for each i, an alternative K_i^*. Those alternatives make up a policy $\mathcal{P}^*$ with the corresponding vector $T(\mathcal{P}^*)$. Now consider the inner product of $T(\mathcal{P}^*)$ and $\Pi(\mathcal{P})$. Since each component of $T(\mathcal{P}^*)$ is greater than the corresponding components of all other $T(\mathcal{P})$ and, since the components of Π are non-negative, it immediately follows that $\mathcal{P}^*$ is the policy that maximizes (6.19). For unconstrained policies, the combinatorial problem of maximizing (6.19) reduces to the N discrete, "uncoupled", maximization problems (6.20). Equation (6.20) is the $\mathcal{P}I$ phase of the VD-$\mathcal{P}I$ algorithm. Thus, that phase is actually the maximization of L with respect to the d_i^k, using the values of Π and V for the current policy. This is represented in Figure 6.1.

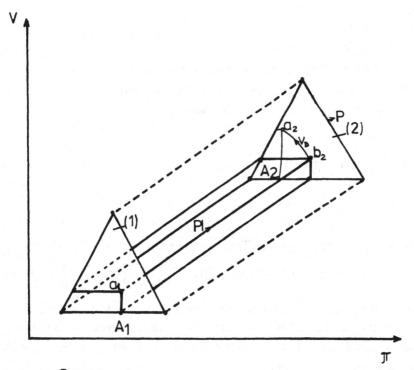

Figure 6.1. A $\mathcal{P}I$-VD iteration.

The variables were grouped into three axes. The $\mathcal{P}$'s are represented by an axis, as are the v 's and π 's. Since the v 's do not exist in the original constrained problem, its set of feasible points lie in the $(\Pi, \mathcal{P})$ plane. Moreover, since the constraints yield unique values for Π, the problem reduces to selecting from a discrete set of points A_i in that plane. Regarding the Lagrangian, for each policy $\mathcal{P}$, it is a function of Π and V. The VD-phase results in points whose Π is identical to that of the given policy, whence the points a_i have the same Π component as the points A_i. The $\mathcal{P}I$-phase consists of moving from a_i, say along the $\mathcal{P}$ direction, holding Π and V constant, to maximize L. This results in point b_2, say. The VD then takes us to a_2, from which we go into the $\mathcal{P}I$ etc.

The very fact that L is maximized along the "ray" emanating from a point a_i does not guarantee that the resultant a_j will give the highest gain possible on the iteration (i.e. the best improvement).

In Figure 6.2, the $\mathcal{P}I$ surveys the points b_2 through b_5 which lie on the "ray" emanating from a_1. It then selects that point b_1 at which L is maximum. For Figure 6.1, b_2 is that point. The VD then gives a_2. However, it cannot be proved that another point b_3, say, at which L is less that b_2, will necessarily yield a point a_3 for which the gain is less than at a_2. In other words, the ordering of the values of L at b_i is not necessarily identical to the ordering of the gain values at the corresponding a_i. Moreover, merely increasing L along the a_i "ray" does not guarantee a better policy. That guarantee is to be provided outside the Lagrangian framework.

The appropriateness of this procedure is justified for two reasons. First, we know that the optimum value of the constrained function we are trying to maximize is identical to that of L, as well as at all points a_i. Second, since we are trying to maximize our original function, it would pay to try to increase L as we go along. This gives us an insight into how to go about policy improvement when there are policy constraints, as long as we bear two things in mind.

REMARK 6.1. *The sole purpose of $\mathcal{P}I$ is to obtain a policy having a higher gain, irrespective of how much higher it is.*

For unconstrained policies, equation (6.19), it reduces to (6.20). which gave us a policy $\mathcal{P}^*$. If the policy constraints do not make $\mathcal{P}^*$ infeasible, then we select $\mathcal{P}^*$. Otherwise we have to solve the combinatorial problem of selecting the vector $T(\mathcal{P})$ from the feasible set of such vectors (corresponding to the feasible set of policies), which maximizes (6.19) and guarantees the increase in gain.

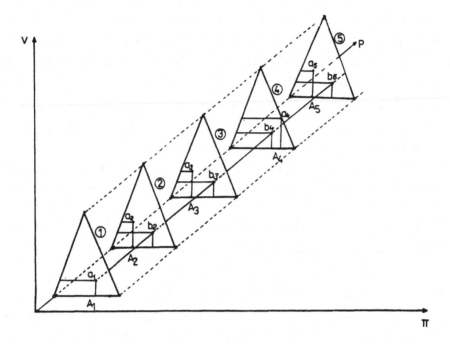

Figure 6.2. The policy improvement routine.

We can have an equivalent representation for (6.19):

$$\sum_{i=1}^{N} \pi_i \sum_{k=1}^{K_i} \left(q_i^k + \sum_{j=1}^{N} p_{ij}^k v_j \right) d_i^k = \sum_{i=1}^{N} \sum_{k=1}^{K_i} \pi_i \left(q_i^k + \sum_{j=1}^{N} p_{ij}^k v_j \right) d_i^k$$

and define:

$$t_i^k = q_i^k + \sum_{j=1}^{N} p_{ij}^k v_j \, .$$

We find that we are trying to maximize:

$$\sum_{i=1}^{N} \sum_{k=1}^{K_i} \pi_i t_i^k d_i^k \, . \tag{6.21}$$

Since d_i^k is zero for all k's but one in any state i, and the t_i^k are the PI test quantities, equation (6.21) tells us that we are trying to maximize the sum of those test quantities, one in each state, weighted by how much time the system spends, on the average, in each state. Maximizing the individual components of a sum, automatically maximizes the whole sum. In the absence of policy constraints, it is possible, as we have shown, to maximize the individual components.

By introducing constraints on the policies, we have to take feasibility into consideration. The states become "coupled" through the constraints, such that it might not be feasible to maximize the individual components independently of each other. It is the "non-coupling" of states which allows the reduction of (6.21) to (6.20). Thus, when policy constraints are present, we have to consider the sum as a whole and seek an efficient method of solving the non-reducible combinatorial problem.

Assume that we have a policy $\mathcal{P}$ for which the VD has been performed. Consider any other policy $\mathcal{P}'$. Each policy has a vector of test quantities T associated with it, where:

$$t_i^k(\mathcal{P}) = q_i^k(\mathcal{P}) + \sum_{j=1}^{N} p_{ij}^k(\mathcal{P}) v_j(\mathcal{P}) \quad i = 1, 2, \dots, N \tag{6.22}$$

$$t_i^k(\mathcal{P}') = q_i^k(\mathcal{P}') + \sum_{j=1}^{N} p_{ij}^k(\mathcal{P}') v_j(\mathcal{P}') \quad i = 1, 2, \dots, N \tag{6.23}$$

where the v_j's are those obtained form the VD policy $\mathcal{P}$. Specifically:

$$q_i^k(\mathcal{P}) + \sum_{j=1}^{N} p_{ij}^k(\mathcal{P}) v_j(\mathcal{P}) = v_i(\mathcal{P}) + g(\mathcal{P}) \quad i = 1, 2, \dots, N \tag{6.24}$$

· Thus, (6.22) reduces to:

$$t_i^k(\mathcal{P}) = v_i(\mathcal{P}) + g(\mathcal{P}). \tag{6.25}$$

Had we solved the policy for $\mathcal{P}'$, we could have had:

$$q_i^k(\mathcal{P}') + \sum_{j=1}^{N} p_{ij}^k(\mathcal{P}') v_j(\mathcal{P}') = v_i(\mathcal{P}') + g(\mathcal{P}') \quad i = 1, 2, \dots, N. \tag{6.26}$$

Combining (6.26) and (6.23), we can write:

$$t_i^k(\mathcal{P}') = v_i(\mathcal{P}') + g(\mathcal{P}') + \sum_{j=1}^{N} p_{ij}^k(\mathcal{P}') \left[v_j(\mathcal{P}) - v_j(\mathcal{P}') \right] \quad i = 1, 2, \dots, N. \tag{6.27}$$

Now compute the components γ_i of the difference vector $\gamma = T(\mathcal{P}') - T(\mathcal{P})$ from (6.24 and (6.25):

$$\gamma_i^k = t_i^k(\mathcal{P}') - t_i^k(\mathcal{P}) =$$

$$= v_i(\mathcal{P}') + g(\mathcal{P}') + \sum_{j=1}^{N} p_{ij}^k(\mathcal{P}') \left[v_j(\mathcal{P}) - v_j(\mathcal{P}') \right] - v_i(\mathcal{P}) - g(\mathcal{P}).$$

Setting

$$\Delta v_i = v_i(\mathcal{P}') - v_i(\mathcal{P})$$

$$\Delta g = g(\mathcal{P}') - g(\mathcal{P})$$

and rearranging terms, we get:

$$\Delta g + \Delta v_i = \gamma_i + \sum_{j=1}^{N} p_{ij}^k(\mathcal{P}') \Delta v_j. \tag{6.28}$$

Equation (6.28) has exactly the same form as (6.17) where the correspondence is:

$$g \leftrightarrow \Delta g, \quad v \leftrightarrow \Delta v, \quad \gamma \leftrightarrow q.$$

The solution for g was $\sum \pi_i q_i$, where the π's are the limiting state probabilities of the policy under consideration. We can immediately write the solution of (6.28) as:

$$g(\mathcal{P}') - g(\mathcal{P}) = \Delta g = \sum_{i=1}^{N} \pi_i(\mathcal{P}') \gamma_i. \tag{6.29}$$

Since the π's are non-negative the above equation states that the sufficient condition for improving the gain is that γ_i be non-negative for all states and strictly positive in at least one state. We do not have to solve the VD for every alternative policy $\mathcal{P}'$. Without knowing $\Pi(\mathcal{P}')$, we can guarantee that $\mathcal{P}'$ is better than $\mathcal{P}$ if at least one γ_i is positive. Other discussions are presented in [103].

The best procedure for guaranteeing a gain improvement is to improve the test quantity in at least one state. In the presence of policy constraints, we have to maximize (6.21), subject to improving at least one state.

6.2.2. DEVELOPMENT AND CONVERGENCE OF THE ALGORITHM

The algorithm consists of the *VD* intact, plus a modified $\mathcal{P}I$ to handle policy constraints, so that:

$$\max_{\mathcal{P}\in F} \sum_{i=1}^{N} \sum_{k=1}^{K_i} \pi_i t_i^k d_i^k \tag{6.30}$$

subject to:

$$\left.\begin{array}{ll} c_l(d_i^k) \leq 0 & l = 1, 2, \dots, m \quad (i, k) \in S_1 \\ c_p(d_i^k) = 0 & p = 1, 2, \dots, q \quad (i, k) \in S_2 \end{array}\right\} \tag{6.31}$$

where all the inequality type of constraints have been grouped together.

One efficient technique for solving combinatorial problems is the branch and bound method. The space over which we are optimizing is divided into sub-sets in such a manner that only a portion of the whole space is examined. We shall develop a method based on these concepts.

We can consider a case with three states, with three alternatives to choose between in each state. Assume, furthermore, that the *VD* has been carried out for a given feasible policy, resulting in the test quantities, t_i^k. We shall consider the t_i^k multiplied by the corresponding π's amd list them, as in Table 6.1, where they are in descending order in each state.

State	Ordered Test Quantities		
1	t_1^2	t_1^3	t_1^1
2	t_2^1	t_2^3	t_2^2
3	t_3^2	t_3^1	t_3^3

Table 6.1. Table of ordered test quantities.

The above table illustrates the policy constraints. Constraints are of the simple mutually exclusive type. Corresponding to the "couplings" in Table 6.1, we can have the constraints:

$$d_1^2 + d_2^1 \le 1 \tag{6.32}$$

$$d_1^1 + d_2^3 \le 1 \tag{6.33}$$

$$d_2^3 + d_3^2 \le 1 \tag{6.34}$$

$$d_2^3 + d_3^1 \le 1 . \tag{6.35}$$

Now we start "branching and bounding" to maximize (6.30), subject to (6.32) to (6.35). The lower bound is initialized to the artificial value of $-\infty$.

Whenever branching gives us a feasible point, the value of L (the function we are trying to maximize) is compared to the current lower bound. If it exceeds that bound, the bound is updated, and any point yielding a value fo L which is lower than the new bound is fathomed. Feasible points are fathomed by definition. The search terminates when no more unfathomed points exist. We start out with the point representing that T vector whose components are given by (6.20). Denote it by T_{01}. Our tree then consists, initially, of one node:

where $T_{01} = (t_1^2, t_2^1, t_3^2)$.

The corresponding policy is $\mathcal{P}_{01} = (2, 1, 2)$; this is the policy the $\mathcal{P}I$ would select. However, in our example, it is infeasible. The lower bound remains at its initial value of $-\infty$, and we have one unfathomed point T_{01} to branch from. Branching consists of selecting each alternative in state 1, in turn, by changing the first component of T_{01}. The remainder of the components are chosen such that they are the largest test quantities in their states, consistent with the constraints imposed by the first component, if any. Thus, selecting t_1^2, for example, would prohibit selecting t_2^1, and hence we have to select t_2^3 instead. For the third component, we can select t_3^3 because it is not "coupled" with t_1^2. The fact that t_2^3 and t_3^3 are coupled is postponed to the next level of branching, if we get there. Our tree becomes:

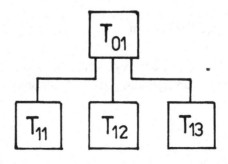

where $T_{11} = (t_1^2, t_2^3, t_3^2)$; $T_{12} = (t_1^3, t_2^1, t_3^2)$; $T_{13} = (t_1^1, t_2^1, t_3^2)$.

The corresponding policies are: $\mathcal{P}_{11} = (2, 3, 2)$; $\mathcal{P}_{12} = (3, 1, 2)$; $\mathcal{P}_{13} = (1, 1, 2)$.

Only $\mathcal{P}_{11}$ is in feasible; the value of the Lagrangian at those two points (T_{12}, T_{13}) is compared to the lower bound $(-\infty)$ and to that at T_{11}. Assume that $L(T_{11}) > L(T_{13}) > L(T_{12})$. In this case the lower bound is updated to $L(T_{13})$ and T_{11} is the only point available for branching. The situation becomes:

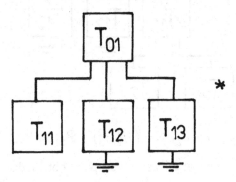

Moreover, $\mathcal{P}_{13}$ is the optimum policy so far. Now we start branching from T_{11}, as we did from T_{01}. Here, we select, in state 2, each alternative in turn. However, the alternatives must be uncoupled from t_1^2. In other words, at each level in the tree, we have a fixed alternative in a number of states. T_{01} represents level 0. No states have fixed alternatives. The next level of T_{11}, T_{12}, and T_{13} has the alternative in state 1 fixed (this is level 1). Only T_{11} goes down one further level (to level 2) to attempt fixing alternatives in state 2, consistent with the constraints imposed by the alternative fixed in the previous level. Since T_{11} has t_2^3 fixed, and t_1^2 is coupled with t_2^1, we cannot fix the latter. Thus, we only have two successor points to T_{11}. The remaining components are selected such that they are the maximum test quantities in their states, under the restriction that they satisfy any constraints imposed by fixing the previous levels. Since we are down to the last level, the points we obtain, if any, have to be feasible. Thus, fixing t_2^3 imposes selecting t_3^3, while fixing t_2^2 enables us to select t_3^2 Note, that, if t_2^3 were also coupled with t_3^3, we would have only one successor point to T_{11}. If, in addition, t_2^2 were coupled with all alternatives in state 3, no branching would be possible from T_{11}, and it would be fathomed. These are interesting cases because such constraints imply that t_2^3 and t_2^2 are not really alternatives at all; they can never be selected in a feasible policy. This can be detected by manipulating the constraints to discover that they impose

$d_2^3 = d_2^2 = 0$. However, we are more inclined to let the branch and bound discover this (along with non-feasible problems). Thus, at this step, our tree would become:

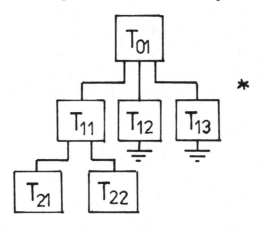

where $T_{21} = (t_1^2, t_2^3, t_3^3)$; $T_{22} = (t_1^2, t_2^2, t_3^2)$; $\mathcal{P}_{21} = (2, 3, 3)$; $\mathcal{P}_{22} = (2, 2, 2)$.

The * next to T_{13} implies that this is the best point obtained so far.

Both $\mathcal{P}_{21}$ and $\mathcal{P}_{22}$ are feasible, whence we compute $L(T_{21})$ and $L(T_{22})$ and compare them to the current bound (the best value of L so far). Also, both points are fathomed. Assume that $L(T_{22}) > L(T_{21}) > L$.

In this case, we update L, set $\mathcal{P}_{22}$ as our optimum so far, and survey the tree for unfathomed points:

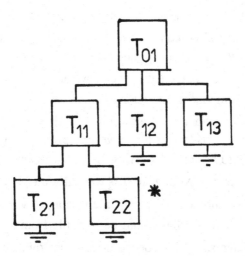

Since no more branching is possible, T_{22} is the optimum. The problem is infeasible if the lower bound is still at its original value of $-\infty$.

The efficiency of branch and bound techniques results from the manner in which branching and fathoming are implemented.

The branch and bound (BB) method maximizes the Lagrangian over the set of feasible policies. A set of propositions will be given without proof (see [103]).

PROPOSITION 6.5. *If branching occurs from some infeasible node I at level k in the tree, the value of L at the resultant nodes cannot exceed that at I.*

PROPOSITION 6.6. *If a given feasible policy $\mathcal{P}$ is not in the BB tree and there exists a fathomed policy $\mathcal{P}'$ in the tree at level k, such that $\mathcal{P}'$ and $\mathcal{P}$ agree in the first k components, then the value of L at $\mathcal{P}$ cannot exceed the optimum value obtained by the BB.*

PROPOSITION 6.7. *If a given feasible policy $\mathcal{P}$ is not in the BB tree and there exists an unfathomed policy I in the tree at level k, such that $\mathcal{P}$ and I agree in the first k components, then there exists a policy $\mathcal{P}'$ in the tree at level k + 1 such that $\mathcal{P}$ and $\mathcal{P}'$ agree in the first k + 1 components.*

PROPOSITION 6.8. *If a feasible policy $\mathcal{P}$ is not in the BB tree and there exists a node N in the tree at level k such that N and $\mathcal{P}$ agree in the first k components, then the value of L at $\mathcal{P}$ cannot exceed the optimum obtained by BB.*

PROPOSITION 6.9. *No feasible policy $\mathcal{P}$ can yield a value of L greater thant the optimum obtained by BB.*

Now that we have a method for solving the combinatorial problem of maximizing the Lagrangian over a set of feasible policies, we have to guarantee an improvement in the gain. To ensure this, we do not allow negative γ's in each iteration.

The obvious question is what if the BB results in the same policy we entered with? What is the characteristic of such a policy? In the following theorem, we show that it maximizes the gain over a sub-set of the feasible policies.

THEOREM 6.2. *If the maximization of the Lagrangian over the set of feasible policies, subject to $\gamma_i \geq 0$ for all i, yields a policy $\mathcal{P}$ for which $\gamma_i = 0$ for all i, then that policy maximizes the gain over the set of all feasible policies that differ with it in exactly one state.*

Proof. Let $\mathcal{P}'$ be a feasible policy differing from $\mathcal{P}$ in the k-th component (i.e. state). Let $\mathcal{P}(k) = l$ and $\mathcal{P}'(k) = m$. Assume that $t_k^m > t_k^l$. Since $T(\mathcal{P})$ differs from $T(\mathcal{P}')$

only in the k-th component, the Lagrangian at $\mathcal{P}'$, is greater than at $\mathcal{P}$. This implies that there exists a feasible policy $\mathcal{P}'$ yielding a value of L greater than the optimum obtained by BB. But this contradicts Proposition 6.9.

Hence:

$$t_k^m \le t_k^l .$$

Thus:

$$\gamma_k \le 0 \quad \text{and} \quad \gamma_i = 0 \quad \text{for } i \ne k$$

where

$$\gamma = T(\mathcal{P}') - T(\mathcal{P}).$$

Hence:

$$g(\mathcal{P}') - g(\mathcal{P}) = \sum_{i=1}^{N} \pi_i (\mathcal{P}') \gamma_i = \pi_k \gamma_k \le 0 ,$$

i.e.

$$g(\mathcal{P}') \le g(\mathcal{P}).$$

Thus, any feasible policy differening with $\mathcal{P}$ in exactly one state cannot have a higher gain then $\mathcal{P}$.

If the policy $\mathcal{P}$ is given by $\mathcal{P}(i) = k$, the constraint is:

$$\sum_{i=1}^{N} d_i^k \le N - 2 . \tag{6.36}$$

In order to increase computational efficiency, we divide the states into two types. The first category is that of "free states" which are not involved in any policy constraints. The states, having alternatives that are "coupled" with each other, we refer to as "coupled states". Given a feasible policy A for which VD has been performed, we first attempt a regular PI. If this does not change A, we have an overall optimum policy. In the case that the resultant policy is feasible, we start a new VD. Otherwise, we maximize over the free states by regular PI, and over coupled states by BB with $\gamma_i > 0$. This results in a policy B. If B differs from A, we enter VD. Otherwise, we know that A maximizes the gain over the set of feasible policies differing with A in exactly one coupled state. In this case, we make policy A, as well as all policies differing from it in exactly one coupled state, infeasible by a (6.36) type constraint (N is the number of coupled states). Then we enter BB again to maximize the Lagrangian over the current feasible policy set, with the $\gamma_i \geq 0$ restriction removed. Basically, we are looking for a feasible policy, irrespective of gain, whence it makes sense to use the latest values of t_i^k since they contain a certain amount of the algorithm's history up to this point.

To get an initial feasible policy, we maximize the sum of the immediate expected rewards q_i^k over the feasible policy set using BB. Schematically, then, our algorithm can be represented in Figure 6.3. The convergence of this algorithm to an optimum feasible policy is readily proved.

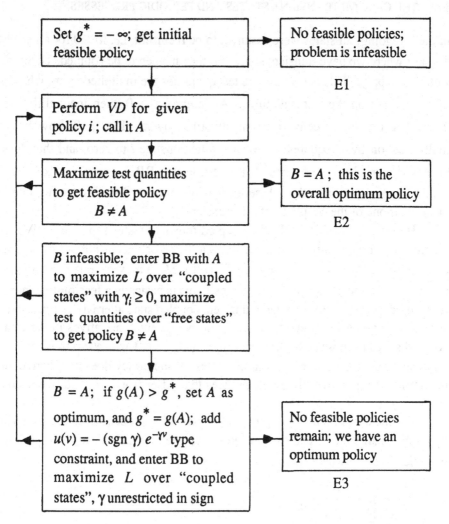

Figure 6.3. Algorithm for risk-indifferent case.

If we exited from the algorithm from E3, then at that point, the given feasible policy $\mathcal{P}$ had become infeasible. The only way this can come about is from (6.36) type constraints.

Since no feasible policy can have a higher gain than that of the policy selected by the algorithm, the latter is the optimal feasible policy.

An extensive example, starting from Howard's taxicab problem, is given in [103].

6.2.3. THE CASE OF TRANSIENT STATES AND PERIODIC PROCESSES

In the foregoing, all states were assumed to be recurrent, i.e. $\pi_i > 0$ for all states. In the case where we have transient states and they happen to be coupled, Theorem 6.1. would not apply. This is because the test quantities are multplied by π_i, whereas the π_i are defined on the test quantities. As long as $\pi_i > 0$, then inequalities of test quantities are not affected by multiplication by π_i. If $\pi_i = 0$, however, then multiplication by π_i equates all test quantities in state i to zero, and that does not necessarily imply γ_i's of zero. It is not true that BB converges to a policy that maximizes the gain over the sub-set of feasible policies differing from it in exactly one state if one of the coupled states is transient.

Hence, we need to deal with coupled transient states separately. We fix the alternatives in the recurrent coupled states to those of the current policy (i.e. set the corresponding $d_i^k = 1$). Then we use BB with $\gamma \ge 0$ to maximize the sum of the test quantities over the transient coupled states. This is basically a feasibility exploration. If this process results in a change in transient coupled state alternatives (in at least one state), we have an improved policy. If not, we fix the alternatives in the transient coupled states to those of the current policy and use BB with $\gamma \ge 0$ to maximize the Lagrangian over the recurrent coupled states. If the policy does not change, then it maximizes the gain over the set of policies that differ from it in exctly one coupled state.

For the case of a periodic process (the transition probability matrix is periodic), $\pi_i = 1/N$ for all states. The case of interest is one in which all feasible policies are periodic. We know that:

$$g = \sum_{i=1}^{N} \pi_i \sum_{k=1}^{K_i} q_i^k d_i^k$$

and it is g we want to maximize. If all policies are periodic, then we want to maximize:

$$1/N \sum_{i=1}^{N} \sum_{k=1}^{K_i} q_i^k d_i^k .$$

In this case, we only need one iteration. Our initial feasible policy is the optimum one, since we use BB to maximize the sum of q_i^k over the feasible policy set to get it. This is an inherently deterministic problem; in the case where we know *a prior* that all feasible policies are periodic, we do not seek to find that out. The only reason for discussing periodic policies, is to shed some light on initial feasible policies and transient coupled states. Maximizing a sum is equivalent to maximizing the average if a uniform probability distribution is assumed. The latter is characteristic of the π_i of periodic policies. In the case of periodic policies, we do not know where we shall find the process if we enter it at a random point in time. This is essentially what we are saying at the start of the algorithm when we do not have any feasible policy available.

6.3 Risk-sensitive Markov decision process with policy constraints

6.3.1. A LAGRANGE MULTIPLIER FORMULATION

Each policy has a disutility contribution matrix Q associated with it. The maximal eigenvalue λ^M of Q determines the certain equivalent gain of the policy $\tilde{g} = -1/\gamma \ln \lambda^M$. We are confronted with the task of selecting that Q which yields the largest value of $\tilde{g}$. For a risk preferring decision maker, γ is negative. Consequently, maximizing $\tilde{g}$ is the same as maximizing λ^M. For a risk averse decision maker, however, γ is positive, and it is the smallest λ^M which gives the largest $\tilde{g}$.

We have, thus, ascertained that our objective function is λ^M, the maximal eigenvalue of Q, whence our constraints are those defining the eigenvalue problem that yields λ^M. To gain more insight into this, we reconsider the risk-indifferent case. If we write the constraints defining the limiting state probabilities π for a transition probability matrix P in vector form, we get:

$$\pi^T \cdot P = \pi^T . \tag{6.37}$$

It is then apparent that π is a "left eigenvalue" of P (or eigenvector of its transpose) with a corresponding eigenvalue of unity. However, that eigenvalue is the maximal eigenvalue of P by virtue of it being a stochastic matrix. This holds for any policy.

Therefore, in the risk-indifferent case, all maximal eigenvalues are unity. The corresponding left eigenvectors then define the objective function through:

$$g = \sum_{i=1}^{N} \pi_i q_i .$$

In the risk-sensitive case, we deal with the Q matrices. To get the constraints, we define vector Z, with components z_i, as the left eigenvector of Q. Thus, Z is the counterpart of π, and it should satisfy

$$Z^T \cdot Q = \lambda^M \cdot Z^T . \tag{6.38}$$

To take alternative selection into account, we use the d_i^k on the rows of Q, as we did in the risk-indifferent case We shall first concentrate on the risk preferring case, i.e. $\gamma < 0$. Here we are dealing with the constrained maximization problem:

$$\max \lambda^M \tag{6.39}$$

subject to:

$$\sum_{i=1}^{N} z_i \sum_{k=1}^{K_i} q_{ij}^k d_j^k - \lambda^M x_j = 0 , \quad j = 1, 2, \dots , N \tag{6.40}$$

$$\sum_{k=1}^{K_i} d_i^k - 1 = 0 , \qquad\qquad i = 1, 2, \dots , N \tag{6.41}$$

$$\left. \begin{array}{ll} c_l(d_i^k) \leq 0 & l = 1, 2, \dots , m \quad (i, k) \in S_1 \\[2mm] c_p(d_i^k) = 0 & p = 1, 2, \dots , q \quad (i, k) \in S_2 \end{array} \right\} \tag{6.42}$$

where λ^M is the largest positive number satisfying (6.40) and S_1 and S_2 are sub-sets of $S = \{(i, k)\}$ the set of all $(i. k)$ pairs defining each alternative in each state.

As noted before, a set of d_i^k describing a policy $\mathcal{P}$ determines a disutility contribution matrix $Q(P)$ and its associated left eigenvector Z defined by:

$$Z^T \cdot Q(P) = \lambda^M \cdot Z^T . \tag{6.43}$$

irreducible primitive matrix Q, the components of the maximal eigenvector all have the same sign. They are either all positive or all negative.

First, we concentrate on unconstrained policies. We showed that the maximizer of the constrained problem is a critical point of the corresponding Lagrangian. We have $2N$ Lagrange multipliers; the first N of those, associated with (6.40), we denote by $u_1, u_2, \ldots, u_N$. The other N, we denote by β_i. Consequently, our Lagrangian is:

$$L = \lambda^M + \sum_{j=1}^{N} \left[\sum_{i=1}^{N} z_i \sum_{k=1}^{K_i} q_i^k d_i^k - \lambda^M z_j \right] u_j + \sum_{i=1}^{N} \beta_i \left[\sum_{i=1}^{N} d_i^k - 1 \right] . \qquad (6.44)$$

To maximize our constrained function, we seek critical points of L. As before, we only set certain partial derivatives to zero to evaluate a policy, then we try to change the policy so as to maximize L (because the values of the constrained function and L are identical whenever a policy is evaluated).

Setting the partial derivatives of L w.r.t. the z_i and λ^M equal to zero, we can write:

$$\frac{\partial L}{\partial z_i} = \sum_{j=1}^{N} u_j \sum_{k=1}^{K_i} q_{ij}^k d_i^k - \lambda^M u_i = 0 \qquad i = 1, 2, \ldots, N \qquad (6.45)$$

$$\frac{\partial L}{\partial \lambda^M} = 1 - \sum_{j=1}^{N} z_j u_j = 0. \qquad (6.46)$$

For a given policy $\mathcal{P}$, the d_i^k are all zero or unity with only one d_i^k equal to unity in each state. In (6.45), the summation over k reduces to one element for each i. This defines the $Q(P)$ of the policy under consideration. Defining U as the vector whose components are u_i, we have:

$$Q(P) \cdot U = \lambda^M U \qquad (6.47)$$

$$Z^T \cdot U = 1. \qquad (6.48)$$

Differentiating L w.r.t. the u_i and equating to zero gives us the original constraints (6.43):

$$Z^T \cdot Q(P) = \lambda^M \cdot Z^T .$$

At this point, there is nothing to indicate that setting u_N to some particular value has any significance. We know, however, that U, as well as Z, has components which are either all positive or all negative. Equation (6.48) tells us that Z and U have the same signs. This will become significant in the PI phase.

In matrix form (6.47) can be written:

$$
\begin{bmatrix}
q_{11}-\lambda^M & q_{12} & \cdot & \cdot & q_{1N} \\
q_{21} & q_{22}-\lambda^M & \cdot & \cdot & q_{2N} \\
\cdot & \cdot & \cdot & \cdot & \cdot \\
\cdot & \cdot & \cdot & \cdot & \cdot \\
q_{N1} & q_{N2} & \cdot & \cdot & q_{NN}-\lambda^M
\end{bmatrix}
\begin{bmatrix}
u_1 \\ u_2 \\ \cdot \\ \cdot \\ u_N
\end{bmatrix}
=
\begin{bmatrix}
0 \\ 0 \\ \cdot \\ \cdot \\ 1
\end{bmatrix}
$$

This is a system of rank $N-1$. If we set $u_N = a$, say, we get:

$$
\begin{bmatrix}
q_{11}-\lambda^M & q_{12} & \cdot & \cdot & q_{1,N-1} \\
q_{21} & q_{22}-\lambda^M & \cdot & \cdot & q_{2,N-1} \\
\cdot & \cdot & \cdot & \cdot & \cdot \\
\cdot & \cdot & \cdot & \cdot & \cdot \\
q_{N-1,1} & q_{N-1,2} & \cdot & \cdot & q_{N-1,N-1}-\lambda^M
\end{bmatrix}
\begin{bmatrix}
u_1 \\ u_2 \\ \cdot \\ \cdot \\ u_{N-1}
\end{bmatrix}
= -a
\begin{bmatrix}
q_{1N} \\ q_{2N} \\ \cdot \\ \cdot \\ q_{N-1,N}
\end{bmatrix}
$$

$$(6.49)$$

where the last line has been omitted.

The resulting system can be written as $A \cdot \underline{U} = V$, where A is the matrix on the left hand side of (6.49), and V is the vector on the right hand side. $\underline{U}$ is the $N-1$ vector composed ot the first $N-1$ components of U. Hence:

$$\underline{U} = A^{-1} \cdot V. \tag{6.50}$$

Now we write (6.43) in matrix form:

$$
\begin{bmatrix}
q_{11}-\lambda^M & q_{21} & \cdot & \cdot & q_{N1} \\
q_{12} & q_{22}-\lambda^M & \cdot & \cdot & q_{N2} \\
\cdot & \cdot & \cdot & \cdot & \cdot \\
\cdot & \cdot & \cdot & \cdot & \cdot \\
q_{1N} & q_{2N} & \cdot & \cdot & q_{NN}-\lambda^M
\end{bmatrix}
\begin{bmatrix}
z_1 \\ z_2 \\ \cdot \\ \cdot \\ z_N
\end{bmatrix}
=
\begin{bmatrix}
0 \\ 0 \\ \cdot \\ \cdot \\ 0
\end{bmatrix}
$$

This being a system of rank $N - 1$, we can set $z_N = b$, say, to get:

$$
\begin{bmatrix}
q_{11} - \lambda^M & q_{21} & \cdot & \cdot & q_{N-1,1} \\
q_{12} & q_{22} - \lambda^M & \cdot & \cdot & q_{N-1,2} \\
\cdot & \cdot & \cdot & \cdot & \cdot \\
\cdot & \cdot & \cdot & \cdot & \cdot \\
q_{1N} & q_{2N} & \cdot & \cdot & q_{N-1,N-1} - \lambda^M
\end{bmatrix}
\begin{bmatrix}
z_1 \\ z_2 \\ \cdot \\ \cdot \\ z_{N-1}
\end{bmatrix}
= -b
\begin{bmatrix}
q_{N1} \\ q_{N2} \\ \cdot \\ \cdot \\ q_{N,N-1}
\end{bmatrix}
$$

$$(6.51)$$

Denoting the $N - 1$ vector whose components are $z_1, \ldots, z_{N-1}$ by $\underline{Z}$ and the right hand side vector of (6.51) by W, we get

$$A^T \cdot \underline{Z} = W$$

where A is the same matrix as in (6.49). Thus:

$$\underline{Z} = (A^T)^{-1} \cdot W. \qquad (6.52)$$

Solving (6.50) implies that (6.52) is solved. All we need to get Z, once we have U, is to transpose the inverse we already have and multiply it by the transpose of the last row of A (and the value of z_N). This is significant for the case of policy constraints.

Now we interpret the Lagrange multipliers u_i and the z_i. Here the Z plays the role that π plays in the risk-indifferent case. Note that the constraints defining the limiting state probabilities are:

$$\pi^T \cdot P = \pi^T$$

$$\pi^T \cdot \mathbf{1} = 1$$

where $\mathbf{1}$ is the vector whose components are all unity. That vector is the eigenvector of P corresponding to π:

$$P \cdot \mathbf{1} = 1.$$

Moreover, the common eigenvalue is unity, the maximal eigenvalue for any transition probability matrix P. Hence, (6.43) represents the "equilibrium of disutility contribution flows" in the limiting case, exactly like the constraints on the limiting state probabilities did for the probabilistic flows. The u_i being Lagrange multipliers, give us the cost of violating the constraints per unit of disequilibrium, just as the v's in the risk-indifferent case.

Once we have evaluated a policy, we want to improve it. We make maximizing the Lagrangian our objective. To do that, we rewrite L to bring out the dependence on d_i^k.

$$L = \sum_{i=1}^{N} z_i \sum_{j=1}^{N} u_j \sum_{k=1}^{K_i} q_{ij}^k d_i^k + \lambda^M \sum_{j=1}^{N} z_j u_j + \sum_{i=1}^{N} \beta_i \left[\sum_{k=1}^{K_i} d_i^k - 1\right].$$

For the given policy $\sum_{j=1}^{N} z_j u_j = 1$, and for any policy $\sum_{k=1}^{K_i} d_i^k = 1$, whence we only have to contend with the first term:

$$\sum_{i=1}^{N} z_i \sum_{j=1}^{N} u_j \sum_{k=1}^{K_i} q_{ij}^k d_i^k. \tag{6.53}$$

The last relation is the inner product of two vectors. The first vector Z is constant. The second is policy dependent. Once a policy $\mathcal{P}$ is selected, it defines a vector $T(\mathcal{P})$ of "test quantities":

$$t_i^k = \sum_{j=1}^{N} q_{ij}^k u_j \tag{6.54}$$

where the q_{ij}^k is the corresponding element of the Q selected by policy $\mathcal{P}$. The sign of t_i^k is the same as that of the U, and hence also Z. Thus, if Z and U are chosen positive ($u_N = -(\text{sgn } \gamma)$ because we are dealing with the case $\gamma < 0$), then to maximize L, we have to select that feasible $T(\mathcal{P})$ with the largest components. Otherwise, we would have to select that with the smallest components (if Z and U are negative). In other words, if the sign of u_N is the same as that of γ, the test quantity has to be *minimized* in order to maximize the Lagrangian. We shall assume, hereafter, that in *PE* we take $u_N = z_N = -(\text{sgn } \gamma)$.

In the absence of policy constraints, we can select:

$$t_i = \max_{k=1,2,\ldots,K_i} \left[\sum_{j=1}^{N} q_{ij}^k u_j\right]. \tag{6.55}$$

If we have policy constraints we redefine the components of T to be:

$$t_i = z_i \sum_{j=1}^{N} q_{ij}^k u_j. \tag{6.56}$$

The original test quantities have been weighted by the corresponding z_i, the variables controlling the equilibrium of disutility contribution flows. Whereas, in the risk-indifferent case, weighting test quantities by the limiting state probabilities was intuitive, this weighting cannot be intuitively derived in the risk-sensitive case. Z encodes the limiting behaviour from both the probabilistic and risk attitude aspects.

In the absence of policy constraints, we can set $u_N = (\text{sgn } \gamma) = -1$ but take care to multiply the test quantities by -1 before maximizing in each state. This is just another way of saying that u_N *has* to be positive when γ is negative.

Equation (6.47) is not a "free" eigenvalue problem, in the sense that we are not free to choose *any* value for one of the u_i. The values *have* to be of different sign than γ in the absence of policy constraints in order to implement the *PI*. Otherwise, the test quantity has to be *minimized*. It was the realization that *PI* is maximization of the Lagrangian which led us to detect this dependence on sign γ.

For the case of policy constraints, we need not worry about the signs. We shall select $z_N = u_N = - (\text{sgn } \gamma)$ for consistency. For constrained policies, we maximize the Lagrangian by BB , as before.

For the case of risk aversion, positive γ, we previously mentioned that we need to minimize λ^M in order to maximize $\tilde{g}$. What applies to constrained maximization, applies to constrained minimization, as far as the Lagrange multiplier rule is concerned.

The minimizer of the constrained problem is still a critical point of the Lagrangian. In the absence of policy constraints, setting $u_N = - (\text{sgn } \gamma)$ makes U and Z both negative. Thus, to minimize the inner product, the variable vector composed of test quantities has to be *maximized*. Setting $u_N = - (\text{sgn } \gamma)$ in the unconstrained policies case, and then maximizing the resultant test quantities, actually *minimizes* the Lagrangian. For constrained policies, we minimize the Lagrangian without worrying about signs, because we multiply the test quantities by z_i. A better approach would be to take the sign of γ into considration, explicity. This could be done by redefining the test quantities as:

$$t_i^k = - (\text{sgn } \gamma) z_i \sum_{j=1}^{N} q_{ij}^k u_j. \tag{6.57}$$

For a risk preference person ($\gamma < 0$), this reduces (6.56), whence all the previous applies. For $\gamma > 0$, (6.57) effectively multiplies the Lagrangian by -1. Since we want to minimize the latter, we can maximize the former. Consequently, *PI* becomes maximization, irrespective of the sign of γ, if we define the test quantities by (6.57).

To summarize, setting $u_N = -(\text{sgn } \gamma)$ makes *PI* maximize test quantities in the absence of policy constraints. While this involves dependence on the sign of γ, it automatically takes into account the fact that in one case we want to maximize L and, in the other, minimize it. When policy constraints are present, we have to explicitly take the sign of γ into account by defining the test quantities as in (6.57), whence we always maximize L in the *PI*.

6.3.2. DEVELOPMENT AND CONVERGENCE OF THE ALGORITHM

We must introduce a sufficient condition for improving $\tilde{g}$, based on a condition derived by Howard and Matheson. If we have a policy A, then the test quantity corresponding to the alternative selected by another policy B, in state i, is defined as:

$$t_i(B) = -(\text{sgn } \gamma) \, z_i^A \sum_{j=1}^{N} q_{ij}^B \, u_j^A \qquad (6.58)$$

where z_i^A and u_j^A are the values obtained by the *PE* for policy A. We shall define Δ_i's analogous to the γ_i's of the risk-indifferent case by:

$$\Delta_i(B, A) = t_i(B) - t_i(A) \qquad (6.59)$$

where the t_i are defined by (6.58).

PROPOSITION 6.10. *Given a policy A for which PE has been performed, and any other policy B, a sufficient condition for $\tilde{g}^B > \tilde{g}^A$ is :*

$$\Delta_i(B, A) \geq 0 \quad i = 1, 2, \dots, N \qquad (6.60)$$

with inequality holding for at least one state i.

Proof. $\tilde{g} = -1/\gamma \ln \lambda$, where λ is the maximal eigenvalue of Q. We need to prove that $\lambda^B > \lambda^A$ for $\gamma < 0$ and vice versa.

From matrix theory, if Q is a non-negative irreducible matrix with maximal eigenvalue λ and x is a vector with positive components x_i, then:

$$\min_i \frac{\sum_{j=1}^{N} q_{ij} x_j}{x_i} \leq \lambda \leq \max_i \frac{\sum_{j=1}^{N} q_{ij} x_j}{x_i} \tag{6.61}$$

with equality holding if, and only if, x is an eigenvector of Q (see also Chapter 4).

Now we consider the implications of (6.60). By virtue of (6.59), we can write $t_i(B) \geq t_i(A)$ with inequalty holding in at least one state. Using (6.58), this reduces to:

$$-(\text{sgn } \gamma) \, z_i^A \sum_{j=1}^{N} q_{ij}^B u_j^A \geq -(\text{sgn } \gamma) \, z_i^A \sum_{j=1}^{N} q_{ij}^A u_j^A \, .$$

Since, from PE, z_i and γ have different signs, the product $-(\text{sgn } \gamma) \, z_i$ is always positive, and we can divide both sides of the inequality by that quantity without reversing its sense. Also, from PE, U is an eigenvector of Q^A, whence:

$$\sum_{j=1}^{N} q_{ij}^B u_j^A \geq \lambda^A u_i^A \tag{6.62}$$

with inequality holding in at least one state i. If U is also an eigenvector of Q^B, (6.62) reduces to $\lambda^B u_i^A \geq \lambda^A u_i^A$ with $\lambda^B u_i^A > \lambda^A u_i^A$ for some i.

For $\gamma < 0$, $u_i > 0$ and $\lambda^B > \lambda^A$, and for $\gamma > 0$, $u_i < 0$ and $\lambda^B < \lambda^A$.

If U is not an eigenvector of Q^B, we first consider $\gamma < 0$, then $\gamma > 0$.

For $\gamma < 0$, $u_j > 0$, and (6.62) can be rewritten as:

$$\frac{\sum_{j=1}^{N} q_{ij}^B u_j^A}{u_i^A} \geq \lambda^A \, .$$

Since this holds for all i, then it is certainly true that:

$$\min_i \sum_{j=1}^{N} \frac{q_{ij}^B u_j^A}{u_i^A} \geq \lambda^A \, .$$

Applying (6.61) to the left hand side of this equality, noting that U is not an eigenvector of Q^B, we get $\lambda^B > \lambda^A$.

For $\gamma < 0$, $u_i < 0$ and (6.62) can be rewritten as:

$$\sum_{j=1}^{N} q_{ij}^B \left(-|u_i^A|\right) \geq \lambda^A \left(-|u_i^A|\right).$$

Hence:

$$\frac{\sum\limits_{j=1}^{N} q_{ij}^B \, |u_j^A|}{|u_i^A|} \leq \lambda^A.$$

Since this holds for all i, it is certainly true that:

$$\max_i \frac{\sum\limits_{j=1}^{N} q_{ij}^B \, |u_j^A|}{|u_i^A|} \leq \lambda^A.$$

Applying (6.61) to the left hand side with u not an eigenvector of Q^B gives $\lambda^B < \lambda^A$.

We thus have a sufficient condition for improving $\tilde{g}$. The foregoing proposition states, in effect, that "improving" the test quantity, as defined in (6.58), in at least one state suffices to improve $\tilde{g}$.

The previous proposition also guarnatees that if, for policy A, each test quantity is maximum in its state, then policy A is optimum. To get an initial feasible policy, we maximize $-(\text{sgn } \gamma) \sum\limits_{j=1}^{N} q_{ij}$ over the feasible policy set by BB. The algorithm can be schematically represented in Figure 6.4.

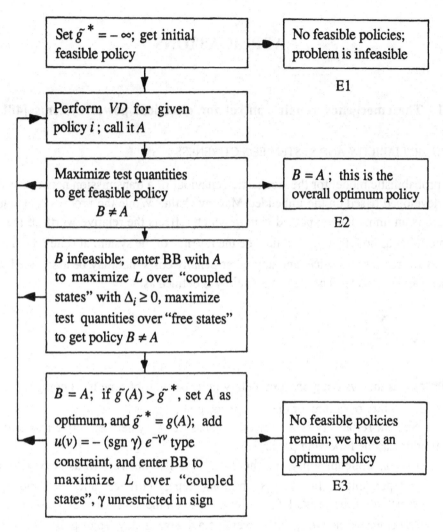

Figure 6.4. Algorithm for risk-sensitive case.

CHAPTER 7

APPLICATIONS

7.1 The emergency repair control for electrical power systems [48]

7.1.1. RELIABILITY AND SYSTEM EFFECTIVENESS

A probabilistic model for the stochastic behaviour of a transmission network system is developed based on the imbedded Markov chain. While the process is at state i, there is an immediate expected gain q_i which reflects the relative worth of the i-th state. Let us note that q_i must include the revenue of the system at state i, the repair costs for restoring service, and any other penalty costs that may be associated with each particular state. The system reliability is defined as:

$$\mathcal{R} = \frac{\sum_{i=0}^{M} q_i \phi_i}{\bar{q}_0} \tag{7.1}$$

where $\bar{q}_0$ is the system gain in monetary units per unit of time if all components had zero probability of failure, or the system could not leave state zero.

System reliability is a subjective measure of system effectiveness and the definition has the following properties:

(a) $\bar{q}_0$ depends only on the load that the network must carry and is independent of failure distributions, repair policies, and the topological structure of the network ($\bar{q}_0$ is invariant with the alternative expansions of the transmission network, the preventive and emergency repair policies). It only depends on the load that must be served and the associated priorities reflected by the prices of the different demand centres.

(b) $g = \sum_{i=0}^{M} q_i \phi_i$ is the expected long-run system gain per unit of time for an ergodic process and it depends on the network configuration, failure and repair distributions and the maintenance and emergency repair policies.

(c) The definition allows the decision maker to trade off between reliability of service and the cost of achieving it.

(d) Maximizing $\mathcal{R}$ with respect to the repair and maintenance policies, we obtain the most effective system performance by achieving the best trade-off between the reliability of the service and the operating cost.

Because $\bar{q}_0$ is independent of maintenance policies and emergency repairs for a given system, maximizing $\Re$ with respect to alternative repair policies is completely equivalent to maximizing the system gain g.

7.1.2. REWARD STRUCTURE

The objective of a transmission network system is to deliver the power from the generating stations to a set of demand centres with requirements $\{d_1^*, d_2^*, \dots, d_m^*\}$. The penalties for not meeting the demand d_i^* at region i may be quite different for different centres. The objective function defining the cost of curtailment should contain priority factors assigned to the different load centres; *the total cost of curtailment must then be minimized subject to the power flow system equations and a set of operational and equipment constraints.*

For the optimal allocation of power, to each state i there corresponds a return R_i *which is the minimum value of the properly defined objective function* (R_i includes penalty costs for not satisfying the demands). The minimum R_i's will reflect the subjective worth of being at state i. *The return rate R_i is assumed to be independent of time.*

Determination of the R_i's requires a model for the physical system which accounts for the induced effects arising from the random failures.

Next we shall assume that there is a penalty function r_j which is a lump sum quantity paid by the system when it just enters a new state j, independently of the last state that it occupied before the transition. *These penalties may be used if we seek a very high degree of reliability.*

Besides the return rate R_i, and the penalty r_j at state i and transition from i to j, the system incurs a cost rate C_i. This is the repair cost per unit of time for trying to restore service and reach normal condition at the zero state (C_i varies with the number of crews employed for the repair). The operational cost per hour is assumed to be proportional to the number of crews employed, and it also varies with the size of the crews.

The repair model is postulated as follows:

(a) For each repair centre the number of crews in reserve is fixed.

(b) To each line there corresponds a finite set of alternative repair policies $k = 1, 2, \dots, K_i$ with operational costs per unit of time $C_i(k)$. A repair policy is the number and size of the crews to send for the repair of each line i. The repair rate $\mu_i(k)$, or, more generally, the repair time distribution, is considered to be a function of the alternative repair policies.

(c) The mean for the repair time distribution is considered to be a non-increasing function of the cost of the repair alternatives.

7.1.3. THE MARKOVIAN DECISION PROCESS FOR EMERGENCY REPAIR

The transmission system operating in time has revenues and costs associated with it, either by remaining in state i, or by making a transition from i to j. Next we shall define the quantities:

$\overline{R}_i(k) = R_i - C_i(k)$ is the yield rate for state i, or the amount that the system will earn per unit of time it remains in state i and the k-th alternative is followed.

r_j is the fixed cost at the instant that the system enters state j.

$v_i(t)$ is the expected cumulative return if the system states at state i and the process operates for a period t following a fixed policy.

$v_i(0)$ is the terminal reward or scrap value of the system being at state i at the end of time t.

Next, the decision process is formulated as a linear programming problem.

The decision maker faces an infinite time horizon (the system operates for a very long time relative to the mean time between transitions) and decisions can be implemented only at the times at which transitions occur. A policy will be determined for each possible state of the system, and, in a given state, a policy choice determines the transition probabilites of the imbedded chain, the distributions of inter-transition times and the returns from the process at that state.

The objective which will be used for choosing a set of policies is to maximize the rate of return from the process when the time value of return is not considered (no discounting). This presentation will mainly be concerned with the case of no discounting since problems in reliability are mainly concerned with penalty costs for which a discount factor has a very vague meaning.

Let the system start at state i, having the accumulated return $v_i(t)$ by time t when a fixed policy is allowed. The probability that the system will not leave state i by time t is ${}^cF_i(t)$, and the contribution to the total expected reward $v_i(t)$ due to not leaving state i is ${}^cF_i(t)\,[\overline{R}_i(t) + v_i(0)]$. The system, however, may leave state i for some other state j at time τ. It will then earn the reward r_j and rate $\overline{R}_i$ for the time τ.

From time τ on, the system will earn the reward $v_j(t - \tau)$. The contribution from the transitions to all possible states j to $v_i(t)$ is given by the relation:

$$\sum_{j=0}^{N} p_{ij} \int_0^t f_{ij}(\tau)\,[r_j + \overline{R}_i\tau + v_j(t - \tau)]\,d\tau$$

and thus:

$$v_i(t) = {}^cF_i(t) [R_i(t) + v_i(0)] + \sum_{j=0}^{N} p_{ij} \int_0^t f_{ij}(\tau) [r_j + \overline{R}_i \tau + v_j(t-\tau)] \, d\tau . \qquad (7.2)$$

Our interest lies in the behaviour of the system for very large t. If the holding time distributions have finite means, then ${}^cF_i(t) \to 0$ and $t.{}^cF_i(t) \to 0$, large t, and any state i. Therefore, when t is large, relation (7.2) can be written as:

$$v_i(t) = \sum_{j=0}^{N} p_{ij} \int_0^t f_{ij}(\tau) [r_j + \overline{R}_i \tau + v_j(t-\tau)] \, d\tau , \qquad (7.3)$$

or:

$$v_i(t) = \sum_{j=0}^{N} p_{ij} r_j + \overline{R}_i \overline{\tau}_i + \sum_{j=0}^{M} p_{ij} \int_0^\infty f_{ij}(\tau) \, v_j(t-\tau) \, d\tau . \qquad (7.4)$$

Now the quantity q_i is defined as the immediate expected earning rate of the system at state i as:

$$q_i = \overline{R}_i + \frac{1}{\overline{\tau}_i} \sum_{j=0}^{N} p_{ij} r_j \qquad i = 0, 1, 2, \dots , N \qquad (7.5)$$

and $v_i(t)$ is now written as:

$$v_i(t) = q_i \overline{\tau}_i + \sum_{j=0}^{N} p_{ij} \int_0^\infty f_{ij}(\tau) \, v_j(t-\tau) \, d\tau . \qquad (7.6)$$

It can be shown that the form of $v_i(t)$ for large values of t is given by:

$$v_i(t) = gt + v_i \quad i = 0, 1, 2, \dots , N \qquad (7.7)$$

where $g = \lim_{t \to \infty} \dfrac{dv_i(t)}{dt}$ and is independent of i if the matrix describing the imbedded chain is ergodic, and the v_i's are the relative values of the policy such that $v_i(t) - v_j(t) = v_i - v_j$ indicates the difference in the long run if the process started in state i rather than state j. After simple substitutions, we can write:

$$v_i + gt = q_i \overline{\tau}_i + \sum_{j=0}^{N} p_{ij} \int_0^\infty f_{ij}(\tau) [v_j + g \cdot (t-\tau)] \, d\tau$$

$$= q_i \bar{\tau}_i + \sum_{j=0}^{N} p_{ij} v_j + gt - g \bar{\tau}_i \qquad (7.8)$$

or for large t :

$$v_i + g \bar{\tau}_i = q_i \bar{\tau}_i + \sum_{j=0}^{N} p_{ij} v_j \qquad i = 0, 1, \dots, N . \qquad (7.9)$$

Since the v_i's are the relative values, the value of v_N is set to zero and the system can now be explicitly solved for g and the remaining v_i's.

The gain g is the same as used in defining the reliability in (7.1). Equation (7.9) is multiplied by π_i, the steady state probability for the i-th state of the imbedded chain, which are summed for all values of i. The result is:

$$\sum_{i=0}^{N} \pi_i v_i + g \sum_{i=0}^{N} \pi_i \bar{\tau}_i = \sum_{i=0}^{N} \pi_i q_i \bar{\tau}_i + \sum_{i=0}^{N} \pi_i \sum_{j=0}^{N} p_{ij} v_j$$

$$= \sum_{i=0}^{N} \pi_i q_i \bar{\tau}_i + \sum_{i=0}^{N} v_j \pi_j \qquad (7.10)$$

and, since the first and last terms of the equation are equal, Equation (7.10) reduces to:

$$g = \frac{\displaystyle\sum_{i=0}^{N} \pi_i \bar{\tau}_i q_i}{\displaystyle\sum_{i=0}^{N} \pi_i \bar{\tau}_i} = \sum_{i=0}^{N} q_i \phi_i \qquad (7.11)$$

where the ϕ's are the limiting probabilities of the original process. Maximizing g is equivalent to maximizing the system reliability. The objective is to choose a set of policies to maximize the system reliability $\mathcal{R}$, or equivalently, the return rate g. Associated with the k-th alternative in state i are all the process parameters: $p_{ij}^k, f_{ij}^k, R_i^k$. The number of alternatives for each state is assumed to be finite, but they may be different in number from one state to another. The optimal steady state policy that maximizes the average reward per unit of time g consists of a selection of one alternative for each state. The value of g is of the form:

$$g = q_i + \frac{1}{\bar{\tau}_i} \left[\sum_{j=0}^{N} (p_{ij} - \delta_{ij}) v_j \right] \qquad i = 0, 1, \dots, N . \qquad (7.12)$$

7.1.4. LINEAR PROGRAMMING FORMULATION FOR REPAIR OPTIMIZATION

As an alternative optimization procedure, the linearity of the simultaneous equations makes it possible to obtain the solution of the decision process by using linear programming.

The basic set of equations which will be used is:

$$v_i + g\bar{\tau}_i = q_i\bar{\tau}_i + \sum_{j=0}^{N} p_{ij} v_j \qquad i = 0, 1, \dots, N.$$ (7.13)

By solving these for g with $v_N = 0$, we can write, as before:

$$g = q_i + \frac{1}{\bar{\tau}_i} \left[\sum_{j=0}^{N-1} (p_{ij} - \delta_{ij}) v_j \right].$$ (7.14)

Now let g and the v_i's be those corresponding to the optimal policy. Then for any non-optimal alternative k in state i we must have:

$$g \geq q_i^k + \frac{1}{\bar{\tau}_i^k} \left[\sum_{j=0}^{N-1} p_{ij}^k v_j - v_i \right] \text{ for all } i$$ (7.15)

and if k_i is the number of alternatives in state i, the gain g must satisfy the following inequality condition:

$$g \geq q_i^k + \frac{1}{\bar{\tau}_i^k} \left[\sum_{j=0}^{N-1} p_{ij}^k - \delta_{ij}) v_j \right]$$ (7.16)

for $((i = 0, 1, 2, \dots, N), k = 1, 2, \dots, K_i)$. The number of inequalities above is equal to K, the number of alternatives in all states such that:

$$K = \sum_{i=0}^{N} K_i.$$

The optimal g is the smallest possible value that satisfies relation (7.16) programming approach to the problem is to minimize g with (7.16) as constraints.

To put the above problem in standard form, we can write:

$$\frac{1}{\bar{\tau}_i} \left[\sum_{j=0}^{N-1} (\delta_{ij} - p_{ij}^k) v_j \right] + g \geq q_i^k$$ (7.17)

and then define the $K \times (N + 1)$ matrix $B = \{b_{nj}\}$ by:

$$b^k_{n(i),j} = (\overline{\tau_i})^{-1} (\delta_{ij} - p^k_{ij}) \quad j = 0, 1, 2, \ldots, N - 1$$

$$b^k_{n(i),N} = 1 \tag{7.18}$$

where:

$$b^k_{m(i)} = k + \sum_{j=0}^{i} k_{j-1} \quad k_{-1} = 0, \quad k_0 = 1 .$$

Next, we shall define the column vectors:

$$V = (v_1, v_1, \ldots, v_{N-1}, g), T = (0, 0, \ldots, 0, 0, 1)$$

and

$$q = (q^1_0, q^1_1, q^2_1, \ldots, q^{k_1}_1, q^1_2, q^2_2, \ldots, q^{k_2}_2, \ldots, q^1_N, q^2_N, \ldots, q^{k_N}_N).$$

The standard linear programming formulation is of the form:

$$\min T'V \tag{7.19}$$

subject to

$$BV \geq q$$

where V is unconstrained in size.

The dual to the above linear programming problem is:

$$\max q'\phi \tag{7.20}$$

subject to

$$B'\phi = T, \quad \phi \geq 0$$

where ϕ is a K-element column vector and is constrained in sign.

It involves the limiting state probabilities for each state i, $(i = 0, 1, \ldots, N)$ and each possible alternative k, $(k = 1, 2, \ldots, k_1)$, and it maximizes the gain g by finding the optimal alternative for each state. The constraints in the relation (7.20) express the fact that:

$$\sum_{i=0}^{N} \sum_{k=1}^{k_i} \phi^k_i = 1 \tag{7.21}$$

and the decision variables ϕ_i^k must satisfy the equation of the state probabilities in relation to the transition matrix given by:

$$\sum_{k=1}^{k_j} \frac{\phi_j^k}{\bar{\tau}_j^k} = \sum_{i=0}^{N} \sum_{k=1}^{k_i} \frac{\phi_i^k P_{ij}}{\bar{\tau}_i^k} \quad j = 0, 1, \dots, N-1. \tag{7.22}$$

The formulation and solution of the decision process presented in this chapter solves the repair control problem. Given a load distribution and the network configuration that carries the power from the generating stations to the demand centres, an optimal steady state repair policy has been found that maximizes the reliability of the system.

7.1.5. THE INVESTMENT PROBLEM

Additional transmission capacity is regularly required in a power system. When an investment is made, the management is interested in the most economical way to expand the system while still retaining high reliability (trade-off and reliability investment).

Let us assume that a set of alternative expansions to a transmission system is given, each one satisfying the demand distribution $D = (d_1^*, d_2^*, \dots, d_m^*)$ under normal operating conditions together with a lower bound on reliability and a budget constraint. Each alternative α_i has a capital cost $C(\alpha_i)$, a maintenance policy, and an emergency repair policy. Therefore, to (α_i, D) there corresponds a $g(\alpha_i, D)$ which is the optimal steady state gain. Since each alternative expansion must meet the same demand D, the return rate $\bar{q}_0(D)$ under normal operating conditions is the same for each α_i, and $\mathcal{R}(\alpha_i, D)$ is given by:

$$\mathcal{R}(\alpha_i, D) = \frac{g(\alpha_i, D)}{\bar{q}_0(D)} \quad \alpha_i \in \{\alpha_1, \alpha_2, \dots, \alpha_n\}.$$

Finding $g(\alpha_i, D)$ will determine which one of the alternatives has the best system reliability.

The decision maker could take into account the time value of the return from his system. Let discounting be exponential at a rate $\alpha > 0$. If x is the rate of return of a continuous flow, the present value of the total return xt will be:

$$u(x, t, \alpha) = (x/\alpha)(1 - e^{-\alpha t}).$$

Since the system α_i earns the amount $g(\alpha_i, D)$ per unit of time, if the planning horizon is of length T, then the continuous flow of $g(\alpha_i, D)$ per unit of time for time T is now worth:

$$g(\alpha_i, D) \cdot \alpha^{-1}(1 - e^{-\alpha T}).$$

The capital cost of the system is $C(\alpha_i)$ and the investment problem can then be stated as:

$$\max_{\alpha_i} \left\{ g(\alpha_i, D) \cdot \alpha^{-1}(1 - e^{-\alpha T}) - C(\alpha_i) \right\}.$$

Dividing the expression to be maximized by the non-zero constant:

$$\alpha^{-1}(1 - e^{-\alpha T})$$

the equivalent statement of the investment problem is:

$$\max_{\alpha_i} \left\{ g(\alpha_i, D) - C(\alpha_i) \left(\alpha + \frac{\alpha}{e^{\alpha T} - 1} \right) \right\} \tag{7.23}$$

where $[\alpha + (\alpha/e^{\alpha T} - 1)]$ is the per unit time cost of a unit of capital or the capital recovery factor, and consists of the continuous interest rate plus the uniform time payment into a T-time sinking fund which will retire the investment after time T.

The solution of the above equation cannot be obtained analytically; it can be easily solved computationally if a reasonably small set of alternatives, α_i, have already been determined. Solution of the investment problem requires the generation of a carefully selected small set of alternatives from which the best is chosen by using the relation (7.23).

7.2 Stochastic models for evaluation of inspection and repair schedules [2]

The discrete-time finite-state Markov chain model is employed for all the optimization problems considered in this paragraph.

The general nature and scope of this model make it applicable to a variety of maintenance scheduling problems. In a multi-unit system, the operating history of each individual unit is modelled by a separate Markov chain.

The single unit transmission matrix may depend on the time and also depend on the repair action taken at that time. The dependence on the repair action is denoted by a sub-script. When the number of alternative repair procedures available for the unit is a_r, the index $r = 1, 2, 3, \ldots, a_r$ is given to the transition matrix when a repair procedure of type r is initiated at time $t : P_r(t)$. The sub-script "0" is used when no repair action is initiated at time $t : P_0(t)$.

	States at Time $(t + 1)$		
	R Repair States $s(t + 1) =$ $1, \ldots, i_r$	M Operating States $s(t + 1) =$ $(i_r+1, \ldots, (i_r+i_m)$	F Failure States $s(t + 1) =$ $(i_r+i_m+1,\ldots, (i_r+i_m+i_f)$
States at Time t — R Repair States $s(t + 1) =$ $1, \ldots, i_r$	$P(M, R \,;\, t)$	$P(M, M \,;\, t)$	$P(R, F \,;\, t)$
M Operating States $s(t + 1) =$ $(i_r+1, \ldots, (i_r+i_m)$	$P(R, R \,;\, t)$	$P(R, M \,;\, t)$	$P(M, F \,;\, t)$
F Failure States $s(t + 1) =$ $(i_r+i_m+1,\ldots, (i_r+i_m+i_f)$	$P(F, R \,;\, t)$	$P(F, M \,;\, t)$	$P(F, F \,;\, t)$

Figure 7.1. Partition of matrrix into nine blocks.

The states in the Markov chain correspond to different modes of operation of the unit. These modes of operation can be divided into three groups corresponding, respectively, to a *unit undergoing repair*, a *unit capable* of performing its *operating* function (although it may be damaged), and a *unit incapable of operation* (a failed unit). The set of all states in the Markov chain model is divided into three mutually exclusive sub-sets, the set of *repair* states, the set of *operating* states, and the set of *failure* states.

The set of repair states (R) contains all states in which the unit is not operating or is undergoing repair. The unit can enter this set only when a repair is initiated. It leaves the set when the repair is completed. When a unit is repaired, the state is immediately reset into an operating set as soon as the repair is completed.

The set of operating states (*M*) contains all states in which the unit is operating or capable of operation. In most practical situations, it is an ordered set, induced by the level of deterioration in the unit.

The set of failure states (*F*) contains all states in which the unit is not capable of operation and is not undergoing repair. In these states, the unit is failed.

The set of all possible states is given by $S = R \cup M \cup F$. No state can be in more than one of the sets *R*, *M* or *F*.

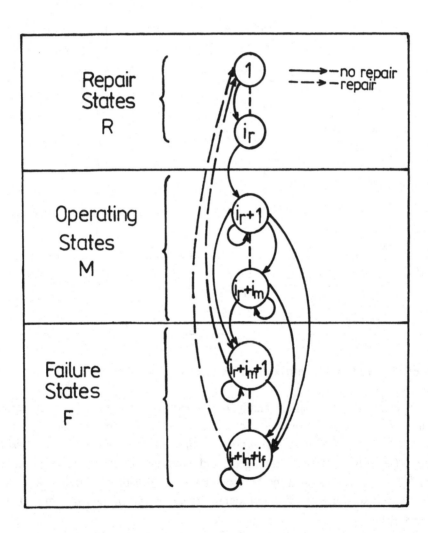

Figure 7.2. Possible state transitions for a Markov chain with non-negative drift and constant non-zero repair time.

The following conventions are used for labelling the *n* states of the chain.

The states in R are labelled $j = 1, 2, 3, ... , i_r$.

The states in M are labelled $j = i_r + 1, i_r + 2, ... , i_r + i_m$, where i_m is number of states in M.

The states in F, are labelled $j = i_r + i_m + 1, i_r + i_m + 2, ... , i_r + i_m + i_f$, where i_f is the number of states in F ($n = i_r + i_m + i_f$).

The matrices $\{P_r(t)|r = 0, 1, ... , a_r\}$ are partitioned into nine blocks (see Figure 7.1).

An example of the possible state transitions is shown in Figure 7.2. Each circle in this figure represents a state of the unit. The lines indicate the allowable state transitions. The dotted arrows indicate the types of transitions possible only when a repair action is initiated.

When a repair action is initiated on the unit at time t according to the r-th available repair procedure, the transition matrix for the Markov chain at that time is $P_r(t)$. As previously mentioned, the transitions matrices $P_r(t)$ are partitioned into nine blocks analagous to the partition in the matrix $P_0(t)$ (see Figure 7.3).

			States at Time ($t + 1$)		
			R 1 2 3 . . . i_r	M (i_r+1) . . . . . (i_r+i_m)	F
States at Time t	R	1 2 . . . i_r	0 1 0 0 . . 0 0 0 1 0 . . . $P_0(R, R ; t)$ . . . . . 1 0 . . . . . 0 1 0 . . . . 0 0	0 . . . . . . 0 0 . . . . $P_0(M, M ; t)$ . . . . 0 . . . . 1 0 0	$P_0(R, F ; t) = 0$
	M	(i_r+1) . .	$P_0(M, R ; t)$	$P_0(R, M ; t)$	$P_0(M, F ; t)$
	F	. .	$P_0(F, R ; t)$	$P_0(F, M ; t)$	$P_0(F, F ; t)$

Figure 7.3. Transition matrix for a unit with error-free repair and constant non-zero repair time.

If a unit is undergoing repair, its state is in set R and initiation of a repair action is superfluous and can have no effect on its transition probabilities.

When the unit is in either an operating or a failure state $s(t) = i$, and the repair procedure initiated requires zero repair time for that particular state, the state of the unit is immediately reset to the "as good as new" state $s(t) = i_0$. The transition probabilities for the state i under this repair procedure are identical to the transition probabilities for the state i_0. For every r, the row in $P_r(t)$ corresponding to the as-good-as-new state is identical to the same row in $P_0(t)$.

When the unit is either in an operating or a failure state, and the repair procedure initiated requires a repair time equal to one or more time steps (non-zero repair time), the state of the unit is moved into R. The row in the matrix $P_r(t)$ corresponding to the state immediately prior to the repair has zeros in all columns corresponding to states in M or F and non-zero elements in columns corresponding to the set of repair states R.

7.2.1. INSPECTION ACTIONS

When an inspection is performed on a unit at time t, its effect is not reflected in a change in the transition probabilities. The inspection does increase the information available to the operator about the state of the unit. The state probability vector is of the form:

$$x_i(t) = \text{Prob. } [s(t) = i|\text{ inspection outcomes and maintenance schedule up to } t],$$

for $i = 1, 2, \ldots , n$.

The effect of the inspection on the sequence of state probability vectors is dependent on the maintenance policy.

The mathematical model describing the possible inspection procedures employs an output process. For an inspection procedure of type i, $i = 1, 2, 3, \ldots , a_i$, where a_i is the total number of alternative inspection procedures available, is given by the elements $h_{i, kj}$ of matrix H_i — "the inspection matrix" — for inspection procedure i :

$$h_{i, kj} = \text{Prob [output is } = j| \text{ unit in state } k, \text{ inspection procedure } i],$$

for $i = 1, 2, \ldots , a_i$; $k = 1, 2, \ldots , n$; $j = 1, 2, \ldots , y_i$.

The output produced can assume one of y_i possible values. The system operator obtains information about the unit by observing the output process (see Figure 7.4).

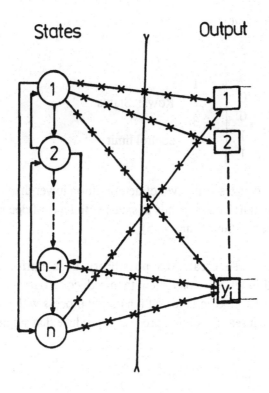

—————▶ : Transitions in the Markov chain, governed by probabilities p_{jk}, elements of the transition matrix P.

✕—✕—✕▶ : Observations, governed by probabilities $h_{i,jk}$ elements of inspection matrix H_i for the i-th inspection alternative.

Figure 7.4. Output process for Markov chains.

7.2.1.1. *Complete Inspection.* In this inspection, the state of the unit is determined exactly. The number of possible outputs in the output process is equal to the number of states in the unit. The matrix H_i is the identity matrix.

7.2.1.2. *Control Limit Inspection.* This procedure applies to units whose states can be ordered according to increasing level of deterioration, up to the control limit. The inspection produces one of two alternative outputs and the matrix H has two columns. For a unit with five states, no repair states, and control limit at state $j^* = 4$:

Columns: Outputs

$$
\begin{array}{cc}
1 & 2
\end{array}
$$

$$
\begin{bmatrix}
1 & 0 \\
1 & 0 \\
1 & 0 \\
0 & 1 \\
0 & 1
\end{bmatrix}
\begin{array}{l}
1 \\
2 \quad \text{Rows: States} \\
3 \\
4 \quad \text{control limit} \\
5
\end{array}
$$

Output 1 corresponds to states with level of deterioration lower than the control limit (states 1, 2, 3) while output 2 corresponds to states with level of deterioration greater than or equal to the control limit (states 4, 5).

7.2.1.3. *Inspection.* To create the possibility of a uniform treatment of inspection/repair at all times, it is convenient to introduce an output at times when no inspection is actually performed on the unit (an output process with only one possible output). Every state "causes" the output process to produce that single output (H_i is a column vector):

$$H_0 = [1, 1, \dots, 1]^T.$$

Each unit in the system is modelled by a Markov chain. Since, in general, a multi-unit system contains various types of units, the Markov chains for the units are not all identical. The inspection and repair procedures are different for various units. The *size* of the inspection and repair problems is given by the sum of the number of states in all Markov chains multiplied by the number of time steps considered in the planning period.

In a system with n units, each having N states, and operating over a period of T time steps, the size of the problem is $n.N.T$.

7.2.2. MARKOV CHAIN MODELS (SEE TABLE 7.1)

a) *A unit with exponentail failure distribution, zero repair time, and a single aggregated failure state.*

The exponential distribution is one of the most used distributions in reliability engineering.

It leads to the simplest possible Markov chain model, with two possible states in the unit. The exponential distribution is a continuous distribution; discretization is required within the framework of the discrete time Markov model.

Failure Distribution	Repair Time	Number of States			Remarks
		R	M	F	
exponential	zero	0	1	1	
gamma	zero	0	α	1	α is integer
general	zero	0	1	1	as-bad-as-old repair
gamma	constant $(4.8t)$	3	3	1	
exponential	random	1	1	1	

Table 7.1. Markov chain models

Under the assumption of no maintenance action, the unit failure distribution:

$$F(\tau) = \text{Prob [failure time is } \leq \tau] = \begin{cases} 1 - \exp(-\lambda\tau), & \tau \geq 0 \\ 0, & \tau < 0 \end{cases} \tag{7.24}$$

and the failure rate is:

$$r(\tau) = \text{Prob [failure in } (\tau, \tau + d\tau) \mid \text{no failure in } (0, \tau)] =$$

$$= \frac{\dfrac{d}{d\tau} F(\tau)}{1 - F(\tau)} = \lambda . \tag{7.25}$$

The geometric distribution with parameter q is:

$$F(t) = \text{Prob [failure time } \leq t] = 1 - q^t, \quad t \geq 0,$$

and the failure rate is:

$$r(t) = \text{Prob [no failure in } t, t + 1) \mid \text{no failure in } (0, t)] =$$

$$= \frac{(1 - q^{t+1}) - (1 - q^t)}{1 - (1 - q^t)} = 1 - q$$

which is a constant. The above formulation of the failure statistics are related by $\tau = \Delta t$ and $q = \exp(-\lambda\Delta t)$ — (Δt is the time interval between successive discrete time steps).

Due to the fact that the failure rate is constant, the unit does not deteriorate during operation, so that:

Prob [no failure prior to time $(\tau + T)$ | unit operating time T] $=$

$$= \frac{1 - F(\tau + T)}{1 - f(t)} = \frac{\exp(-\lambda\,(\tau + T))}{\exp(-\lambda\,T)} = \exp(-\lambda\,\tau) =$$

$= $ Prob [no failure prior to τ | unit operating at time $= 0$].   (7.26)

The unit has only a single operating state $(s = 1)$. Because of the zero repair time, the set of repair states R is empty, and, by assumption, the set of failure states contains only a single state $(s = 2)$: $(R = \varnothing, M = \{1\}, F = \{2\})$.

The transition matrix $P_0(t)$ is independent of time:

$$P_0(t) = \begin{bmatrix} q & p \\ 0 & 1 \end{bmatrix}, \text{ where } q = \exp(-\lambda\,\Delta t)$$

and $p = 1 - q$. The four elements of the matrix are obtained as follows:

$q = $ Prob [state at time $(t + 1) = 1$ | state at time $t = 1$]

$\quad = 1 - $ Prob [state at time $(t + 1) = 2$ | state at time $t = 1$]

$\quad = 1 - $ Prob [failure in $(t, t + 1)$ | not failed at time t]

$\quad = 1 - r(t)$,   (7.27)

$p = $ Prob [state at time $(rt + 1) = 2$ | state at time $t = 1$]

$\quad = r(t) = 1 - q$.   (7.28)

The number of alternative repair procedures is one; every repair resets the unit into the operating state. The transition matrix

$$P_1(t) = \begin{bmatrix} q & p \\ q & p \end{bmatrix}$$ (7.29)

assumes that the repair procedure is also independent of time.

b) *A unit with zero repair time, single aggregated failure state, good-as-new repair, and a gamma failure distribution* (a gradually increasing damage mechanism that may be present in a unit).

The continuous two-parameter (α, λ) gamma distribution has the probability density function:

$$f(\tau, \alpha, \lambda) = \begin{cases} 0, & t < 0 \\ \dfrac{\lambda^\alpha \tau^{\alpha-1}}{\Gamma(\alpha)} \exp(-\lambda.\tau), & \tau \geq 0. \end{cases} \tag{7.30}$$

For $\alpha > 1$, the failure rate $r(\tau)$ is an increasing function of time with $r(0) = 0$, and $r(\tau)$ increasing to a limit λ. This increasing failure rate reflects the presence of a damage mechanism leading to deterioration of the unit.

A discrete-time finite state Markov chain that generates failure probabilities at times $\tau = 0$, $\tau = \Delta t$, $\tau = 2 \Delta t$, ..., $\tau = t$. ΔT for a unit whose failure distribution is a gamma distribution is obtained from the relationship:

$$x(\tau + \Delta t) = \exp(A \Delta t) \cdot x(\tau). \tag{7.31}$$

In this Markov chain, state 1 is the "good-as-new" state, states 2, 3, ... are operating states, and $(\alpha + 1)$ is the failure state. The transition matrix is:

$$P_0(t) = [\exp(A \Delta t)]^T, \tag{7.32}$$

where

$$A = \begin{bmatrix} -\lambda & 0 & . & . & . \\ +\lambda & -\lambda & 0 & . & . \\ 0 & +\lambda & -\lambda & 0 & . \\ . & . & . & . & . \\ . & . & 0 & +\lambda & 0 \end{bmatrix} \tag{7.33}$$

The transition matrix for this approximation becomes:

$$P_0(t) = P_0 = \begin{bmatrix} q & p & 0 & 0 & . & . & 0 \\ 0 & q & p & 0 & . & . & . \\ . & 0 & q & p & 0 & . & . \\ . & . & . & . & . & . & . \\ . & . & . & . & . & . & 0 \\ . & . & . & . & . & q & p \\ 0 & . & . & . & . & 0 & 1 \end{bmatrix}$$

(7.34)

The transition matrix P_0 has non-negative drift. The operating states 1, 2, ... , α are arranged in the order of increasing deterioration. In the case of as-good-as-new repair with zero repair time, the effect of a repair action is to reset the unit immediately into state 1. The transition matrix is:

$$P_1(t) = P_1 = \begin{bmatrix} q & p & 0 & . & 0 \\ q & p & 0 & . & . \\ . & . & . & . & . \\ . & . & . & . & . \\ q & p & 0 & . & 0 \end{bmatrix}$$

(7.35)

c) *A unit with a failure distribution F, as-bad-as-old repair and zero repair time.*

The repair action has no effect on the level of deterioration but only removes the causes of failure that may be present.

The time-dependent transition matrix $P_0(t)$ is:

$$P_0(t) = \begin{bmatrix} [1 - r(t)] & r(t) \\ 0 & r(t) \end{bmatrix}$$

(7.36)

The transition matrix $P_1(t)$ is:

$$P_1(t) = \begin{bmatrix} [1 - r(t)] & r(t) \\ [1 - r(t)] & r(t) \end{bmatrix}$$

(7.37)

State at time $(t+1)$

-A- No Repair Action $P_0(t) = P_0$

States at time t		Repair States R			Operating States M			Failure State F
		1	2	3	4	5	6	7
Repair States R	1	0	1	0	0	0	0	0
	2	0	0	1	0	0	0	0
	3	0	0	0	1	0	0	0
Operating States M	4	0	0	0	q	p	0	0
	5	0	0	0	0	q	p	0
	6	0	0	0	0	0	q	p
Failure State F	7	0	0	0	0	0	0	1

-B- Repair Action $P_1(t) = P_1$

States at time t		Repair States R			Operating States M			Failure State F
		1	2	3	4	5	6	7
Repair States R	1	0	1	0	0	0	0	0
	2	0	0	1	0	0	0	0
	3	0	0	1	0	0	0	0
Operating States M	4	0	q	0	0	p	0	0
	5	0	0	0	1	0	0	0
	6	0	1	0	0	0	0	0
Failure State F	7	1	0	0	0	0	0	0

Figure 7.5. Transition matrices for the unit of example (d).

d) *Units with non-zero repair time.*

The transition matrices for such units are given in Figure 7.5.

7.2.3. COST STRUCTURES AND OPERATING REQUIREMENTS

The total cost incurred in the planning period is composed of inspection costs, repair costs, and operating costs. Operating requirements are discussed in conjunction with operating costs.

7.2.3.1. *Inspection Costs.* Inspection costs are incurred whenever one or more units are inspected irrespective of the operating condition of these units. These costs depend on the set of units that are inspected. The total cost for inspection of a particular set of units may be smaller than the sum of the inspection costs incurred when each units is inspected separately. In a two unit system, typical values for inspection costs may be:

C = 10, when only unit 1 is inspected,
C = 5, when only unit 2 is inspected,
C = 11, when units 1 and 2 are inspected, simultaneously.

For a general multi-unit system, the inspection costs are:

$$C_I(t) = \sum_{l=1}^{N_c} K_l \cdot p[J_l \cap I(t)] \tag{7.38}$$

U = $\{1, 2, 3, \dots, N\}$: the set of indices of all units,
$I(t)$: the set of all units inspected at time t .

$$p(S) = \begin{cases} 0, \text{ if the set } S \text{ is empty} \\ 1, \text{ if the set } S \text{ is not empty} \end{cases}$$

$J_l, \quad l = 1, 2, \dots, N_c$: sub-sets of U characterizing the inspection cost group structure

$K_l, \quad l = 1, 2, \dots, N_c$: inspection cost incurred if one or more units in set $J_l, \quad l = 1, 2, \dots, N_c$, are inspected.

7.2.3.2. *Repair costs.* Repair costs are incurred when a unit makes a transition from the set of failure states into the set of repair states, or, for the zero repair time units, into a more advantageous operating state. The repair costs depend on the unit repaired, the state this unit occupies, and the repair procedures employed.

The total expected value of the repair cost is:

$$E\{C_r(t)\} = \sum_{j=1}^{N} \sum_{i=1}^{n_j} \sum_{r=1}^{a_r^j} d_{ir}^j \cdot \alpha_{ir}^j(t) \cdot x_i^j(t) \qquad (7.39)$$

where:

x_i^j = Prob [unit j is in state i at time t]

$\alpha_{ir}^j(t)$ = Prob [repair action r chosen for unit j at time step t | unit j in state i at time t]

$d_{ir}^j(t)$ = repair cost incurred for repair of unit j from state i, using repair action

r :

for

$$j = 1, 2, 3, \dots, N,$$
$$i = 1, 2, 3, \dots, n^j,$$
$$r = 1, 2, 3, \dots, a_r^j,$$

where $\alpha_{ir}(t)$ are either equal to zero or equal to one.

7.2.3.3. *Operating Costs and Requirements.* Operating costs and requirements are formulated in terms of a structure function ψ which espresses the system operating states in terms of the unit operating states. This structure function is described as [38]:

$\psi[\sigma(t)]$ = a Boolean variable (= 0 or 1) specifying the operating state of the system as a function of the operating states of the units,

$$\psi(\cdot) = \begin{cases} 0 & \text{if the system is operating} \\ 1 & \text{if the system is not operating} \end{cases}$$

$\sigma(t)$ = an N -vector of Boolean variables with j-th component $\sigma^j(t)$, for $j =$ 1, 2, ... , N,

$\sigma^j(t)$ = a Boolean variable associated with the unit j defined by:

$$\sigma^j(t) = \begin{cases} 0 & \text{if unit } j \text{ (in state } M) \text{ is operating at time } t \\ 1 & \text{if unit } j \text{ (in state } R \text{ or } F) \text{ is not operating at time } t \end{cases}$$

For a system with simple structure, $\psi(\sigma)$ can be obtained by inspection.

Associated with the structure function $\psi(\sigma)$ is a *reliability function* $h(z)$ which gives the probability of a system failure in terms of the probabilities of individual unit failure. This function is defined as follows:

$h[z(t)] = \text{Prob } [\psi[\sigma(t)] = 1 = $ the probability that the system is not operating at time t.

The total expected cost due to loss of system operating capability is:

$E\{C_F(t)\} = H \cdot \text{Prob }$ [system not functioning at time step t]

$\qquad = H \cdot \text{Prob } [\psi(s) = 1]$

$\qquad = H \cdot h[z(t)],$ \hfill (7.40)

where H is the cost incurred when the system is not functioning.

For the second cost criterion, the cost is given by the inspection and repair costs, and the constraint on the probability of a loss of operating capability is:

$h[z(t)] \leq \gamma$ \hfill (7.41)

where, here, γ is the maximum allowable value of system failure probability.

The total expected cost during the planning period is:

$$E\{C_A(t)\} = \sum_{t=0}^{T=1} H \cdot g[z(t), z(t+1)] \qquad (7.42)$$

which is proportional to the total expected number of accidents.

System operating costs consist of one component reflecting loss of functional capability, and a second, reflecting occurrence of accidents.

7.2.3.4. *Inspection and Repair Policies.* During the planning period, the system operator is confronted with a sequence of maintenance action decision alternatives. The flow-chart in Figure 7.6 depicts the decisions to be made at each time.

The figure illustrates the sequential approach that the operator can adopt in determining these maintenance actions at time t. With this type of policy, the operator makes use of all the information contained in the sequence of inspection outcomes. The maintenance schedule is then a closed-loop policy.

7.2.3.5. *Closed Loop Policies.* Closed loop policies require a means of expressing the information content of inspection and maintenance histories for the units in the system. The information content can be summarized by an N-vector π (N being the number of states in a unit) which contains the conditional state probabilities for the unit:

$\pi(t)$ = an N-vector of conditional state probabilities at time t for the unit, given all previous maintenance actions and inspection outcomes *up to but not including* time t.

$\pi[t, \theta(t)]$ = an N-vector of conditional state probabilities at time t for the unit, given all previous maintenance adtions and inspection outcomes up to but not including time t, and the inspection outcomes $\theta(t)$ observed at time t.

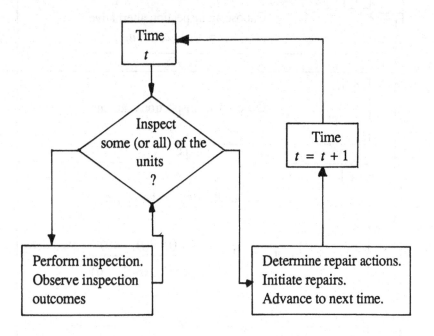

Figure 7.6. Maintenance decision alternatives at time t for the inspection and repair problem.

The information required for calculating the vector $\pi[t, \theta(t)]$ is:

(i) the vector $\pi(t)$,

(ii) the inspection procedure $i(t)$ used at time t. $i(t)$ may be one of the $(a_i + 1)$ possible alternatives $i(t) = 0, 1, \dots , a_i$,

(iii) the inspection outcome $\theta(t)$ observed after application of inspection procedure $i(t)$.

The information required for calculating the vector $\pi(t + 1)$ is:

(i) the vector $\pi[t, \theta(t)]$,

(ii) the transition matrix $P_r(t$ chosen at time t, i.e. the repair alternative $r(t)$ for time t. $r(t)$ may be one of the $(a_r + 1)$ possible alternatives $r(t) = 0, 1, \dots , a_r$ (see Figure 7.7).

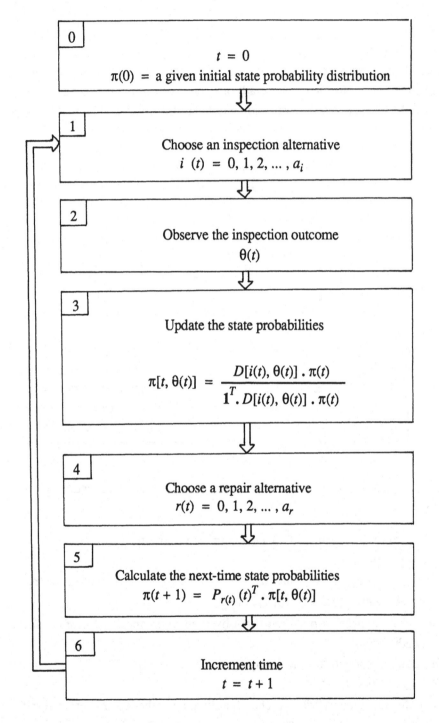

Figure 7.7. Solution strategy for the single unit inspection and repair problem.

7.2.3.6. *Updating State Probabilities after an Inspection.* Assume that the output $\theta(t)$ observed at inspection at time t is $\sigma(t) = m$. The k-th component of $\pi[t, \theta(t)]$ is:

$$\pi_k[t, \theta(t)] = \text{Prob}\ [s(t) = k \mid \theta(t) = m,\ i(t),\ \varepsilon(t)] =$$

$$= \frac{\text{Prob}\ [s(t) = k, \theta(t) = m \mid i(t),\ \varepsilon(t)]}{\text{Prob}\ [\theta(t) = m \mid i(t),\ \varepsilon(t)]} =$$

$$= \frac{\text{Prob}\ [\sigma(t) = m \mid s(t) = k, i(t)] \cdot \text{Prob}\ [s(t) = k \mid \varepsilon(t)]}{\displaystyle\sum_{l=1}^{n} \text{Prob}\ [s(t) = l, \theta(t) = m \mid i(t),\ \varepsilon(t)]} =$$

$$= \frac{H_{i(t),\ km} \cdot \pi_k(t)}{\displaystyle\sum_{l=1}^{n} H_{i(t),\ lm} \cdot \pi_l(t)}. \qquad\qquad (7.43)$$

7.2.3.7. *Obtaining Next-Time State Probabilities using Transition Matrix.* Using transition matrix $P_{r(t)}(t)$, the transition relationship for the Markov chain is:

$$\pi(t + 1) = P_{r(t)}(t)^{\text{T}} \cdot \pi[t, \theta(t)] \qquad\qquad (7.44)$$

The sequence of vectors $\pi(0)$, $\pi[0, \theta(0)]$, $\pi(1)$, $\pi[1, \theta(1)]$, ... , $\pi[t - 1, \theta(t - 1)]$, $\pi(t)$, $\pi[t, \theta(t)]$ ... constitutes a Markov process.

The π-process is a completely observable Markov process, and can be determined exactly after observation. The problem of optimization of closed-loop strategy for the π-process can be approached with the aid of the theory of dynamic programming in completely observable processes.

In a multi-unit system there is a π-process corresponding to each individual Markov chain. The sequential decision process for the multi-unit system is shown in Figure 7.8.

START

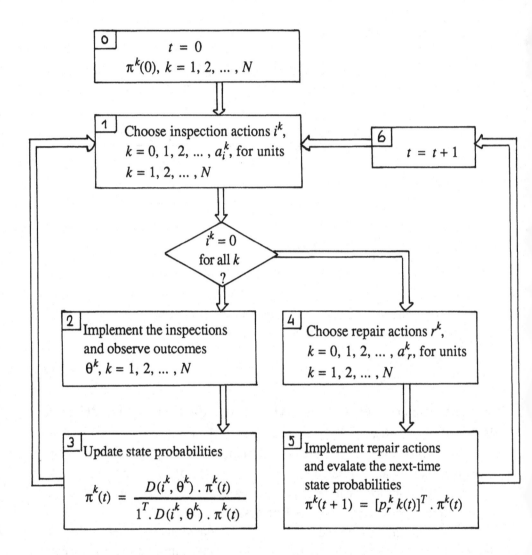

Figure 7.8. Solution strategy for the multi-unit inspection and repair problem.

A closed loop inspection and repair policy for an n-unit system in a planning period extending over T time intervals is a set of rules:

$$i^k (\pi^1, \pi^2, \dots, \pi^n) \text{ and } r^k ((\pi^1, \pi^2, \dots, \pi^n)$$

for each unit $k = 1, 2, \dots, n$, and $t = 0, 1, \dots, T-1$.

7.2.3.8. *Open Loop Policies.* Open loop polices are obtained when the maintenance actions at time t are not dependent on the conditional probability $\pi^k(t)$. To determine an optimal open-loop policy it is not necessary to solve a sequential decision process. Instead, the probability distribution of importance is:

$$x_k^j(t) = \text{Prob } [s^j(t) = k \mid \mu(t)]$$

$$= \text{Prob } [\text{unit } j \text{ in state } k \text{ at time step } t \mid \text{the initial state probabilities for unit } k \text{ and the maintenance schedule up to time step } t]$$

for $k = 1, 2, \dots , N^j$; $j = 1, 2, \dots , n$; $t = 0, 1, \dots , T$.

The symbol $\mu(t)$ is used to denote the information contained in the sequence of maintenance actions up to but not including time t, and in the initial state probabilities.

The succession of vectors $x(t)$, $t = 0, 1, 2, \dots , T$ for a unit is obtained by application of the rule of total probability:

$$x_k (t + 1) = \text{Prob } [s(t + 1) = k \mid \mu(t + 1)]$$

$$= \sum_{i=1}^{n} \sum_{r=0}^{q_r} \text{Prob } [s(t + 1) = k \mid s(t) = i, \ r(t) = r]. \tag{7.45}$$

$$\text{Prob } [r(t) = r \mid s(t) = i] \cdot \text{Prob } [s(t) = i \mid \mu(t)]$$

$$= \sum_{i=1}^{n} \sum_{r=0}^{q_r} P_{r(t), \ ik}(t) \cdot \alpha_{ir}(t) \cdot x_i(t),$$

$$k = 1, 2, \dots , N ; \ t = 0, 1, 2, \dots , T . \tag{7.46}$$

and $x_k(0) = $ the initial probability that the unit is in state k, $k = 1, 2, \dots , N$,

where $\alpha_{ir}(t) = \text{Prob } [r(t) = r \mid s(t) = i]$; $A(t) = aN[Nx(a_r + 1)]$ matrix with elements $\alpha_{ir}(t)$, N is the number of states in the unit and a_r is the number of repair alternatives, $r = 0$ denotes the "no repair" decision.

An open-loop maintenance schedule for the unit is a sequence of matrices $A(t)$, $t = 0, 1, \dots , T - 1$, specifying the repair actions in terms of the state of the unit. The maintenance policy is pure (all elements of $A(t)$ are either zero or one). Otherwise, it is called a random policy.

To implement the maintenance policy, an inspection must be performed at each time t where the schedule prescribes a non-zero probability of repair for at least one of the states in the unit. Following the inspection, a repair must be performed in accordance with the outcome of the inspection.

We can write that:

$$u_{jr}(t) = \alpha_{jr}(t) \cdot x_j(t) = \text{Prob}\,[r(t) = r, s(t) = j], \tag{7.47}$$

which leads to:

$$\sum_{x=0}^{a_r} u_{jr}(t) = x_j(t). \tag{7.48}$$

Finally, we have:

$$\sum_{x=0}^{a_r} u_{kr}(t+1) = \sum_{j=1}^{n} \sum_{r=0}^{a_r} P_{rjk}(t) \cdot u_{jr}(t), \text{ for } k = 1, 2, \dots, n \tag{7.49}$$

with the condition:

$$\sum_{x=0}^{a_r} u_{kr}(0) = x_k(0) \text{ for } k = 1, 2, \dots, n \tag{7.50}$$

and the constraints:

$$u_{kr}(t) \geq 0 \text{ for all } k, r \text{ and } t. \tag{7.51}$$

In multi-unit systems with two-state units, the maintenance policies, called here, open loop policies, are equivalent to closed loop policies [3], [11], [37], [46], [48], [80], [81], [82], [84].

7.3 A Markovian decision model for clinical diagnosis and treatment applied to the respiratory system

The complex process of medical diagnosis has traditionally relied on the experience and judgement of the clinician. Decision theory is well established and several decision models have been applied to medical problems. Next, a unified approach to clinical decision making is presented. This approach combines partially observable Markov decision processes with cause-effect models as a probabilistic representation of the diagnostic process. This class of model has a direct application to medical diagnosis and treatment, and a respiratory system example is presented. The methodology is given for combining the patient's state of health, the clinician's state of knowledge of the cause-effect representation from the observation space and,

finally, the treatment decisions with which to restore the patient to a more desirable
state of health [32].

A pictorial representation is given in Figure 7.9.

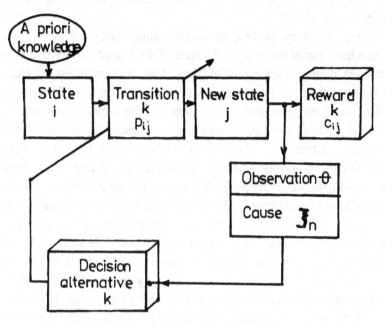

Figure 7.9. Markovian model representation for clinical diagnosis and treatment.

Clearly, many different effects are involved in the whole system operation, e.g.
blood flow rate, rate of breathing and the well recognized feedback control of
breathing. Furthermore, the plant efficiency depends on blood circulation time and
the efficiencies of lung and tissue gas exchange systems; also mechanical factors
such as rib-cage mechanics, hardening of the arteries and heart pump defects. even
without obvious clinical defect, the dynamic performance of the system deteriorates
with age. Clinical decisions regarding diagnosis and treatment are rendered more
difficult by this dynamic complexity and the fact that the system is only partially
observable.

In presenting our unified approach to clinical decision making, it is assumed
that the clinician wishes to choose an action from a finite set of treatment alternatives.
The states of the patient (and, in this case, his respiratory states) are considered to
follow Markovian dynamics describable in terms of a performance index based upon
deterioration of the system with time or due to a fault such as infection. The system
is complex and not completely observable, so any change of patient state can only be
assessed on the basis of clinical obsrvations. These may be cheap or expensive to
perform. In addition, care must be taken to ensure that any measurements performed

on the patient do not lead to a deterioration in his state of health. It is on the basis of these observations that decisions are made.

In carrying out the diagnosis, the clinician aims to ensure that the outcome of the decision is satisfactory both to the patient and himslef. During this process, three stages have to be considered:

(i)   taking measurements (e.g. pulse rate, temperature),

(ii) reaching a conclusion as to the state of the process; whether it is a case of heart disease, lung disease, lack of blood flow, poor environment, etc.,

(iii) taking a decision regarding any course of action (e.g. do nothing, apply medication, change environment, perform additional measurements).

For the respiratory system, the measurement space may consist of heart rate, respiratory rate, temperature, tidal volume, noise level in lungs and oxygen and carbon dioxide concentrations in venous and arterial blood (these last observations being time-consuming and expensive). A given set of observations could result from a number of different malfunctions. For example, a high ventialtion rate could be due to low perfusion ventilation or a slow rate of blood flow through the system.

The clinician has also to decide at what point to stop making further observations and start making decisions about medication or some alternative treatment. So decision making concerns not only the final treatment, but also, whether a given set of observations reveals the cause of the ailment with any degree of certainty. Traditionally, all this has been done in an implicit semi-intuitive fashion.

The model developed in this section provides a description, in terms of dynamic decision making, for the complex process of diagnosis and treatment which form a part of the clinical health care process. The proposals described incorporate Markov decision processes and the cause-effect models. This new class of model has a direct application to medical diagnosis and treatment, and a respiratory system example is presented. The methodology is given for combining the patient's state of health, the clinician's state of knowledge of the cause-effect representation from the observation space and, finally, the treatment decisions with which to restore the patient to a more desirable state of health.

The development of decision models depends on the availability of probabilistic information. The assignment of numerical values to probabilitites is a subjective process and as such depends on the particular decision maker, his experience and his state of knowledge regarding the process under consideration.

7.3.1. CONCEPT OF STATE IN THE RESPIRATORY SYSTEM

The class of Markovian models presented here has considerable advantage over other such models described in the literature. For large scale processes, Markov models suffer "the curse of dimensionality" and, therefore, an aggregation in the process state space is needed if appropriate computational optimization techniques such as "policy iteration" or linear programming are to be used. Consider the building up of a Markov model on the level of causes in the diagnosis problem. If there are n causes (primary events for the health states of the patient) where each of them could take a dichotomic value $(0, 1)$ (see Figure 7.10) then it is easy to see that the process state will have 2^n states, and, even for a relatively small number of causes, the state space for the Markov process increases exponentially $(2^{10} > 10^3, 2^{20} > 10^6)$.

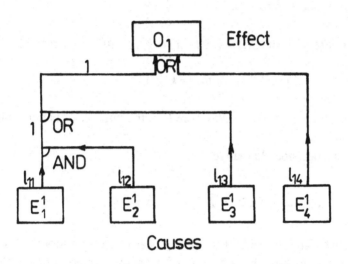

Figure 7.10. Cause-effect model.

Therefore, by defining an aggregate state of health for the "core process", and using an observation space and a diagnosis description for such observations (cause-effect models), it is possible, so to speak, to look inside the patient's body without using any attempt at a completely observable process (e.g. this can happen with surgery when the surgeon has the possibility of observing the state of any "component" of the respiratory system).

The Markov decision model, however, with logical considerations for the diagnosis process given here, uses two levels.

First it aggregates the state of any "component" in the respiratory system on the level of "state of health" and then relaxes the uncertainty about the state of these by using a "fixed scheme" — cause-effect models — which is probabilistically correlated with the "core process" dynamics given by probabilities p^k_{ij}, for any pertinent treatment alternative, k.

To assign numerical values to each different state of the system would be extremely difficult. It is considered adequate, however, to create a state space where each state such as "well", "ill", or "very ill" describes the performance index for the respiratory system.

The formulation is given in terms of the quadruplex $\{S, \mathcal{D}, \mathcal{O}, \mathcal{E}\}$ for the patient-state process S, the control process $\mathcal{D}$, observation $\mathcal{O}$ and $\mathcal{E}$ the state of knowledge.

Our assumption is that the system has Markov behaviour. The system dynamics are therefore given by probabilities describing the transitions between patient states (see also Figure 7.9).

$$p^k_{ij} = \text{Pr} \{S(t + 1) = j \mid S(t) = i, S(t - 1) = 1, \dots , S(0) = m, \ u(t) = k, \ \mathcal{E}(t)\}$$

$$= \text{Pr} \{S(t + 1) = j \mid S(t) = i, u(t) = k, \mathcal{E}(t)\} \tag{7.52}$$

$i = 1, 2, \dots , N ; \ j = 1, 2, \dots , N$ where N is the number of patient states.

$u (t) = k$ is a decision alternative.

$$k = 1, 2, \dots , K_i \text{ and } p^k_{ij} \geq 0, \ \sum_{j=1}^{N} p^k_{ij} = 1; \ i = 1, 2, \dots , N ; \ k = 1, 2, \dots , K_i .$$

This probabilistic measure of the patient states in respiratory illness provides sufficient information about the behaviour of the respiratory system and shows that the current state of the system $S(t + 1)$, depends only on the immediate past transition and the current state of knowledge, $\mathcal{E}(t)$, about the system, that is to say which treatment alternative led to state $S(t) = i$.

For a time dependent Markov process, the state vector $\pi(t)$, with components $\pi_i(t)$ — the probability that state i is occupied at time t — can be written as:

$$\pi(t) = [\pi_i(t), i = 1, 2, \dots , N]$$

$$\pi_i(t) = \text{Pr} \{S(t) = i \mid \mathcal{E}(t)\}, \ i = 1, 2, \dots , N]. \tag{7.53}$$

The process states at time zero, therefore $\pi_i(0) = \text{Pr} \{S(0) = i \mid \mathcal{E}(0)\}$ for any $i = 1, 2, \dots , N$.

Time $t = 0$ corresponds to the instant at which the systems investigation begins, for example, at the first presentation of the disease symptoms to the clinician.

In equation (7.53), $\mathcal{E}(t)$ contains all past information and history of treatment and patient health evolution whereas $\pi(t)$ is only a partial encoding of the representation of that information. Vector $\pi(t)$ represents the state of knowledge on the process at time t rather than that through time t, $\mathcal{E}(t)$. $\mathcal{E}(t)$ is a function of $u(0)$, $u(1), \ldots, u(t-1)$.

The state vector depends on past outputs and the history of treatment decisions, and measures the probability of the patient being in each state as a result of a particular treatment decision and patient health evolution. A set of all possible state vectors can be defined by:

$$\Pi = \text{set} \, [\pi_i : \pi_i \geq 0 \, ; \, \sum_{i=1}^{N} \pi_i = 1] \tag{7.54}$$

$\pi = \lim_{t \to \infty} \pi(t)$ for the case in which the process is allowed to make a large number of transitions.

7.3.2. THE CLINICAL OBSERVATION SPACE

By performing a set of independent observations, $\theta = 1, 2, \ldots, M$, the clinician obtains information about the patient's state. These observations may be classified as cheap or expensive (Figure 7.9). The new state of knowledge regarding the state of the respiratory system at the $(t + 1)$-th transition is given by:

$$\mathcal{E}(t+1) = [\mathcal{E}(t), Z(t+1) = \theta, Y(t+1) = \zeta_n] \tag{7.55}$$

where $Z(t + 1)$ is an event representing an observable output at time $(t + 1)$, $Y(t + 1)$ is an event representing a cause in a cause-effect model and ζ_n is a primary event in a cause-effect model; the cause to which the observation θ is attributed.

The random inspection (clinical observation) is described by the probability:

$$r_{j\theta}^{k} = \Pr \{Z(t+1) = \theta \mid S(t+1) = j, \, S(t) = i, \, u(t) = k, \, \mathcal{E}(t)\} \tag{7.56}$$

$$j = 1, 2, \ldots, N \, ; \, \theta = 1, 2, \ldots, M \, ; \, k = 1, 2, \ldots, K_j$$

and

$$0 \leq r_{j\theta}^{k} \leq 1, \, \sum_{\theta=1}^{M} r_{j\theta}^{k} = 1; \, j = 1, 2, \ldots, N$$
$$k = 1, 2, \ldots, K_j \tag{7.57}$$

This probability defines the probability of achieving observations θ, given that state j results from a particular treatment history.

The functional form of the decision variable $u(t)$ is precisely specified e.g. $u(t) = f(t, Z(1), \ldots, Z(t), u(1), \ldots, u(t-1))$.

Following an observation, a cause-effect model can be built in which all the logical combinations (such as AND, OR, TRUE, FALSE) of the causes are correlated with the observation outputs. A cause-effect model is then characterized by:

$$l_{\theta\tau_n} = \text{Pr}\,\{Y(t+1) = \tau_n,\ S(t+1) = j\,|\,Z(t+1) = \theta,\ S(t) = i, \mathcal{E}(t)\}$$

$$= \text{Pr}\,\{Y(t+1) = \tau_n\,|\,Z(t+1) = \theta,\ S(t) = i, \mathcal{E}(t)\}$$

$$\theta = 1, 2, \ldots, M\ ;\ \tau_n = 1, 2, \ldots, N_{A_n} \tag{7.58}$$

where N_{A_n} is the number of cases which could give rise to a given observation, and

$$0 \leq l_{\theta\tau_n} \leq 1\ ;\ l_{\theta\tau_n} + \overline{l}_{\theta\tau_n} = 1 \tag{7.59}$$

where $l_{\theta\tau_n}$ specifies the probability of a particular case, given an observation θ and the present state being achieved as a result of a specific treatment history. A sample cause-effect model for the respiratory system is given in Figure (7.10).

7.3.3. COMPUTING PROBABILITIES IN CAUSE-EFFECT MODELS AND OVERALL SYSTEM DYNAMICS

It is necessary to have methods of calculating probabilities $l_{\theta\tau_n}$ in cause-effect models which involve logical combinations such as AND/OR gates. Examples of calculating these probabilities for simple cause-effect models are given in Table 7.2.

The state space of the process described above for the respiratory system is no longer Markov; its form is more complicated even than that given by Smallwood and Sondik [19]. The dynamics of the new state-space (see Figures 7.11, 7.12, 7.13 and 7.14) are described below.

Let

$$X^k = (R_\theta^k \nabla L_{\zeta_n}^k) \tag{7.60}$$

where ∇ is in effect a logical multiplication operator. X^k therefore measures the probability of a particular cause having given rise to a specific state which was achieved as a result of decision alternative (treatment history) k.

	Cause effect model	Probability Evaluation
1.		$\Pr\{B \mid A\} = p \; ; \; \Pr\{B \mid C\} = q$ $\Pr\{B \mid A, \text{NOT } C\} = p$ $\Pr\{B \mid \text{NOT } A, C\} = q$ $\Pr\{B \mid \text{NOT } A, \text{NOT } C\} = 0$ $\Pr\{B \mid A, C\} = p + q - pq$ $B = (A : p) \text{ OR } (C : q)$
2.		$\Pr\{B \mid A\} = 0; \; \Pr\{B \mid C\} = 0$ $\Pr\{B \mid A, \text{NOT } C\} = 0$ $\Pr\{B \mid \text{NOT } A, C\} = 0$ $\Pr\{B \mid \text{NOT } A, \text{NOT } C\} = 0$ $\Pr\{B \mid A, C\} = p \cdot q$ $B = (A : p) \text{ AND } (C : q)$
3.		$D = ((A : p) \text{ OR } (B : q)) \text{ OR } (C : r)$

Table 7.2. Probabilities for simple cause-effect models.

For an example in which there are three patient states, two observations and two causes, X^k has 12 elements. Considering the patient in state j as a result of treatment history k :

$$X_j^k = R_j^k \nabla L_{\zeta_n}^k = [R_{j1}^k \cdot L_{11}^k \; R_{j1}^k \cdot L_{12}^k \; R_{j2}^k \cdot L_{21}^k \; R_{j2}^k \cdot L_{22}^k]$$

where L_{uv} refers to observation u and the v-th element in the L vector; and R_{jw} refers to the w-th element in a row.

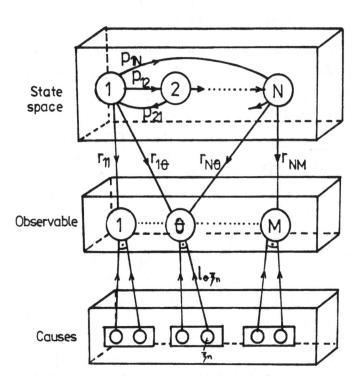

Figure 7.11. Markov model for clinical diagnosis for the respiratory system.

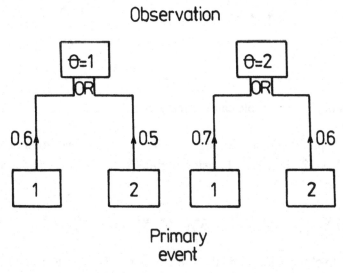

Figure 7.12. Cause-effect model for the simple respiratory example.

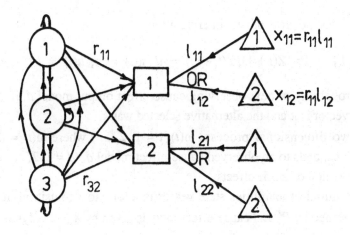

State   Observation Cause - Effect
                              Model

Figure 7.13. 'Logical multiplication' for the respiratory example

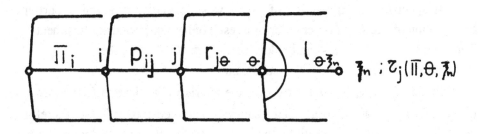

Figure 7.14. State dynamics.

Let d be an index for the cause-effect model. In the example with two causes and two observations, d has four elements corresponding to each combination of cause and observation.

As X^k is a non-square matrix, a square matrix X_d^k is defined as diag X^k for constant d, $d = 1, 2, \ldots, \mathcal{M}$ and every other element of X_d^k is zero, then:

$$\{d \mid \pi, k\} = \pi P^k X_d^k \mathbf{1} \tag{7.61}$$

where $\mathbf{1} = [1 \ldots . 1]^T$

and $\{d \mid \pi, k\}$ has the probabilistic interpretation:

$$\{d \mid \pi, k\} = \Pr \{Z(t+1), Y(t+1)) = d \mid \pi(t) = \pi, \ u(t) = k\},$$

that is the probability of next observing cause-effect output model d, if the current information vector is π and the alternative selected was k.

The two dimensional process, $\omega(t), \ t = 1, 2, \ldots)$ where $\omega(t) = (Z(t+1), Y(t+1))$, corresponds to an observed cause-effect model d, $(d = 1, 2, \ldots, \mathcal{M})$ rather than an observation on some effects.

As the initial probabilities state vector is π, due to a transition in the "core process" governed by P^k and cause-effect models given by X_d^k, the dynamics of the new state space, in matrix form, become:

$$T(\pi \mid d, k) = \frac{\pi P^k X_d^k}{\{d \mid \pi, k\}} \tag{7.62}$$

where k represents treatment decision alternative k at time t in the triplex $D = \{S, \mathcal{O}, \mathcal{D}\}$ such that $k \in \mathcal{D}$ and $d \in \mathcal{O}$.

It specifies the dynamics of the new state vector π given a particular combination of cause and observation as a result of having carried out treatment k. Thus, the system dynamics can be described by $\pi, \pi', \pi'', \ldots$ where

$\pi' = T(\pi \mid d, k), \ \pi'' = T(\pi' \mid d, k)$, etc.

Dimensionally, $\{\pi \mid d, k\}$ involves the matrix multiplication $(1 \times N) \ (N \times N) \ (N \times N) \ (N \times 1)$ leading to a scalar quantity. The elements of T are, therefore, normalised by dividing through by the index d, so that the sum of the probabilities constituting the elements of T equal unity.

7.3.4. DECISION ALTERNATIVES IN RESPIRATORY DISORDERS

To improve the patient's state, or even to maintain him in an acceptable state, it is necessary to carry out a set of treatment decisions. From the viewpoint of the model, we can ascribe rewards to the implementing of alternative treatments to patients in each state, given the current observations.

When the clinician is able to choose amongst alternative treatments, a Markov decision process can be built up. Since the treatment and observation (measurement) time is short in comparison with the lifetime of the patient, an infinite horizon $(t \to \infty)$ model will be considered. In effect, the problem becomes one of evaluating the available information.

Let us define:

$g(\pi)$ – expected return on treatment in the case of a completely observable process (after a transition in the systems state space, the new process state can be identified precisely),

$\bar{g}(\pi)$ – expected return on treatment for a partially observable process (only a probabilistic observation of the system's states).

$g^*(\pi)$ – expected return on treatment for probabilistic observation of the patient's states and auxiliary cause-efeect models.

The representation of such models is given in Figure 7.15.

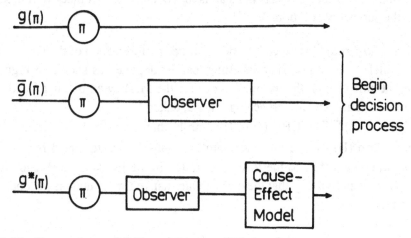

Figure 7.15. Comparison of different classes of Markov processes.

It can be shown that:

$$g(\pi) \le g(\bar{\pi}) \le g^*(\pi).$$ (7.63)

A Branch and Bound algorithm is now developed for evaluating alternative treatments. In the case of the respiratory system these decisions might be drug therapy, intensive care nursing, surgery, other therapies, or, even 'do nothing'. The computational advantage of the Branch and Bound optimization technique is that it can handle relatively large processes without recourse to the complicated approximation techniques normally described in the literature for infinite horizon models, such as dynamic programming (policy iteration technique using finite transient policies). Nevertheless, the Branch and Bound algorithm given here is not the only suitable optimization technique for the class of problems treated in this work and further effort is required in the search for "efficient" model formulation and adequate computational techniques [26], [29], [36], [50], [51], [83], [93], [96], [100].

7.3.4.1. *Branch and Bound Algorithm.* The Branch and Bound algorithm is applied to this Markov decision process, specified by the patient state dynamics, cause-effect models, and treatment decision alternatives. As was already emphasized, in essence this is an enumerative scheme for solving optimization problems. By applying the procedures of this algorithm, the optimal treatment decision alternative corresponding to each state of the core process can be identified out of the set of possible solutions.

The Branch and Bound method deals with:

"*Branching* which consists of dividing collections of sets of solutions into sub-sets.
Bounding which consists of establishing bounds of the objective function over the sub-set of solutions".

The *branching rule* $\mathcal{B}$ is given by a decision tree representation of the process (Figure 7.16) which, after appropriate labelling, can be depicted as shown in Figure 7.17, where, at time $(t + 1)$, the terminal nodes for the Markovian tree can be identified, and used as value function branching criteria, either:

(a) DIFF = $(\beta)^{level} \times$ (UPB $(I) - (I)) \times$ prob. that the node is maximum for that branch (Prob (I)), where DIFF is simply a branching criterion, β is the discount factor, and level refers to the level of iteration in the Branch and Bound algorithm Prob (I) is the probability that the decision alternative for the node is the optimum for that branch;

  or

(b) the lower bound for the decision k is greater than the upper bound for any decision $j \neq k$ for that particular branch.

It can be shown that an upper bound of the cost functional exists and is given by:

$$g(\pi) \leq \sum_{i,\zeta_n} \pi_i \, v_i \, l_{\zeta_n} \tag{7.67}$$

where:

$$v_i(t) = \max_k \left\{ \sum_{j=1}^{N} p_{ij}^k \, (c_{ij}^k + \beta \, v_j \, (t - 1)) \right\} \tag{7.68}$$

$$i = 1, 2, \dots, N .$$

c_{ij}^k is the treatment reward structure for a completely observable process which makes the transition form state i to state j, $v_i(t)$ is the maximum expected return if the process finishes in state i as a result of decision alternative k :

$$v_i = \lim_{t \to \infty} v_i(t) \quad \text{i.e. the steady state value of infinite horizon.}$$

It can be shown that a lower bound of the reward functional exists and is given by:

$$g(\pi) > [I - \beta P^k]^{-1} c^k \qquad (7.66)$$

where the expected return for one transition $c_i^k = \sum_j p_{ij}^k c_{ij}^k$; $i = 1, 2, \ldots, N$.

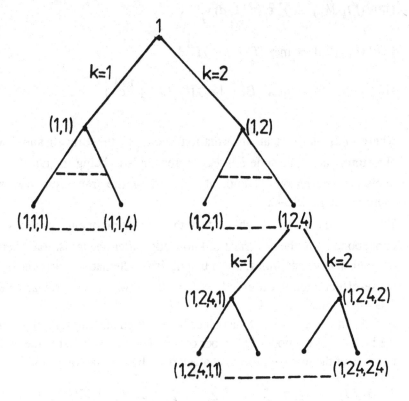

Figure 7.16. Decision tree for the Markov process.

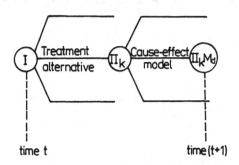

Figure 7.17. Modified decision tree.

7.3.4.2. *Steps in the Branch and Bound Algorithm*

Step 1 : Perform the branching rule $\mathcal{B}$ and identify all nodes in the tree (Figure 7.16),

Step 2 : Calculate the upper and lower bounds for the terminal nodes in the decision tree:

$$\text{UPB } (I\,I_k\, \mathcal{M}_d) = \sum_j \tau_j\, (\pi \mid k, d)\, v_j$$

$$\text{LOB } (I\,I_k\, \mathcal{M}_d) = \max_A\, [T(\, \cdot \mid \cdot\,)\, L^A\,]$$

$$= \max_A\, \{T\, (\, \cdot \mid \cdot\,)\, [I - \beta P^A\,]^{-1}\, c^A\, \}$$

where $(I\,I_k\, \mathcal{M}_d)$ is an information vector in the Branch and Bound algorithm such that node $I\,I_k\, \mathcal{M}_d$ corresponds to being in state I, taking decision alternative k, and, at the end of the transition, observing cause-effect model $\mathcal{M}_d$.

$T(\pi \mid \cdot\,) = [\tau_j\, (\pi \mid \cdot\,)]$ where τ_j is the j-th component of the vector corresponding to the new state of knowledge after one transition where the original state vector was π, given that decision alternative k was chosen and cause-effect model observed. L^A is the lower bound reward vector following some policy A.

Step 3 : For any initial node which has other branches emanating from it, calculate the bounds for the nodes I (number of initial nodes I is equal to the number of states of the internal process characterized by the Markov model).

$$\text{UPB } (I\,I_k) = \sum_j \pi_j\, p_{ij}^k\, c_{ij}^k + \beta \sum_{\mathcal{M}_d}\, \sum_j\, \sum_i\, \pi_j\, p_{ij}^k\, x_{jd}^k \cdot \text{UPB } (II_k\, \mathcal{M}_d)$$

$$\text{LOB } (I\,I_k) = \sum_j \pi_j\, p_{ij}^k\, c_{ij}^k + \beta \sum_{\mathcal{M}_d}\, \sum_j\, \sum_i\, \pi_j\, p_{ij}^k\, x_{jd}^k \cdot \text{LOB } (II_k\, \mathcal{M}_d)$$

where $\sum_j p_{ij}^k\, c_{ij}^k$ is the expected immediate reward resulting from the transition.

$$\text{UPB } (I) = \max_k\, \{\text{UPB } (II_k)\}$$

$$\text{LOB } (I) = \max_k\, \{\text{LOB } (II_k)\}$$

Step 4 : Select nodes for further branching in accordance with some criteria given above (Figure 7.16 and Table 7.3).

Step 5 : Scan all the nodes in the decision tree from the terminal nodes (which now correspond to the highest level in the tree).

Step 6 : If, at a given level, a dominating decision exists, (LOB (a, l) > UPB (b, l) $a \neq b$), where a and b are decision alternatives, then go to Step 7. If not, repeat the procedure with the new tree and go to Step 2.

Step 7 : Stop and print the results.

The algorithm finds an optimal solution for the problem in a finite number of iterations.

7.3.5. A NUMERICAL EXAMPLE FOR THE RESPIRATORY SYSTEM

The Markovian decision modelling techniques are now applied to the respiratory system. In order to limit the computational complexity, only a small model is considered. The data used fall within physiological limits and are based on current clinical knowledge. The methods, however, can be readily applied to more realistic complex situations.

Let us consider the respiratory system with $N = 3$ states (ill, satisfactory and good), and $M = 2$ observable outputs (low peak ventilation, i.e. less than normal but greater than 100 l.min^{-1}, and very low peak ventilation, i.e. less than 100 l.min^{-1}.) Let us suppose that there are two primary events (causes) for each observation as shown in Figure 7.12. Observation $\theta = 1$ is very low peak ventilation; $\theta = 2 =$ low peak ventilation: cause 1 = emphysema; cause 2 = pneumonia.

The other process parameters are listed below. Consider that the number of treatment alternatives $k = 2$. Alternative 1 ($k = 1$) is to apply medication and alternative 2 is to 'do nothing'. The following represent the state transition probabilities.

Decision alternative 1	Decision alternative 2

$$[p_{ij}^1] = \begin{bmatrix} 0.1 & 0.1 & 0.8 \\ 0.2 & 0.5 & 0.3 \\ 0.7 & 0.1 & 0.2 \end{bmatrix} \qquad [p_{ij}^2] = \begin{bmatrix} 0.1 & 0.8 & 0.1 \\ 0.7 & 0.1 & 0.2 \\ 0.1 & 0.9 & 0 \end{bmatrix}$$

where 1 = ill; 2 = satisfactory; 3 = good.

For example, if as a result of 'doing nothing' the patient is in state 3, there is an 0.9 probability that his next transition will be to state 2.

Initial State Vector

$$\pi = [\,0.3 \quad 0.2 \quad 0.5\,]$$

Random Inspection (Clinical Observation) Probabilities

$$k = 1 \qquad\qquad\qquad\qquad\qquad k = 2$$

$$[r^1_{j\theta}] = \begin{bmatrix} 0.7 & 0.3 \\ 0.1 & 0.9 \\ 0.4 & 0.6 \end{bmatrix} \qquad\qquad [r^2_{j\theta}] = \begin{bmatrix} 0.2 & 0.8 \\ 0.4 & 0.6 \\ 0.3 & 0.7 \end{bmatrix}$$

For example, the probability of achieving observation $\theta = 1$, given that state 2 has resulted from treatment alternative $k = 1$ is 0.1

Cause-Effect Model Probabilities

$$k = 1 \qquad\qquad\qquad k = 2$$

$$L^1_1 = \begin{bmatrix} 0.6 \\ 0.5 \end{bmatrix} \qquad\qquad L^2_1 = \begin{bmatrix} 0.6 \\ 0.5 \end{bmatrix}$$

$$L^1_2 = \begin{bmatrix} 0.7 \\ 0.6 \end{bmatrix} \qquad\qquad L^2_2 = \begin{bmatrix} 0.7 \\ 0.2 \end{bmatrix}$$

For example, there is a probability of 0.7 that observation $\theta = 2$ results from cause 1 and a probability of 0.2 that the same observation results from cause 2, where in both cases the patient was in his present state as a result of treatment alternative $k = 2$.

Return on Termination of the Process

$$k = 1 \qquad\qquad\qquad k = 2$$

$$v^1 = \begin{bmatrix} 6 \\ 4 \\ 5 \end{bmatrix} \qquad\qquad v^2 = \begin{bmatrix} 7 \\ 9 \\ 6 \end{bmatrix}$$

Reward Function for Change of State

$$k = 1 \qquad\qquad\qquad k = 2$$

$$c^1 = \begin{bmatrix} 3 & 5 & 7 \\ 2 & 4 & 3 \\ 6 & 4 & 2 \end{bmatrix} \qquad c^2 = \begin{bmatrix} 2 & 4 & 6 \\ 3 & 5 & 7 \\ 2 & 2 & 5 \end{bmatrix}$$

Using the cause-effect model representation and relations derived above, causes may be related to states as follows:

$$k = 1 \qquad\qquad\qquad k = 2$$

$$c^1 = \begin{bmatrix} 0.42 & 0.35 & 0.21 & 0.18 \\ 0.06 & 0.05 & 0.63 & 0.54 \\ 0.24 & 0.20 & 0.42 & 0.36 \end{bmatrix} \qquad c^2 = \begin{bmatrix} 0.12 & 0.10 & 0.56 & 0.16 \\ 0.24 & 0.20 & 0.42 & 0.12 \\ 0.18 & 0.15 & 0.49 & 0.14 \end{bmatrix}$$

This 'logical multiplication' is depicted in Figure 7.13. the dynamics of the new state space can then be calculated using Equation (7.62). The results for the two decision alternatives are shown in Table 7.3.

$\tau_j(\pi \,|d, 1)$	State j		
	1	2	3
$d = 1$	0.622	0.038	0.340
$d = 2$	0.622	0.039	0.339
$d = 3$	0.239	0.307	0.454
$d = 4$	0.239	0.307	0.454

where cause-effect model $d = 1$ corresponds to $\theta = 1$, $\zeta_n = 1$

$$d = 2 \qquad\qquad \theta = 1, \zeta_n = 2$$
$$d = 3 \qquad\qquad \theta = 2, \zeta_n = 1$$
$$d = 4 \qquad\qquad \theta = 2, \zeta_n = 2$$

Table 7.3. (a) First Iteration: Decision Alternative $k = 1$.

| | State j | | |
$\tau_i(\pi \mid d, 2)$	1	2	3
$d = 1$ 0.126	0.814	0.060	
$d = 2$	0.126	0.814	0.060
$d = 3$	0.270	0.654	0.076
$d = 4$	0.270	0.654	0.076

Table 7.3. (b) First Iteration: Decision Alternative $k = 2$.

The Branch and Bound algorithm is now applied leading to the results shown in Table 7.4.

Node	Upper Bound	Lower Bound
1, 1	6.06572	6.2288184
1, 2	5.0517224	6.0051232

Table 7.4. First Iteration.

Since LOB (1, 1) is greater than UPB (1, 2), decision 1 dominates decision 2, and, therefore, decision 1 is optimal. That is to say, for a patient who is currently in state 1, the optimal strategy is to apply medication ($k = 1$). In a similar manner optimal decision strategies can be calculated for other patient states.

With this particular example, the solution has emerged at the first iteration. A typical node distribution for a more general case requiring two iterations is shown in Figure 7.16, with the nodes starting from the corresponding state 1 (ill). The general node nomenclature follows the pattern ($I \, I_k \, M_d$) so, for example, node (1, 2, 3) corresponds to being in state 1, taking decision alternative 2 and observing cause-effect model 3.

The bounding procedure is carried out once a new branching has been implemented using the branching rule $\mathcal{B}$. Typical values adopted would be a discount factor $\beta = 0.5$, and Pr $\{I\} = 0.8$ (probability of realizing a segment history of events such as states, observations, decisions and cause-effect models), enabling the value of the branching criterion DIFF to be evaluated. For the case shown in Figure 7.16 the terminal node (1, 2, 4) yields the value of DIFF which is nearest to that calculated for the initial node 1 and this node, therefore, becomes node 1 for the second iteration which would be expanded in a manner identical to that carried out in the first iteration.

7.3.6. CONCLUSIONS

In applying decision modelling techniques to clinical problems, many of the difficulties are associated with data acquisition. First, it has been assumed that the process parameters remain stationary during the period of investigation. The whole problem of confidence in the data needs to be examined, considering factors such as process parameter sensitivity. The problems of encoding need to be examined; that is to say, at what point do we stop making observations and begin the decision process? What is the cost of additional observations and what benefit in terms of more reliable probabilities is achieved? These questions have been highlighted by this discussion, and clearly much more work remains to be done.

BIBLIOGRAPHY

1. Blackwell, D. : 'Discounted dynamic programming', *Ann. Math. Stat.* **36** (1965), 226-235.

2. Bonhomme, N.M. : *Evaluation and optimization of inspection and repair schedules*, Ph.D. Dissertation, Carnegie-Mellon University, May (1972).

3. Gheorghe, A.V. : 'On risk sensitive Markovian decision models for complex system maintenance', *Economic Computation and Economic Cybernetics Studies and Research,* **1** (1976).

4. Grinold, R.C. : 'Elimination of sub-optimal actions in Markov decision problems', *Oper. Research,* **2** (3) (1973), 848-852.

5. Gantmacher, F.R. : *The Theory of Matrices,* **1** and **2**, Chelsea (1960).

6. Hastings, N.A.J., Mello, J.M.C. : 'Tests for sub-optimal actions in discounted Markov programming', *Management Science,* **19** (1973), 1019-1023.

7. Howard, R.A. : *Dynamic Programming and Markov Processes*, M.I.T. Press, Cambridge, Mass, 1960.

8. Howard, R.A. : 'The foundations of decision analysis', *IEEE Trans.,* **SSC-4** (3), September (1968), 211-220.

9. Howard, R.A. : *Dynamic Probabilistic Systems*, John Wiley, New York, (1970).

10. Howard, R.A., Matheson, J.E. : 'Risk-sensitive Markov decision processes', *Management Science,* **18** (7) (1972), 356-370.

11. Kao, E.P.C. : 'Optimal replacement rules when changes of state are semi-Markovian', *Oper. Research,* **21** (6) (1973), 1231-1250.

12. Kao, E.P.C. : 'A semi-Markovian population model with application to hospital planning', *IEEE Trans.,* **SMC-4** (3), July (1973), 327-336.

13. Lawler, E., Wood, D.E. : 'Branch and Bound methods - a survey', *Oper. Research,* **14** (4) (1970), 699.

14. Leonte, A., *et al.* : *Elemente de calcul matriceal cu aplicatiif*, Editura Tehnica, Bucuresti, (1975).

15. Janssen, J. : *Semi-Markov Models. Theory and Applications*, Plenum Press, New York (1986).

16. MacQueen, J.B. : 'A test for sub-optimal actions in Markovian decision problems', *Oper. Research,* **15** (4) (1967), 559-561.

17. Mine, H., Osaki, S. : *Markovian Decision Processes*, Elsevier, New York, (1970).

18. Satia, J.K., Lave, R.E. : 'Markovian decision processes with probabilistic observation of states', *Management Science,* **20** (1) (1973), 1-14.

19. Smallwood, R.D., Sondik, E.J. : 'The optimum control of partially observable Markov processes over a finite horizon', *Oper. Research*, **21** (5) (1973), 1071-1088.

20. Sondik, E.J. : *The optimum control of partially observable Markov processes,* Ph.D. Dissertation, Stanford University, EES Dept., California (1971)

21. Takahashi, Y. : 'On the effects of small deviations in the transition matrix of a finite Markov chain', *J. Oper. Research Soc. of Japan*, **16** (2) (1973), 104-130.

22. Porteus, E.L. : 'Some bounds for discounted sequential decision processes', *Management Science*, **18** (1971), 7-11.

23. Jaquette, S.C. : *Utility criteria for Markov decision processes*, Technical Report No. 200, Dept. of Operations Research, Cornell University (1973).

24. McPherson, P.K.M. : 'Systems science and interdisciplinary education', *J. Systems Engineering*, **4** (1) August (1974), 20-39.

25. Onicescu, O., Oprisan, Gh., Popescu, Gh. : *Renewal Processes with Complete Connections*, Institutul National pentru Creatie Stiintifica si Tehnica, Institutul de Matematica, nr. 27, (1982), ISSN 0250-2638.

26. Iosifescu, M. : 'Processus aleatoires a liaisons completes purement discontinus', *C.R. Acad. Sci. Paris*, **266** (1968), A1159-A1161.

27. Kesten, H. : 'Occupation times for Markov and semi-Markov chains'. *Trans. Amer. Math. Soc.*, **103** (1962), 82-112.

28. Cinlar, E. : 'Markov renewal theory', *Advances Appl. Probab.*, **1** (1969), 123-187.

29. Norman, M.F. : *Markov Processes and Learning*, Academic Press, New York (1972).

30. Onicescu, O., Mihoc, G. : 'Sur les chaines statistiques', *C.R. Acad. Sci. Paris*, **200** (1935), 511-512.

31. Popescu, G. : 'Functional limit theorems for random systems with complete connections', *Proc. fifth Conf. on Probab. Theory*, Brasov, Romania, Sept. 1-6 (1974).

32. Gheorghe, A. : *Applied Systems Engineering*, John Wiley, New York, 1982.

33. Mihoc, Gh., Niculescu, St. P. : *Procese stohastice de reinnoire*, Editura Academiei R.S. Romania, Bucuresti (1982).

34. Iosifescu, M. : 'An extension of the renewal equation', *Z Wahrschein-lichkeitstheorie und Verw. Gebiete*, **23** (1972), 148-152.

35. Pyke, R. : 'The existence and uniqueness of stationary measures for Markov renewal processes', *Ann. Math. Stat.*, **37** (1966), 1439-1462.

36. Iosifescu, M., Theodorescu, R. : *Random Processes and Learning*. Springer Verlag, Berlin (1969).

37. Iosifescu, M. : *Finite Markov Processes and their Application*, Wiley, New York, (1980).

38. Gheorghe, A. : *Ingineria Sistemelor*, Editura Academiei R.S. Romania, Bucuresti, (1979).

39. Pyke, R. : 'Markov renewal processes: definitions and preliminary properties', *Ann. Math. Stat.*, **32** (1961), 1231-1242.

40   Pyke, R. : 'Markov renewal processes with finitely many states', *Ann. Math. Stat.*, **32** (1961), 1231-1242.

41. Pyke, R., Schaufele, R. : 'Limit theorems for Markov renewal processes', *Ann. Math. Stat.*, **35** (1964), 1746-1764.

42. Yackel, J. : 'Limit theorems for semi-Markov processes ', *Trans. Amer. Math. Soc.*, **123** (1966), 74-104.

43. Yackel, J. : 'A random time change relating semi-Markov and Markov processes', *Ann. Math. Stat.*, **93** (1967), 358-364.

44. Lippman, S. : 'Semi-Markov decision processes with unbounded rewards', *Management Science,* **19** (7), March (1973), 717-731.

45. Dobryden, V.A. ; 'Optimal observation of a semi-Markov process', *Engineering Cybernetics*, No. 4, (1971), 631-633.

46. Kredentser, B.P. : 'Reliability of systems with equipment and time surpluses and instantaneous failure detection', *Engineering Cybernetics*, No. 4, (1971), 640-650..

47. Voskoboyev, V.F. : 'Allowance for incomplete renewal in control of the state of a technological system', *Engineering Cybernetics*, No. 4, (1971), 650-657..

48. Branson, M.H., Shah, B. : 'Reliability analysis of systems comprised of units with arbitrary repair-time distributions', *IEEE Trans.* **R-20** (4) (1971), 217-222.

49. Hinomato, H. : 'Selective control of independent activites: linear programming of Markovian decisions', *Management Science*, **18** (1), September (1971), 88-96.

50. de Ghellink, G.T., Eppen, G. : 'Linear programming solutions for separable Markovian decision problems', *Management Science*, **13** (5), January (1967), 371-393.

51. Veinott, Jr., A. :'Discrete dynamic programming with sensitive discount optimality criteria ', *Ann. Math. Stat.*, **40** (5) (1969), 1635-1660.

52. Jaquette, S.C. : *Markov decision processes with a new optimality criterion*, Technical Report No. 15, May (1971), Dept. of Operations Research, Stanford University, Stanford, California.

53. Tahara, A., Nishida, T. : 'Optimal replacement policies for a repairable system with Markovian transition of states', *J. Operations Research Soc. of Japan,* **16** (2) June (1973).

54. Satia, J.K. : *Markovian Decision Processes with Uncertain Transition Matrices or/and Probabilistic Observation of State*, Ph.D Dissertation, Dept. of Industrial Engineering, Stanford University, Stanford, California, (1968).

55. Holmes, D.S. : *Markov Chains and PERT*, Institute of Administration and Management Monograph IAM-7414, Schenectady, N.Y. (1975).

56. Lave, Jr., R.E. : *Markov Models for Management Control*, Technical Report No. 64-5, Dept. of Industrial Engineering, Stanford University, Stanford, California, December (1971).

57. Eagle, II, J.N. : *A Utility Criterion for the Markov Decision Process*, Ph.D. Dissertation, Dept. of Engineering - Economic Systems, Stanford University, Stanford, California, August (1975).

58. Britney, R. : *The Reliability of Complex Systems with Dependent Sub-system Failures: an Absorbing Markov Chain Model.* Working Paper Series, No. 26R, School of Business Administration, the University of Western Ontario, London, Canada (1973).

59. Bather, J. : *Optimal Decision Procedures for Finite Markov Chains*, Technical Report No. 43, Dept. of Statistics, Stanford University, Stanford, California, July (1972).

60. Howard, R.A. : 'Comments on the origin and application of Markov decision processes'. In: *Dynamic Programming and its Applications*, Academic Press, Inc. (1978).

61. Stone, J.D. : 'Necessary and sufficient conditions for optimal control of semi-Markov jump processes', *SIAM J. Control*, **11** (2) May (1973), 187-201.

62. Cinlar, E. : 'Markov renewal theory: a survey', *Management Science*, **21** (7) March (1975), 727-752.

63. Klein, M. : 'Inspection–Maintenance–Replacement schedules under Markovian deterioration', *Management Science*, **9** (1962), 25-32.

64. Osaki, S., Mine, H. : 'Linear programming algorithms for semi-Markovian decision processes', *J. Math. Anal. Appl.*, **22** (1968), 356-381.

65. Ross, S.M. : *Applied Probability Models with Optimization Applications*, Holden-Day, San Francisco, (1970).

66. Derman, C. : *Finite State Markovian Decision Processes*, Academic Press, New York (1970).

67. Zidaroifu, C. : *Programarea dinamica discreta*, Editura Tehnica, Bucuresti, (1975).

68. Huang Tui : 'Concave programming with linear constraints', *Soviet Math. Dokl.*, **5** (1964), 1437-1440.

69. Gheorghe, A. : 'Partially observable Markov processes with a risk-sensitive decision-maker', *Revue Roumaine Math. Pur. Appl.*, **22** (1977), 461-482.

70. Gheorghe, A., *et al* : 'Dynamic decision models for clinical diagnosis', *Int. J. Bio-Medical Computing*, **7** (2) (1976), 81-93.

71. Smallwood, R.D. : Private communication (1973).

72. Sato, M., Abe, K, Takeda, H. :'Learning control of finite Markov chains with an explicit trade-off between estimation and control', *IEEE Trans.* **SMC 18** (5) Sept/Oct (1988).

73. Doshi, B., Shreve, S. : 'Strong consistency of a modified maximum likelihood estimator for controlled Markov chains', *J. Appl.Prob.* **17** (3) Sept (1980).

74. El-Fattah, Y.M. : 'Recursive algorithms for adaptive control of finite Markov chains', *IEEE Trans.* **SMC 11** (2) February (1981).

75. Kumar, P.R., Becker, A. : 'A new family of optimal adaptive controllers for Markov chains', *IEEE Trans.* **AC 27** (2) February (1982).

76. Sato, M., Abe, K, Takeda, H. :'Learning control of finite Markov chains with unknown transition probabilities', *IEEE Trans.* **AC 27** (4) April (1982).

77. Sato, M., Abe, K, Takeda, H. :'An asymptotically optimal learning controller for finite Markov chains with unknown transition probabilities', *IEEE Trans.* **AC 30** (11) November (1985).

78. Keilson, J., Sumito, U., Zachmann, M. : 'Row–continous finite Markov chains. Structure and algorithms', *J. Operations Research Society of Japan*, **30** (3) September (1987).

79. Kawai, H., Katoh, N. : 'Variance constrained Markov decision process, *J. Operations Research Society of Japan*, **30** (1) March (1987).

80. Ohashi, M. : 'Replacement policy for components in a Markovian deteriorating system with minimal repair', *J. Operations Research Society of Japan,* **29** (4) December (1986).

81. Mine, H., Kawai, H. : 'An optimal inspection and maintenance policy deteriorating system', *J. Operations Research Society of Japan,* **25** (1) March (1982).

82. Tatsumo, K., Ohi, F. : 'Optimal repair for stochastically failing system', *J. Operations Research Society of Japan,* **25** (3) September (1982).

83. Wakuta, K. : 'Semi-Markov decision process with incomplete state observation – discounted cost criterion', *J. Operations Research Society of Japan,* **25** (4) December (1982).

84. Kurano, M. : 'Semi-Markov decision processes and the applications in replacement models', *J. Operations Research Society of Japan,* **28** (1) March (1985).

85. Kurano, M. : 'Average–optimal adaptive policies in semi-Markov decision processes including an unknown parameter', *J. Operations Research Society of Japan*, **28** (3) September (1985).

86. Smith, A.R., Bartholomew, D.J. : 'Manpower planning in the United Kingdom: an historical review', *J. Operational Research Soc.*, **39** (3) (1988).

87. Taggras, G. : 'An integrated cost model for the joint optimization control and maintenance', *J. Operational Research Soc.*, **39** (8) (1988).

88. Sztrik, J. : 'The ⟨G/M/r/FiFo⟩ machine interference model with state-dependent speeds', *J. Operational Research Soc.*, **39** (2) (1988).

89. White III, C. : 'Note on "A partially observable Markov decision process with lagged information" ', *J. Operational Research Soc.*, **39** (2) (1988).

90. Kim, S.H., Jeong, B.H. : 'A partially observable Markov decision process with lagged information', *J. Operational Research Soc.*, **38** (1987), 439-446.

91. Kizima, M. : 'Upper bounds of a measure of dependence and the relaxation time for finite state Markov chains', *J. Operations Research Society of Japan*, **32** (1) March (1989).

92. Cavazos-Cadena, R. : 'Weak conditions for the existence of optimal stationary policies in avaerage Markov decision chains with unbounded costs', *Kybernetika* **25** (3) (1989).

93. Hordijk, A., Puterman, M.L. : 'On the convergence of policy iteration in finite state undiscounted Markov decision processes: The Unichain Case', *Mathematics of Operations Research*, **12** (1) February (1987).

94. Tsitsiklis, J.N. : 'Markov chains with rare transitions and simulated annealing', *Mathematics of Operations Research*, **14** (1) February (1989).

95. Ozekici, S. : 'Optimal periodic replacement of multi-component reliablity systems', *J. Operational Research Soc.*, **36** (4) July/August (1985).

96. Hopp, W.J., Bean, J.C., Smith, R.L. : 'A new optimality criterion for non-homogeneous Markov decision processes', *Operations Research*, **35** (6) Nov/Dec (1987).

97. Lovejoy, W.S. : 'Some monotonicity results for partially observed Markov decision processes', *Operations Research*, **35** (5) Sept/Oct (1987).

98. Sheskin, Th.J. : 'Successive approximations in value determination for a Markov decision process', *Operations Research*, **35** (5) Sept/Oct (1987).

99. Feinberg, B.N., Chui, S.S. : 'A method to calculate steady-state disruptions of large Markov chains by aggregating states', *Operations Research*, **35** (2) March/April (1987).

100. Schweitzer, P.J. : 'On undiscounted Markovian decision process with compact action spaces', *RAIRO*, **19** (1) February (1985).

101. Mattson, A., Thorburn, D. : 'A simple check of the time homogeneity of Markov chains', *J. Forecasting*, **8** (1), Jan-March (1989).

102. Porteus, E.L. : 'Improved iterative computation of the expected discounted return in Markov and semi-Markov chains', *Z. Oper. Res.*, Ser.A, **24** (1980), 155-170.

103. Nafeh, D. : *Markov Decision Processes with Policy Constraints*, Ph.D. Dissertation, Stanford University, Stanford, California (1978).

104. Arvanitidis, N.V. : *A generalized reliability model with applications in electrical power systems*, Systan Inc., SRR - 67 - 12, December (1967).

105. Szetzler, K., *et al* : 'Encoding probabilities'. In: Howard, R. (ed) *Readings in Decision Analysis*, SRI, Menlo Park, California (1980).

106. Gheorghe, A. :'Markovian models with logical conditions for fault isolation with insufficient observations', *Revue Roumaine Math. Pur. Appl.*, **24** (1979), 1041-1064.

107. Aranda, G.J. : Some relations between Markovian models of netwoks of queues', *Trab. Estab. Invest. Oper.*, **33** (1982), 3-29 (in Spanish).

108. Arjas, E., *et al* : 'Semi-Markov processes on a general state space: alpha-theory and quasi-stationarity', *J. Aust. Math Soc.*, Ser. A, **30** (1980), 187-200.

109. Barbour, A.D., Schassberger, R. : 'Insensitive average residence times in generalized semi-Markov processes', *Adv. Appl. Probab.*, **13** (1981), 720-735.

110. Boel, R. : 'Stochastic models of computer networks', *Proc. NATO ASI: Stochastic Systems*, Les Arcs, 1980, (1981), 141-167.

111. Carravetta, M., Dominicis (de) R., D'Esposito, M.R. : 'Semi-Markov processes in modelling and reliability: a new approach', *Appl. Math. Modelling* **5** (1981), 269-274.

112. Chayny, F. *et al*: 'Nursing care demand prediction based on a decomposed semi-Markov population model', *Oper. Res. Lett.* **2**, (1984), 279-284.

113. Cullmann, G. : 'Les chaines de Markov multiples' - *Programmation dynamique*. Masson et Cie, Paris (1980).

114. Dominicis (de) R. : 'Estimation of average waiting times and asymptotic behaviour of the semi-Markov process corresponding to a Markov renewal process', *Statistica* **40** (1980), 467-475.

115. Doshi, B.T. : 'Generalized semi-Markov decision processes', *J. Appl. Probab.*, **16** (1979), 618-630.

116. Duyn Schouten (van der), F.A. : 'Markov decision processes with continuous time parameters', *Mathematical Centre Tracts*, 167, Amsterdam (1983).

117. Federgruen, A., Spreen, A.D. : 'A new specification of the multi-chain policy iteration algorithm in undiscounted Markov renewal programs', *Management Science* **26** (1980), 1211-1217.

118. Gupta, A. : 'Markovian approach to determine optimal maintenance policies for production systems'. *Indian J. Technol.* **19** (1981), 353-356.

119. Harmalov, B.P. : 'Property of "correct exit" and one limit theorem for semi-Markov processes', *J. Soviet Math.* **23** (1983), 2352-2362.

120. Jaiswal, N.K., Ramamurthy, K.G. : 'Some results for a pseudo semi-Markov process', *Math. Operationsforsch. Stat. ser. Optimization* **13** (1982), 123-132.

121. Kim Kwang, W., White C.C III. : 'Solution procedure for vector criterion Markov decision processes', *Large Scale Syst.* **1** (1980), 129-140.

121. Rosberg, Z. : 'Semi-Markov decision processes with polynomial reward', *J. Appl. Probab.* **19** (1982), 301-309.

123. Yasuda, M. : 'Semi-Markov decision processes with countable state space and compact action space', *Bull. Math. Statist.* **18** (1978), 35-54.

SUBJECT INDEX